COVID-19 in Alzheimer's Disease and Dementia

T0396200

COVID-19 in Alzheimer's Disease and Dementia

P. Hemachandra Reddy

Albin John

ELSEVIER

ACADEMIC PRESS

An imprint of Elsevier

Academic Press is an imprint of Elsevier
125 London Wall, London EC2Y 5AS, United Kingdom
525 B Street, Suite 1650, San Diego, CA 92101, United States
50 Hampshire Street, 5th Floor, Cambridge, MA 02139, United States
The Boulevard, Langford Lane, Kidlington, Oxford OX5 1GB, United Kingdom

ISBN: 978-0-443-15256-6

For information on all Academic Press publications visit our website at https://www.elsevier.com/books-and-journals

Publisher: Nikki Levy
Acquisitions Editor: Anna Valutkevich
Editorial Project Manager: Michaela Realiza
Production Project Manager: Swapna Srinivasan
Cover Designer: Miles Hitchen

Typeset by TNQ Technologies

Contents

List of contributors

Kiran Ali School of Medicine, Texas University Health Sciences Center, Lubbock, TX, United States

Sanjay Awasthi Department of Internal Medicine, Division of Hematology and Oncology, Texas Tech University Health Sciences Center, Lubbock, TX, United States

Chhanda Bose Department of Internal Medicine, Division of Hematology and Oncology, Texas Tech University Health Sciences Center, Lubbock, TX, United States

Rathika Chinniah Madurai HLA Centre, Madurai Kidney Centre & Transplantation Research Institute, Madurai, Tamil Nadu, India

Sonam Deshwal Department of Biochemistry, Panjab University, Chandigarh, Punjab, India

Neha Dhiman Department of Biochemistry, Panjab University, Chandigarh, Punjab, India

Tochi Eboh School of Medicine, Texas Tech University Health Sciences Center, Lubbock, TX, United States

Bernardo Galvan School of Medicine, Texas Tech University Health Sciences Center, Lubbock, TX, United States

Alec Giakas School of Medicine, Texas Tech University Health Sciences Center, Lubbock, TX, United States

Giridhar Goudar Food Quality Analysis and Biochemistry Division, Biochem Research and Testing Laboratory, Dharwad, Karnataka, India

Prashanth Gowda Department of Internal Medicine, Texas Tech University Health Sciences Center, Lubbock, TX, United States

Zachery C. Gray Department of Internal Medicine, School of Medicine, Texas Tech University Health Sciences Center, Lubbock, TX, United States

Cameron Griffith Department of Anthropology, Texas Tech University, Lubbock, TX, United States

Madison Hanson Department of Internal Medicine, Texas Tech University Health Sciences Center, Lubbock, TX, United States

P. Hemachandra Reddy Department of Internal Medicine, Texas Tech University Health Sciences Center, Lubbock, TX, United States; Nutritional Sciences Department, Texas Tech University, Lubbock, TX, United States; Department of Pharmacology and Neuroscience, School of Medicine, Texas Tech University Health Sciences Center, Lubbock, TX, United States; Department of Neurology, Texas Tech University Health Sciences Center, Lubbock, TX, United States; Neurology, Departments of School of Medicine, School of Medicine, Texas Tech University Health Sciences Center, Lubbock, TX, United States; Public Health Department of Graduate School of Biomedical Sciences, School of Medicine, Texas Tech University Health Sciences Center, Lubbock, TX, United States; Department of Speech, Language and Hearing Sciences, School of Health Professions, School of Medicine, Texas Tech University Health Sciences Center, Lubbock, TX, United States

Ashly Hindle Department of Internal Medicine, Division of Hematology and Oncology, Texas Tech University Health Sciences Center, Lubbock, TX, United States

Katherine G. Holder School of Medicine, Texas Tech University Health Sciences Center, Lubbock, TX, United States

Albin John Department of Neurology, Texas Tech University Health Sciences Center, Lubbock, TX, United States

Balakrishnan Karuppiah Madurai HLA Centre, Madurai Kidney Centre & Transplantation Research Institute, Madurai, Tamil Nadu, India

Sonia Y. Khan School of Medicine, Texas University Health Sciences Center, Lubbock, TX, United States

Shazma Khan School of Medicine, Texas University Health Sciences Center, Lubbock, TX, United States

Jonathan Kopel School of Medicine, Texas Tech University Health Sciences Center, Lubbock, TX, United States

Vivek Kumar Department of Biotechnology, IMS Engineering College, Ghaziabad, Uttar Pradesh, India

Subodh Kumar Center of Emphasis in Neuroscience, Department of Molecular and Translational Medicine, Paul L. Foster School of Medicine, Texas Tech University Health Sciences Center El Paso, El Paso, TX, United States

Pulak R. Manna Department of Internal Medicine, School of Medicine, Texas Tech University Health Sciences Center, Lubbock, TX, United States

Munikumar Manne Clinical Division, ICMR-National Institute of Nutrition, Hyderabad, Telangana, India

Harrison Marsh Department of Ophthalmology, School of Medicine, Texas Tech University Health Sciences Center, Lubbock, TX, United States

Hallie Morton Department of Internal Medicine, Texas Tech University Health Sciences Center, Lubbock, TX, United States

Sasiharan Pandi Madurai HLA Centre, Madurai Kidney Centre & Transplantation Research Institute, Madurai, Tamil Nadu, India

Jangampalli Adi Pradeepkiran Department of Internal Medicine, Texas Tech University Health Sciences Center, Lubbock, TX, United States

Bhagavathi Ramasubramanian Department of Internal Medicine, Texas Tech University Health Sciences Center, Lubbock, TX, United States; Department of Neurology, UT Southwestern Medical Center, Dallas, TX, United States

Padma Malini Ravi Department of Immunology, School of Biological Sciences, Madurai Kamaraj University, Madurai, Tamil Nadu, India

Sparsh Ray School of Medicine, Texas University Health Sciences Center, Lubbock, TX, United States

Stephen Rossettie School of Medicine, Texas Tech University Health Sciences Center, Lubbock, TX, United States

Rajat Sandhir Department of Biochemistry, Panjab University, Chandigarh, Punjab, India

Ashish Sarangi University of Missouri, Columbia, MO, United States; Department of Psychiatry and Behavioral Sciences Baylor College of Medicine, Houston, TX, United States

Emma Schindler University of Miami Miller School of Medicine, Miami, FL, United States

Vandit Sevak Madurai HLA Centre, Madurai Kidney Centre & Transplantation Research Institute, Madurai, Tamil Nadu, India

Shyam Sheladia Department of Internal Medicine, Texas Tech University Health Sciences Center, Lubbock, TX, United States

Shivam Sheladia Department of Internal Medicine, Texas Tech University Health Sciences Center, Lubbock, TX, United States

Sharda P. Singh Department of Internal Medicine, Division of Hematology and Oncology, Texas Tech University Health Sciences Center, Lubbock, TX, United States

Dhinakaran Thadakanathan Madurai HLA Centre, Madurai Kidney Centre & Transplantation Research Institute, Madurai, Tamil Nadu, India

Murali Vijayan Department of Internal Medicine, Texas Tech University Health Sciences Center, Lubbock, TX, United States

Rishi Virani Department of Internal Medicine, Texas Tech University Health Sciences Center, Lubbock, TX, United States; Department of Biomedical Engineering, The University of Texas at Austin, Austin, TX, United States

Preface

P. Hemachandra Reddy, PhD

Ever since the outbreak of the severe acute respiratory syndrome coronavirus 2 (SARS-CoV-2) in 2019, many researchers who study aging and Alzheimer's disease have become growingly interested in the effects of the disease on their fields of interest. Our book summarizes the current status of coronavirus disease 2019 (COVID-19) in aging populations with comorbidities, including hypertension, diabetes, obesity, kidney disease, respiratory illnesses, and various infectious diseases, as well as in those suffering from dementia and Alzheimer's disease.

Our book is split into two sections. The first section includes 10 chapters that provide a general description of COVID-19, including SARS-CoV-2 structure, function, and biology, and its impact on the elderly with chronic conditions such as hypertension, diabetes, obesity, kidney disease, respiratory illnesses, and infectious diseases. This section also discusses the effects of the virus on the immune system. The second section of eight chapters shifts to the impact of COVID-19 on those with dementia or Alzheimer's disease, with special emphasis on age, gender, ethnic background, and lifestyle. By bringing this focus on neurodegenerative disease in one comprehensive resource, this volume is an essential reference for neuroscientists, clinicians, biomedical scientists, virologists, immunologists, and most importantly to our communities, friends, and families.

Albin John, MBA

The key points of our book are (1) a description of SARS-CoV-2 structure, function, and biology; (2) an examination of its impact on the elderly with chronic conditions including hypertension, diabetes, obesity, kidney disease, respiratory, illnesses, and infectious diseases; (3) a description of Alzheimer's disease and other dementias as comorbidities for COVID-19; (4) the role of the blood−brain barrier disruption following COVID-19 infection; (5) the role of age-related oxidative stress and mitochondrial damage as factors of COVID-19; and (6) a discussion of the effects of race, gender, and sex as additional risk factors.

We sincerely thank all the contributors for their outstanding chapters. We also thank Texas Tech University Health Sciences Center, leaders Dr Scott Shurmur, Chair of Internal Medicine, School of Medicine Dean Dr. Steven Berk, Provost Dr. Darrin Dagostino, Sr VPR and Innovation Dr. Lance MacMahon and President, Dr Lori Rice-Spearman for their encouragement and support. Our heartfelt thanks to Ms. Nikki Levy, Ms. Anna Valutkevich, Ms. Swapna Srinivasan, Mr. Mohan Raj Rajendran, and Ms. Michaela Realiza at Elsevier, for their support and help in assembling this volume.

Section One

COVID-19

COVID-19 and immunity: an overview

1

Pulak R. Manna[1], *Zachery C. Gray*[1] *and P. Hemachandra Reddy*[2,3,4,5]
[1]Department of Internal Medicine, School of Medicine, Texas Tech University Health Sciences Center, Lubbock, TX, United States; [2]Department of Pharmacology and Neuroscience, School of Medicine, Texas Tech University Health Sciences Center, School of Medicine, Lubbock, TX, United States; [3]Neurology, Department of School of Medicine, School of Medicine, Texas Tech University Health Sciences Center, Lubbock, TX, United States; [4]Public Health Department of Graduate School of Biomedical Sciences, School of Medicine, Texas Tech University Health Sciences Center, Lubbock, TX, United States; [5]Department of Speech, Language and Hearing Sciences, School Health Professions, School of Medicine, Texas Tech University Health Sciences Center, Lubbock, TX, United States

Abstract

Coronavirus disease-19 (COVID-19), caused by a β-coronavirus and its genomic variants, is associated with substantial morbidities and mortalities globally. The COVID-19 virus enters host cells upon binding to the angiotensin converting enzyme two receptors. Patients afflicted with COVID-19 may be asymptomatic or present with critical symptoms possibly due to diverse lifestyles, immune responses, aging, and underlying medical conditions. Geriatric populations, especially men in comparison to women, with immunocompromized conditions, are the most vulnerable to severe COVID-19-associated infections, complications, and mortalities. Notably, whereas immunomodulation, involving nutritional consumption, is essential to protecting an individual from COVID-19, immunosuppression is detrimental to the host with this hostile disease. As such, immune health is inversely correlated to COVID-19 severity and resulting consequences. Advances in genomic and proteomic technologies have helped us to understand the molecular events underlying symptomatology, transmission, and pathogenesis of COVID-19 and its genomic variants. Accordingly, there has been development of a variety of therapeutic interventions, ranging from mask wearing to vaccination to medication. Regardless of various measures, a strengthened immune system can be considered as a high priority of preventive medicine for combating this highly contagious disease. This chapter provides an overview of pathogenesis, effects of comorbidities on COVID-19 and their correlation to immunity, and prospective therapeutic strategies for the prevention and treatment of COVID-19.

Keywords: Aging; COVID-19; Immunomodulation; Prevention and treatment of COVID-19; Underlying medical conditions

COVID-19 in Alzheimer's Disease and Dementia. https://doi.org/10.1016/B978-0-443-15256-6.00013-1

1. Introduction

COVID-19, a very contagious disease, is caused by severe acute respiratory syndrome coronavirus 2 that was first identified in December 2019 at Wuhan, China.[1,2] This new virus has since spread globally, leading to a severe health crisis. COVID-19 is a member of the family of viruses known as Coronaviridae (order, *Nidovirales*; subfamily, *Orthocoronavirinae*), which displays similar clinical features as those of severe acute respiratory syndrome (SARS) and Middle East respiratory syndrome (MERS).[1,3,4] Airborne transmission is the primary mode of infection in the spread of the COVID-19 virus that enters host cells upon binding to the angiotensin converting enzyme 2 (ACE2) receptors that expressed in a variety of tissues.[5,6] The pathophysiological manifestations of COVID-19 include moderate to life-threatening symptoms that are frequently associated with fever, headache, respiratory distress, hypoxia, lung injury, inflammation, and cardiovascular diseases (CVDs). COVID-19 can lead to grave outcomes by affecting the immune system and damaging multiple organ systems through a plethora of pathophysiological events.[7,8]

A large body of epidemiological evidence indicates a strong correlation between the intake of vitamins, minerals, antioxidants, and a reduction and/or prevention of COVID-19 and other pathogens.[9–11] Nutritional status can influence COVID-19 infections to variable degrees, with implications for duration, harshness, and overall consequences. Deficiency of nutrients, involving impaired immunity, is more susceptible to severe COVID-19-related infections and fatal outcomes.[12–14] It is noteworthy that people with immunocompromized conditions are more inclined to develop multi-organ complications and deadly consequences from COVID-19.[15,16]

COVID-19 placed an immense burden on nearly every country in the world. A wide variety of measures to control this virus were implemented including lockdowns, social distancing, mask wearing, vaccinations, and many emergency use authorization (EUA) drugs and/or antibodies.[11] However, genetic variants of the disease have prolonged the fight against COVID-19 and pushed healthcare systems to their limits. Herein, we summarize current literature that helps comprehend disease pathogenesis, its relevance to healthy immunity, and therapeutic potentials, for the management of COVID-19.

2. COVID-19: risk factors and pathogenesis

COVID-19 is an acute respiratory disease that attacks alveolar epithelial cells of the lungs, with clinical features essentially similar to SARS and MERS, which are essentially spherical (60–200 nm) and single-stranded RNA viruses.[2,4] Stemming from the same family of coronaviruses, COVID-19 was thought to emerge as a zoonotic transmission from bats, which have been the major evolutionary reservoirs of coronavirus diversity. Upon emergence, this virus made human to human transmission and spread rapidly throughout the globe with a high morbidity and mortality.[3,4] The World Health Organization (WHO) declared COVID-19 a global pandemic in March 2020.

While the majority of COVID-19 patients display mild to moderate symptoms, a small number of cases develop critical signs, including pneumonia, acute respiratory distress syndrome (ARDS), sepsis, multi-organ failure, and death.[17–19] Of note, the median incubation period for COVID-19 was estimated to be between 5 and 6 days. Despite various pathophysiological conditions, COVID-19-associated infections and mortalities are considerably higher among men in comparison to women.[11]

COVID-19 is spread from human to human via both direct contact and airborne transmission. A wide variety of risk factors for COVID-19 include activities, procedures, products, and events, ranging from low to very high.[1,11,20] Noteworthy, however, that the contribution of numerous risk factors to COVID-19 infections and resultant complications are dependent on immune health, aging, and underlying medical conditions. There is increasing evidence that children and adolescents are generally asymptomatic to COVID-19 or exhibit mild symptoms such as fever, headache, fatigue, and nasal congestion, then recover from the infection by their healthy immune system.[6,21] On the other hand, a subset of patients, possessing impaired immunity or immunocompromized conditions, display severe clinical manifestations, requiring hospitalizations and life supporting treatments, along with mortalities.[11,22]

The genomic configuration of COVID-19 ($\sim$30 kb) is highly conserved with previously identified SARS and MERS, all of which possess large positive sense RNA (++RNA) genomes. Among the four different isoforms (α, β, γ, and δ), COVID-19 is categorized as a β-virus that enters host cells through endocytosis involving three steps: binding, cleavage, and fusion. It encompasses four structural proteins; spike, membrane, envelope, and nucleocapsid (Fig. 1.1). Additionally, COVID-19 possesses hemagglutinin esterase, a glyocoprotein that is utilized for an invading mechanism. The spike protein is composed of two functional subunits, S1 and S2, in which the former binds to the ACE2 receptors.[6,23] The binding of S2 allows for insertion of the RNA genome into the host cells, which then undergoes cleavages by host proteases (e.g., furin and trypsin), and translation to form polyproteins that are then assembled to make replication–transcription complexes. Once the complex is formed, a copy of the RNA genome is made; different structural proteins are synthesized in the cytoplasm, and all parts are assembled with help from the endoplasmic reticulum and Golgi apparatus.[6,23,24] These viral particles are then released from the cell by exocytosis and have the ability to infect other cells to continue the replication process (Fig. 1.2).

3. COVID-19 variants and their impact on global health tragedy

The COVID-19 pandemic is associated with a number of genomic variants that are consistently occurring. The WHO has declared certain strains of COVID-19 as "variants of concerns" (VOCs). These variants have increased COVID-19 transmissibility, severity, epidemiology, clinical disease presentation, or have decreased the effectiveness of current treatment options.[25,26] Notably, COVID-19 α, β, γ, δ, and Omicron variants have shown significant effects in different parts of the world, in which both

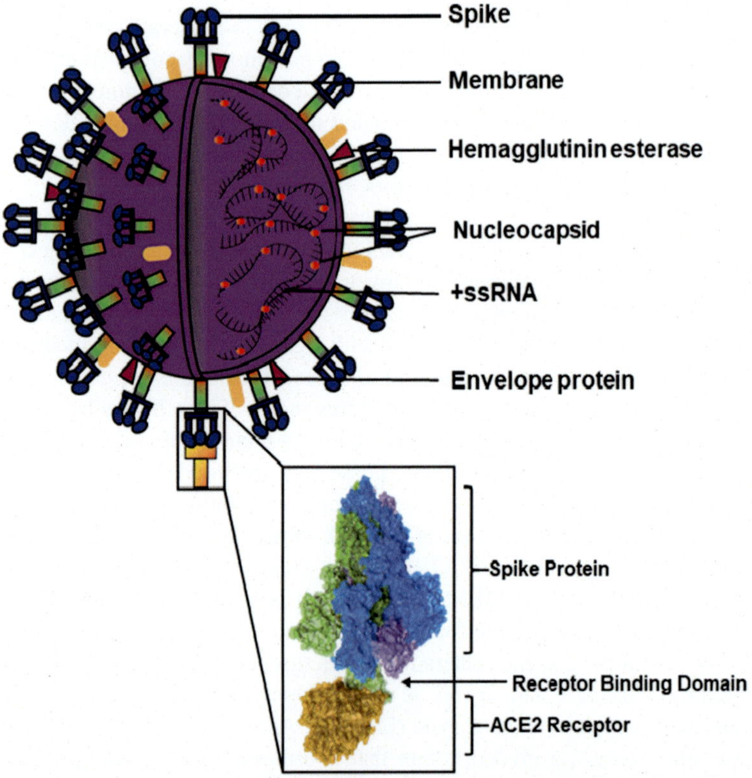

Figure 1.1 Schematic representation of a COVID-19 virus and its different components including the spike protein (*red-to-green box with blue crown*), plasma membrane (*dark purple*), hemagglutinin-esterase enzyme (*light purple triangle*), nucleocapsid proteins (*red circles*), positive sense single stranded RNA (*black lines*), and the envelope protein (*orange pill shape*). The spike protein is further magnified to show various parts including the receptor binding domain and its binding to the ACE2 receptors.

δ and Omicron variants are responsible for the majority of COVID-19-associated complications and mortalities.[25]

The COVID-19 α-variant (B.1.1.7) was first discovered in September 2020, in the United Kingdom, and was initially considered to have higher rates of transmissibility than the original COVID-19 virus with the most notable mutation being N501Y (asparagine to tyrosine substitution) and a deletion of amino acid 69/70 on the spike protein causing increased binding affinity.[27] It was shown to have a 43%–90% higher transmission rate than the original virus in England but did not increase disease severity.[28]

The β-variant of COVID-19 (B.1.351) was initially identified in October 2020, South Africa, and contains the same N501Y mutation along with many others but lacks the 69/70 deletion.[29] This variant is not associated with increased transmission or disease severity but showed resistance to neutralizing antibodies due to changes of the

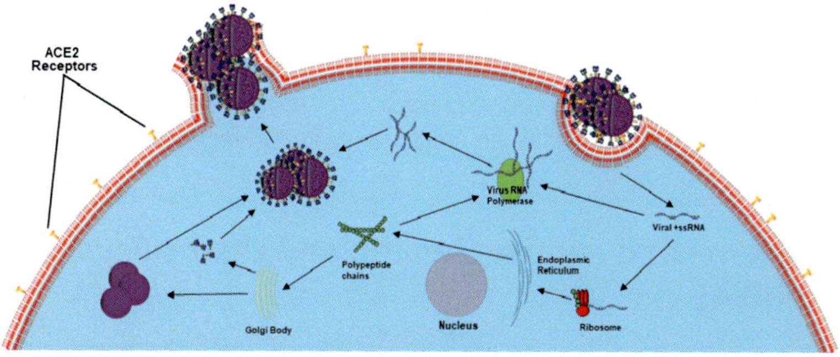

Figure 1.2 Entry of the COVID-19 virus into a body cell upon binding to the ACE2 receptors, and subsequent replication process using cellular machinery. The viral genome is expulsed and first travels to the cellular ribosomes to be translated into the virus specific RNA polymerase and various viral components. The RNA polymerase makes copies of the + ssRNA and meets up with the viral components after their respective posttranscriptional modifications have been completed in the endoplasmic reticulum and Golgi body. The + ssRNA is assembled along with the various components and the newly formed viruses undergo exocytosis to leave the cell and infect new cells.

spike protein's primary structure, thus, demonstrated possibilities for increased reinfection rates and certain treatment resistance.[30]

Brazil was thought to be the initial location (November 2020) of the COVID-19 γ-variant (P.1) and consists of the same N501Y spike protein mutation without the 69/70 deletion but with additional mutations on other components such as the nucleocapsid protein.[31,32] The P.1 strain was the most common strain in Brazil during a significant COVID-19 outbreak showing greater disease severity against younger populations, as compared with a previous COVID-19 outbreak mostly consisting of other variants.[33] It is thought that mutations to the spike protein account for its ability to evade certain monoclonal antibody treatments.[34]

The COVID-19 δ-variant (B.1.617.2) is the most predominant strain causing the disease in the United States and accounts for nearly all cases (99%).[35,36] It was first identified in May 2021, India, and shows many novel mutations such as L452R (leucine to arginine) and P681R (proline to arginine) of the spike protein that can inhibit antibody binding and increase its affinity for the ACE2 receptor expressed in somatic cells.[37] The δ-variant has a higher transmissibility rate (40%−60% increase over α-variant) and has alluded to more severe complications, including hospitalizations and mortalities, than that of the original COVID-19 virus.[35,38]

The Omicron variant (B.1.1.529) is thought to have originated in South Africa, Botswana, or Hong Kong.[39] This COVID-19 variant was first reported in November 2021 and shows signs of considerable reinfection capability, increased transmissibility, disease severity, and augmented antibody evasion.[40] This new variant has extensive spike protein mutations, including N501Y and an EPE (glutamate, proline, and glutamate) insertion at amino acid 214, nucleocapsid mutations, and other deletions similar to previous VOCs.[39] While genomic and proteomic analyses have identified more than 30 mutations in Omicron, from the original COVID-19 virus, much is unknown surrounding this variant.[40]

It is noteworthy that the COVID-19 virus and its genomic variants enter host cells upon binding to the ACE2 receptors expressed in a variety of tissues, but higher prevalences of this receptor are within the lungs, heart, and kidneys (Fig. 1.3).

4. Nutrients and immune health, and their relevance to COVID-19

Nutrients are required for normal growth, reproduction, and various physiological activities. Vitamins, minerals, and antioxidants are fundamental to proper functioning of the immune system, which prevent an organism from contracting COVID-19 and other pathogens. Deficiency of nutrients, involving impaired immunity, is associated with a variety of health issues and increased susceptibility to severe COVID-19-associated complications and mortalities.[12–14]

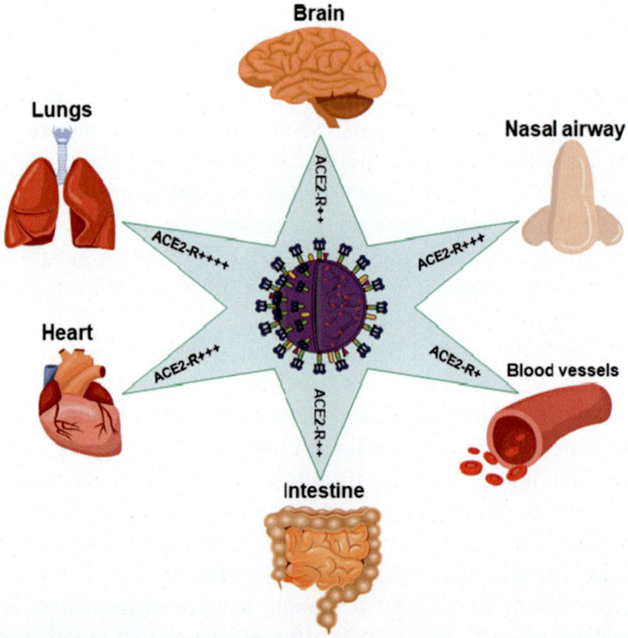

Figure 1.3 Schematic representation showing the potential target of the COVID-19 virus or its variants to various organs expressing the ACE2 receptors. The organ/organ systems shown are the brain, lungs, heart, intestine, blood vessels, and nasal airways. Different organs with the ACE2 receptor (ACE2-R) expression are arbitrarily depicted as ACE2-R+ (low), ACE2-R++ (medium), and ACE2-R+++/++++ (high), thus, demonstrating the diverse abilities of these viruses to infect respective organs.

It is unquestionable that nutritional status plays an essential role in maintaining bodily homeostasis, as well as overall healthy physiology. Vitamins are divided into two groups: water-soluble (C and 8 B vitamins, i.e., B1, B2, B3, B5, B6, B7, B9, and B12) and fat-soluble (A, D, E, and K). All of these vitamins are obtained mostly from various food sources that people consume regularly and exert diverse effects on biological activities. However, the recommended daily amount (RDA) varies between adult men and women (Table 1.1). B vitamins play integral roles in many important processes in the body such as immune cell proliferation, hormonal equilibrium, energy production, heart and neurological health, oxygen transportation, decreasing the risk of comorbidities, cytokine formation, and antibody production.[11] These processes help strengthen the immune system for the recognition and neutralization of COVID-19 and other harmful environmental factors. Vitamins A, C, D, E, and K serve in many cellular functions, including anti-inflammation, antioxidation, immunomodulation, and antithrombotic states.[11] All of these diverse processes aid in the de-escalation or protection from severe inflammation and tissue damages inflicted by COVID-19.

Macronutrients such as carbohydrates, fats, and proteins provide various building blocks and energy sources for an organism to develop and/or repair immune system function. Certain fats such as omega-3 fatty acids provide anti-inflammatory effects contribute to appropriate immune responses and have shown to assist patients with COVID-19 infections and complications.[41,42]

Various micronutrients are instrumental to the appropriate functioning of the immune system and, therefore, prevent individuals from COVID-19 and other relevant diseases. In accordance, zinc, iron, selenium, copper, and magnesium are essential for the inhibition of viral replication, proliferation of various immune cells and components, anti-inflammation, antioxidation, immunomodulation, and serve as cofactors with enzymes in many necessary reactions.[43]

Anti-inflammatory substances have the ability to assist and/or regulate the body's natural immune response associated with the cytokine storm influenced by severe COVID-19 infections that causes low oxygen saturation, lung damage, multiorgan failure, and ultimately death. Antioxidants provide safe mechanisms to avoid tissue damage and provide an avenue to neutralize reactive oxygen species (ROS) used throughout the immune system. The correct usage of inflammation and ROS provides the immune system with an accurate and efficient response to pathogens thus increasing its chance of survival with minimal damages. Therefore, nutrients, by strengthening and modulating the immune system, serve as the primary defense against COVID-19 and other invading pathogens.

The immune system is a collection of biological processes, including various organs and cellular structures, which prevent organisms affected by a variety of environmental toxins, bacteria, and viruses.[44,45] Briefly, the immune system, involving innate and adaptive/acquired responses, is vital to proper functioning of many important physiological processes, as it serves as a barrier between pathogens and the internal milieu.[46,47] Noteworthy, healthy immunity has recently been reported as a high priority of preventive medicine for combating COVID-19.[11] The innate immune system is the culmination of physical barriers and literal gene expressions of an organism that are present at birth such as skin, epithelial tissue linings, respiratory

Table 1.1 Vitamins, their major sources, RDA values for adults, and various functions.

		RDA for adults		
Vitamins	**Sources**	**Women**	**Men**	**Functions**
Vitamin A (retinol)	Leafy greens, apricots, cantaloupe, carrots, squash, sweet potatoes, pumpkin, dairy, liver, and eggs.	700 μg	900 μg	Vision, skin, bones, immune health, and mucous membranes.
Vitamin B1 (thiamine)	Whole-grain, seeds, nuts, legumes, and pork.	1.1 mg	1.2 mg	Nerve function, metabolism, cell growth, kidney function, nerve health, and immune health.
Vitamin B2 (riboflavin)	Leafy greens, whole-grain, almonds dairy, pork, liver, chicken, and salmon.	1.1 mg	1.3 mg	Metabolism, neurological health, immune health, vision, and skin health.
Vitamin B3 (niacin)	Whole-grains, breads and cereals, mushrooms, asparagus, leafy greens, legumes, peanuts, banana red meats, poultry, and fish.	14 mg	16 mg	Metabolism, nerve function, digestion, and skin.
Vitamin B5 (pantothenic acid)	Nearly all foods.	5 mg	5 mg	Metabolism and immune health.
Vitamin B6 (pyridoxine)	Chickpeas, leafy greens, bananas, papayas, oranges, cantaloupe, liver, tuna, salmon, poultry, pork, and beef.	1.2 mg	1.3 mg	Metabolism, skin health, vision, neurological health, immune health, and nerve function.
Vitamin B7 (biotin)	Avocados, sweet potatoes, nuts, seeds, cereals, liver, eggs, salmon, tuna, and pork.	20–30 mg	20–30 mg	Metabolism, hair, nail, and skin health.
Vitamin B9 (folic acid)	Leafy greens, legumes, seeds, oranges, peanuts, whole-grain, liver, seafood, eggs, meats, berries, melons, and asparagus.	400 μg	400 μg	DNA synthesis, growth and development, and neurological health.

Vitamin B12 (cyanocobalamin)	Fish, shellfish, liver, red meat, eggs, poultry, and dairy.	2.4 μg	2.4 μg	Nerve function, immune cell development, metabolism, and red blood cell function.
Vitamin C (ascorbic acid)	Citrus fruit, bell peppers, strawberries, tomatoes, broccoli, cabbage, cauliflower, potatoes, kiwi, and kale.	75 mg	90 mg	Antioxidant, metabolism, immune health, iron uptake, gene expression, nerve and muscle health.
Vitamin D (ergocalciferol)	Cod liver oil, fish, beef liver, egg yolk, and dairy.	15 μg	15 μg	Calcium, magnesium, and phosphate absorption, neurological health, immune health, and bone health.
Vitamin E (tocopherol)	Wheat, sunflower oil, soybean, olive oil, almonds, peanuts, leafy greens, pumpkin, bell pepper, asparagus, mango, avocado, and dairy.	15 mg	15 mg	Antioxidant, immune health, anti-clotting, and cell signaling.
Vitamin K (phylloquinone, menaquinone)	Leafy greens, soybean, canola oil, broccoli, asparagus, kale, cauliflower, spinach, meat, dairy, and seafood.	90 μg	120 μg	Aids in clotting factors, bone health, and neurological health.

tract, and genitourinary tract, as well as mucus layers that coat these tissues. The cells and other components specific to the innate system are the neutrophils, monocytes, macrophages, cytokines, and specific proteins (such as antimicrobial peptides) that work to broadly attack pathogens and invaders.[48–50] Neutrophils are the body's first cellular line of defense for external pathogens that are ingested through phagocytosis and subsequently metabolized.[48,51] The adaptive immune system is thought to have evolved alongside the innate system in complex vertebrates to identify and recognize explicit threats that have been presented to an organism. Both T- and B-lymphocytes generated in the thymus and bone morrow comprise the cellular components of the adaptive immune system, in which mature T-cells are responsible for cytokine production, antigen destruction, and immunomodulation.[47,52,53] Macrophages serve in pathogen recognition, attaching to them, and escorting them to T- and B-lymphocytes for destruction.[48] Initially, B-cells recognize a pathogen and develop antibodies against it, which then respond rapidly to recognize and contain infections.[53–55] This acquired memory naturally adapts to each individual and the pathogens they come into contact with.

An effective immune response provides its host with the greatest possible chance to weather and protect against various diseases including COVID-19 without undue harm. Whereas immunomodulation, influenced by a variety of health promoting factors including nutrients, contributes to protecting an organism from pathogens and overactive immune responses, immunosuppression is unable to adequately recognize and neutralize those invaders (Fig. 1.4). COVID-19 is a nonsevere disease in population majorities especially among the young and healthy; however, it has shown severe effects among a subset of the population such as elderly and obese individuals and people with other underlying medical conditions.[22] As well, an overreactive cytokine storm is associated with severe COVID-19-linked complications and mortalities.[8] Ranging from underlying complications to the inadequate intake of nutrients and cofactors, the ability of an organism, possessing impaired immunity, to fend off infection is especially relevant within the confines of COVID-19.

5. Immunocompromised conditions and COVID-19

Whereas a boosted immune system helps keep foreign bodies away, an impaired immune response is incapable of protecting individuals from invading pathogens (Fig. 1.4). Specifically, for COVID-19, immunosuppression plays a predominant role toward infections, resulting hospitalizations, and mortalities.[56] Certain conditions generate immunocompromised states that include the human immunodeficiency virus, hepatitis B, and acquired immunodeficiency syndrome/sexually transmitted diseases, all of which destroy T-cells and attack the immune system. Immunocompromised conditions also occur with chronic lung and kidney diseases, dementia, and/or neurological and inherited diseases.[14,57] Long-term use of high-dose steroids, immunosuppressant drugs, organ transplants, chemotherapy, hematological malignancies, and autoimmune diseases can also lead to immunosuppression.[58] All of these

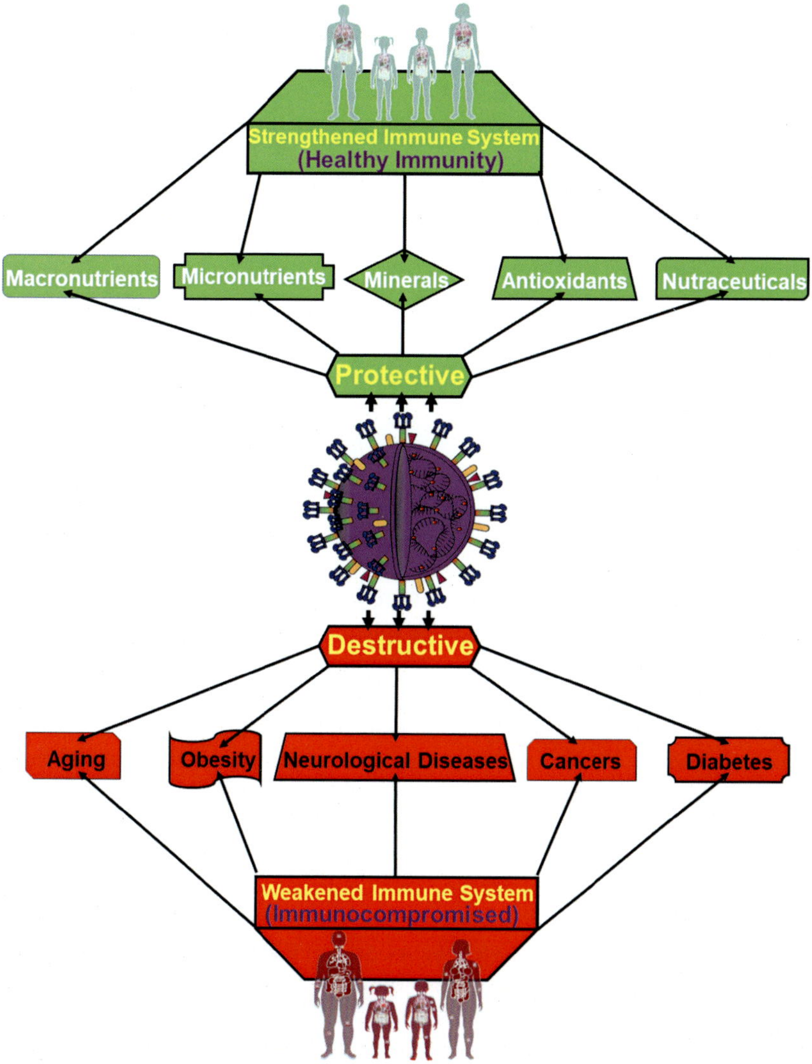

Figure 1.4 Graphical representation illustrating the impact of COVID-19 to men and women possessing either a strengthened immune system or an impaired immune system. Individuals maintaining healthy immunity/immunomodulation (upper section), affected by vitamins, minerals, and nutrients, are generally protective against COVID-19 (middle) linked infections and resulting consequences. On the other hand, people with immunocompromised conditions (lower section), affected by underlying medical situations, including aging, obesity, cancers, neurological disease, and diabetes are most vulnerable to severe COVID-19-associated infections and complications, with fatal outcomes.

immunocompromised conditions are liable to be susceptible not only by COVID-19 infections but also with other pathogens and can lead to fatal consequences.

Epidemiological evidence suggests that COVID-19-associated infections and complications are moderate in children and adolescents who generally possess a strengthened immune system.[6,21] However, a small group of patients, owning deteriorated immunity or underlying medical conditions, develop critical symptoms such as pneumonia, ARDS, and multiorgan failure, involving deadly outcomes.[17–19] Underlying medical conditions can be genetic and acquired, can dampen the immune system, and predispose individuals to a higher risk of developing severe complications in conjunction with COVID-19. Additionally, chronic inflammation can lead to increased ACE2 receptors and cytokine overproduction that can render this subset of patients more vulnerable to severe COVID-19-related complications and mortalities.[59–61]

5.1 Aging

An overwhelming amount of evidence indicates aging is the most significant risk factor for COVID-19.[9,62] This inevitable phenomenon, connecting dampened immunity, is associated with many health complications, including memory and cognitive function, decreased muscle mass and bone density, diminished eyesight, sexual dysfunction and depression, increased risk of CVDs, and skin disorders.[9,63–66] Aging is also connected with the accumulation of dysfunctional mitochondria and oxidative damage, which modulate the immune system and contribute to increased morbidity and mortality.[67–70] It has been reported that COVID-19-related mortalities are higher among males in all age groups 20 years or older, with a mortality rate two times higher in men than women.[71,72] During aging, excessive production of free radicals and ROS occurs in the mitochondria, and an increase in free radicals/ROS is inversely correlated with antioxidant capacity in the central nervous system and its associated glands.[68,73] Studies have shown that an imbalance between production of free radicals/ROS and protective antioxidant systems, affecting cellular damage, induces age-associated complications and diseases.[68,69,73] In addition, ROS disrupts mitochondrial function and decreases the steroidogenic acute regulatory protein (StAR), a key mitochondrial factor that is implicated in age-related decline of steroid hormones.[67,74] As such, oxidative damage induced by ROS is deleterious to the functional efficiency of various cellular processes, including the immune system, leading to severe COVID-19 allied complications and mortalities.

5.2 Diabetes

Diabetes is a group of metabolic disorders in which the pancreas produces either little to no insulin or the body has built resistance to the insulin thus affecting the breakdown of foods to sugar/glucose, an important process for energy metabolism and proper functioning of various physiological activities. Diabetes is one of the most common comorbidities of COVID-19 infections.[75–77] Studies have demonstrated that the expression of ACE2 receptors is upregulated in diabetes allowing higher levels of the virus to bind to host cells and result in progressive COVID-19-associated complications.[78,79] The chronic condition in diabetes increases circulating levels of furin, which is known to assist host–pathogen interactions and the entry of COVID-19 as well.[80] Diabetes inhibits the function of neutrophils in chemotaxis, phagocytosis, and intracellular microbe neutralization, leading to a hyperinflammatory state.[76,81] Impaired immunity, linked with diabetes, is recognized as a potential mechanism for increased morbidity and susceptibility to severe COVID-19. It is plausible that hyperinflammation is connected with an increased ability of the COVID-19 virus to bind to upregulated ACE2, involving higher proliferation rates and subsequent infections of different organs, resulting in fatal outcomes.

5.3 Obesity

Obesity is a chronic disease characterized by the excessive accumulation of body fat and is commonly associated with various health problems. Obesity increases the risk for diabetes, CVDs, strokes, and certain cancers.[82] It is often described as a result of high fat and sugar intake, whereby the food eaten is often processed and devoid of many essential nutrients for healthy growth and development.[83,84] Excessive consumption of these unhealthy foods is an important risk factor of CVDs and cerebrovascular diseases. Obesity is a known risk factor toward other diseases, including hypertension, heart failure, and type 2 diabetes.[85,86] In obesity, the immune system function in the adipose tissues become considerably altered as the cells responsible for the regulation of systemic metabolism and bodily homeostasis are exchanged for cells responsible for inflammatory responses.[82,87] The immune system generates a proinflammatory response in the body through higher than average cytokine production,[88] and an increased level of the latter leaves an obvious pathway to develop severe COVID-19-related complications. Furthermore, the overabundance of adipose tissue affects the respiratory system and limits the ability of an individual to effectively provide oxygen to various tissues and exchange CO_2 in adequate amounts for proper cellular function.[88,89] Importantly, cells that are not able to function at optimal levels display lesser responses to pathogenic intrusions and are more susceptible to various infections, including COVID-19.

5.4 Cancers

Cancer is a leading cause of death stemming from the aberrant growth of cells. Most common cancers are connected with breast, ovary, colon, and prostate tissues; however, metastasis is the key cause for cancer deaths.[90] Alterations in gene expression result in the uncontrolled growth of cells involving tumor progression.[90,91] Underlying medical conditions such as cancers show worse outcomes from COVID-19 infections.[92,93] Malfunction in the immune system, connecting impaired immunity, caused by diverse factors and/or processes, often leads cancer patients to develop various comorbidities such as diabetes, obesity, and CVDs, which are detrimental to COVID-19.[94,95] Specific cancer therapies (chemotherapy, radiotherapy, and surgical recoveries) can also suppress an individual's immune response to fight off COVID-19 and other pathogens.[96,97] It has been reported that the expression of ACE2 receptors is upregulated in various cancerous tissues, which allow the COVID-19 virus to bind and enter host cells effectively and results in higher problems and mortalities.[98,99] Consequently, the effect of COVID-19 has been harsh in cancer patients, as they frequently possess weakened immune systems either by cancer itself or due to associated therapies and therefore develop severe complications with lower incidences of survival.

5.5 *Neurological diseases*

Neurological disorders are pathologies of the brain, spine and the nerves such as Alzheimer's disease (AD), stroke, Huntington's disease (HD), epilepsy, and Parkinson's disease (PD), all of which affect the cognitive, behavioral, and social skills of a person's ability to act independently. Inflammation in the brain can bring about seizures, delirium, coma, and other neurological manifestations.[100–103] Patients afflicted with neurological diseases, for example AD, may be more susceptible to COVID-19. AD is the most common cause of dementia, which results in neuronal cell death due to cerebrovascular dysfunction.[104,105] Accumulation of amyloid-β precursor protein (APP) and Tau in the brain is the pathological hallmark of AD, a condition most prevalent in elderly men and women.[106,107] Our current data reveal that co-expression of StAR, with either mutant APP or Tau, in hippocampal neuronal HT22 cells partially reinstates APP/Tau mediated cell survival and increased pregnenolone production, pointing to a neuroprotective role of StAR in AD (Manna PR et al., manuscript in preparation). It has been reported that AD patients express higher levels of ACE2 receptors in the brain, in comparison to individuals without AD, implicating that AD patients are more prone to COVID-19 infections.[108] Additionally, increased levels of interleukin-6 in serum, reflecting aberrant cellular immunity, are associated with higher fatality rates in COVID-19 patients.[109,110]

Overall, it is unambiguous that people with immunocompromised conditions and comorbidities are the most vulnerable to severe COVID-19 infections, serious complications, and result in higher mortalities (Fig. 1.4).

6. Diverse measures for preventing COVID-19

Disease prevention is paramount to modern healthcare, and the COVID-19 pandemic needs special attention as no unique measure is available to control this hostile disease. As a consequence, the WHO and the governments of numerous countries and their disease prevention and control centers have advocated several actions and/or practices for limiting the spread of COVID-19 infections and its deadly consequences.[111] These preventive procedures include frequent hand washing, sanitization, face coverings, avoidance of parties and/or gatherings, physical distancing, and cleaning of commonly touched surfaces, in addition to available COVID-19 vaccination.[11,112] Many of these practices also involved shutting down nonessential activities such as travel, school, the workplace, recreation, and meetings/parties (Table 1.2).

Table 1.2 Influence of a variety of practices and/or measures for the prevention of COVID-19 and its variants.

Practices/measures	Usefulness	Effectiveness (+, low; + +, medium; + + +, high-very high)	References
Hand washing	Very high	+++	112,113
Face covering	Very high	+++	114,115
Physical distancing	Very high	++	112,116
Avoidance of large gatherings	Medium	++	111,112
Avoid travel	Low/medium	++	117
Cleaning of common touch areas	High	++	112,118
Vaccination/ immunization	Very high	+++	119,120
Micro/macronutrients	Medium	++	11
Vitamin A supplementation	Low	+	57,121
B Vitamins	High	+++	122–129
Vitamin C supplementation	Medium	++	130,131
Vitamin D supplementation	High	+++	57,132,133
Vitamin E supplementation	Medium	++	122,134
Vitamin K supplementation	Medium	++	135–137
Zinc supplementation	Low/medium	+	138
Iron supplementation	Medium	++	139
Selenium supplementation	Low/medium	+	140,141
Melatonin supplementation	Medium	++	142
Magnesium supplementation	Low	+	143
Omega-3 fatty acid supplementation	High	+++	41,42
Copper supplementation	High	+	139

One of these preventive approaches is hand washing, which is known to lower various infectious disease rates, including COVID-19, influenza/seasonal flus, whooping cough, and the common cold caused by viruses/pathogens.[118] It has been reported that a substantial amount of hospital-acquired illnesses can be mitigated through the proper hand washing of healthcare workers.[144] While hand washing is effective, the

COVID-19 virus spreads through aerosol droplets from infected individuals reaching the nose, eyes, and mouth after being expelled through coughing or sneezing. Consequently, the use of face masks has been employed to prevent the spread of COVID-19.[114–116] Another implemented protective measure for COVID-19 is physical and social distancing. Studies of viral infections such as influenza have reported that physical distancing of at least 1 m is effective at limiting the spread of disease, a scenario certainly influential for protection against COVID-19 infections.[116] Additionally, a number of agents (with pharmacological effects) have been postulated to play fundamental roles in the prophylaxis and/or improvement of COVID-19 associated symptoms. These agents display antioxidative (e.g., vitamin C, trans-resveratrol, kale, and pecans), anti-inflammatory (e.g., miodesin, berries, nuts, and curcumin), and immunomodulatory with either endogenous (e.g., hormones, cytokines, and growth factors) or exogenous (e.g., nutritional supplements such as vitamins, zinc, selenium, and spirulina) effects in preventing and/or ameliorating the severity of COVID-19 allied infections and complications.[142,145] Many of these compounds, including vitamins, phytochemicals, and nutraceuticals, strengthen the immune system for defending against invading pathogens (Table 1.2). Additionally, melatonin, a bioactive compound with many health benefits, along with anti-inflammatory, antioxidative, and immunomodulatory properties, has been reported to regress/limit severe symptoms and complications in COVID-19 patients.[11,142]

The most effective preventive measure for COVID-19 and other contagious diseases is vaccination/immunization, which essentially prepares and boosts the immune system, especially in a susceptible population.[119,120,146] Vaccines allow the preemptive development of memory B- and T-cells to neutralize pathogens. However, the efficacies of vaccines are dependent on many factors, including age, immune response disorders, and underlying medical conditions.

Three vaccines that are currently available include those developed by Pfizer-BioNTech, Moderna, and Johnson and Johnson in the United States.[147] These vaccines are highly efficacious, at ages 12 years and older, and are capable of reducing the severity of COVID-19-associated complications and hospitalizations. Of note, the Pfizer-BioNTech vaccine is available to individuals of age 5 years and above currently.

7. Potential therapies for the treatment of COVID-19

Certain treatment plans have been urgently approved and implemented to manage the severity of COVID-19-associated complications, hospitalizations, and mortalities. Since respiratory complications are commonly observed in COVID patients, the first line of intervention involves artificial oxygen delivery via multiple methods, including noninvasive positive pressure ventilation, intubation and invasive mechanical ventilation, and extracorporeal membrane oxygenation.[6,23,148] Generally, severe COVID-19 infections are associated with excessive and persistent inflammation in the lungs and other tissues, causing multiorgan failure and death. Thus, anti-inflammatory interventions have been introduced in combating COVID-19 complications utilizing different antagonists.[7] It has been reported that the introduction of mesenchymal stem cells containing the ACE2 receptors into COVID-19 patients suffering from critical illnesses reduces the inflammatory response and disease progression.[149] Small interfering RNA is another treatment option against COVID-19 patients by introducing RNA molecules that regulate viral gene expression and inhibit replications.[150]

The COVID-19 virus uses a number of enzymes, including RNA polymerase, proteases, methyltransferase, and exoribonuclease, for its replication, and this process aligns with SARS and MERS viruses that have been extensively studied.[151] Accordingly, several antiviral and antiretroviral drugs being used to combat COVID-19-related complications are based upon other virus treatment regimens.[152–154] These drugs include Ribavirin (Tribavirin), Ritonavir (Lopinavir/Norvir), Remdesivir (Veklury), Nelfinavir (Viracept), Umifenovir (Arbidol), and chloroquine/hydroxychloroquine (Table 1.3), which have been used to treat COVID-19 patients with varying degrees of effectiveness.[152,178] Among these drugs, both Nelfinavir and Lopinavir have worked effectively against previously known SARS and MERS viruses; however, the results are neither satisfactory nor conclusive with COVID-19.[178,195,196] Inversely, both remdesivir and chloroquine had received EUA by the FDA for treatment of COVID-19 patients, and these drugs, especially Remdesivir, were found to effectively diminish COVID-19-related complications in certain age groups.[175,197] Similarly, umifenovir was reported to have moderate effects in the management of COVID-19 severity.[180] Recently, Merck and Co., in collaboration with Ridgeback Therapeutics, has developed an antiviral oral drug, named molnupiravir, for the treatment of COVID-19 patients, which shows promising effects against infections and reduces hospitalizations by 50%. It should be noted that the European Medicines Agency has issued emergency authorization of Molnupiravir (Lagevrio or MK4482) for adult COVID-19 patients suffering with increased complications and illnesses.[182,183] In addition, a phase two trial of molnupiravir is currently underway with an estimated completion date of May 5, 2022 (https://clinicaltrials.gov/ct2/show/NCT04575597). Pfizer Inc. has also recently developed an antiviral oral drug named paxlovid (nirmatrelvir/ritonavir), and this drug therapy (with a low dose of Ritonavir that is used in treating HIV) has been reported to reduce the risk of hospitalizations and mortalities by 89% when administered within 3 days of the first symptom(s).[184,185]

A number of monoclonal antibodies have also been used for the treatment of COVID-19 patients (Table 1.3). These antibodies attach to the spike protein of COVID-19 and limit its ability to bind the ACE2 receptors and subsequent replication. The FDA has approved an EUA for antibody-based treatments generated by different pharmaceutical companies, i.e., bamlanivimab + estesevimab (Eli Lilly and Co.), casirivimab + imdevimab (Regeneron), sotrovimab (GlaxoSmithKline plc), and tixagevimab + cilgavimab (Astrazena), respectively.[186,187,189] A combination of the latter antibodies, called Evusheld, has recently been approved in the United Kingdom for

Table 1.3 Potential therapeutic interventions for the treatment of COVID-19 and its variants.

Therapeutic approaches	Common name/(company)	Effectiveness (+, low; + +, medium; + + +, high)	References
Artificial oxygenation	Multiple	++	23,148
Anti-inflammatories (TNF-α inhibitors)	Multiple	+	155,156
Immunomodulators (colchicine, fluvoxamine)	Multiple	++	157–159
IL-1 inhibitors (anakinra, canakinumab)	Anakinra-Kineret (Swedish Orphan Biovitrum), Canakinumab-Ilaris (Novartis), respectively	+	160,161
IL-6 inhibitors (sarliumab, tocilizumab	Sarliumab-Kevzara (regeneron), Tocilizumab-Actemra (Genentech), respectively	++	162–164
Janus kinase inhibitors (baricitinib, tofacitinib, and ruxolitinib)	Baricitinib-Olumiant (Eli Lilly), Tofacitinib-Xeljanz (Pfizer), and Ruxolitinib-Jakafi (Incyte), respectively	+	165–167
Corticosteroids (Dexamethasone, Budesonide, Ciclesonide)	Dexamethasone-multiple, Budesonide-Pulmicort respules (AstraZeneca), Ciclesonide-Alvesco (AstraZeneca), respectively	++	168–171
Mesenchymal stem cells	Multiple	+	149
Ribavirin (tribavirin)	Multiple	+	172
Ritonavir	Norvir (AbbVie)	++	173,174
Remdesivir	Veklury (Gilead)	+++	175,176
Nelfinavir	Viracept (multiple)	+	177
Lopinavir/Norvir	Kaletra, Aluvia (AbbVie)	+	178,179
Umifenovir	Arbidol (Pharmstandard)	++	180
Chloroquine/ hydroxychloroquine	Multiple	+	178,181
Molnupiravir	Lagevrio or MK4482 (Merck/Ridgeback)	+++	182,183

Nirmatrelvir/ritonavir	Paxlovid (Pfizer)	+++	184,185
Balanivimab + Estesvimab	(Eli Lilly)	+++	186,187
Casirivimab + Imdevimab	Regen-Cov (regeneron)	+++	186,188
Evusheld	Rixagevimab + Cilgavimab (Astrazeneca)	+++	189
Sotrovimab	Xevudy (GlaxoSmithKline, vir Biotechnology)	++	186,190
Siltuximab	Sylvant (Johnson and Johnson)	+	191,192
Small interfering RNAs	Multiple	++	193,194

COVID-19 prevention by the Medicines and Healthcare Products Regulatory Agency for immunocompromized people.[189] Treatment regimens for these antibodies and/or cocktails include intravenous infusion at the onset of infection to lower the viral load by limiting its initial replication process. Noteworthy, the Regeneron's antibody treatment has also received a sponsorship by the WHO to be used in people who are not developing natural immunity to COVID-19 and may be at a high risk for severe COVID-19-associated illnesses.

8. Summary and conclusions

COVID-19 is an ever-emerging multiorgan system disorder, which represents a serious health crisis all over the world. The manifestations of this disease include aberrant respiratory distress, hypoxia, lung injury, inflammation, and a cytokine storm.[7,8] Patients afflicted with COVID-19 display either mild to moderate, or critical, symptoms with severe complications involving deadly outcomes. Notably, COVID-19-associated morbidities and mortalities are relatively higher with geriatric populations and people with underlying medical conditions such as obesity, diabetes, kidney diseases, autoimmune, inherited diseases, cancers, and neurological disorders, in which the function of immune responses are strikingly impaired.[198] Regardless of various immunocompromized situations, men in comparison to women, are drastically more affected by severe COVID-19-associated complications, along with mortalities. As such, maintenance of a strengthened immune system is the primary and natural preventive measure for combating COVID-19 and its genomic variants.

Despite the significance of the immune system, technological advances have provided insights into the molecular events that facilitate a better understanding of COVID-19 pathogenesis. The pathophysiological analyses of COVID-19 have developed various therapeutic interventions for the management of this hostile disease. In accordance with this, a variety of measures have been implemented toward targeting the prevention (e.g., human-to-human transmission) and treatment (e.g., antiviral, antibodies, and others) of COVID-19. Even so, there is no dynamic measure currently available that can effectively prevent and/or treat COVID-19. As emergence of additional patient/clinical data, along with more discoveries, become available, the precise understanding of molecular mechanisms of COVID-19 will lead to develop novel therapeutic strategies in the prevention and treatment of this aggressive disease.

Acknowledgments

The authors would like to thank many co-workers, collaborators, and the studies of several research groups whose contributions helped in preparing this chapter. This work was supported in part by National Institutes of Health Grants AG042178, AG047812, NS105473, AG060767, AG069333, AG066347, and R41 AG060836 to PHR, and funds from the Department of Internal Medicine to PRM.

References

1. Liu YC, Kuo RL, Shih SR. COVID-19: the first documented coronavirus pandemic in history. *Biomed J*. 2020;43:328–333.
2. Berekaa MM. Insights into the COVID-19 pandemic: origin, pathogenesis, diagnosis, and therapeutic interventions. *Front Biosci*. 2021;13:117–139.
3. Sheervalilou R, Shirvaliloo M, Dadashzadeh N, et al. COVID-19 under spotlight: a close look at the origin, transmission, diagnosis, and treatment of the 2019-nCoV disease. *J Cell Physiol*. 2020;235:8873–8924.
4. Umakanthan S, Sahu P, Ranade AV, et al. Origin, transmission, diagnosis and management of coronavirus disease 2019 (COVID-19). *Postgrad Med*. 2020;96:753–758.
5. Yuki K, Fujiogi M, Koutsogiannaki S. COVID-19 pathophysiology: a review. *Clin Immunol*. 2020;215:108427.
6. Rando HM, MacLean AL, Lee AJ, et al. Pathogenesis, symptomatology, and transmission of SARS-CoV-2 through analysis of viral genomics and structure. *mSystems*. 2021;6: e0009521.
7. Ye Q, Wang B, Mao J. The pathogenesis and treatment of the 'Cytokine Storm' in COVID-19. *J Infect*. 2020;80:607–613.
8. Hu B, Huang S, Yin L. The cytokine storm and COVID-19. *J Med Virol*. 2021;93: 250–256.
9. Manna PR, Stetson CL, Slominski AT, Pruitt K. Role of the steroidogenic acute regulatory protein in health and disease. *Endocrine*. 2016;51:7–21.
10. Pecora F, Persico F, Argentiero A, Neglia C, Esposito S. The role of micronutrients in support of the immune response against viral infections. *Nutrients*. 2020;12.
11. Manna PR, Gray ZC, Reddy PH. Healthy immunity on preventive medicine for combating COVID-19. *Nutrients*. 2022;14(5).
12. Calder PC. Nutrition, immunity and COVID-19. *BMJ Nutr Prev Health*. 2020;3:74–92.
13. Jayawardena R, Misra A. Balanced diet is a major casualty in COVID-19. *Diabetes Metabol Syndr*. 2020;14:1085–1086.
14. Gorji A, Khaleghi Ghadiri M. Potential roles of micronutrient deficiency and immune system dysfunction in the coronavirus disease 2019 (COVID-19) pandemic. *Nutrition*. 2021;82:111047.
15. Kalra RS, Dhanjal JK, Meena AS, et al. COVID-19, neuropathology, and aging: SARS-CoV-2 neurological infection, mechanism, and associated complications. *Front Aging Neurosci*. 2021;13:662786.
16. Mainali S, Darsie ME. Neurologic and neuroscientific evidence in aged COVID-19 patients. *Front Aging Neurosci*. 2021;13:648662.
17. Domingo JL, Marques M, Rovira J. Influence of airborne transmission of SARS-CoV-2 on COVID-19 pandemic. A review. *Environ Res*. 2020;188:109861.
18. Xu L, Mao Y, Chen G. Risk factors for 2019 novel coronavirus disease (COVID-19) patients progressing to critical illness: a systematic review and meta-analysis. *Aging*. 2020; 12:12410–12421.
19. Rong Y, Wang F, Liu J, et al. Clinical characteristics and risk factors of mild-to-moderate COVID-19 patients with false-negative SARS-CoV-2 nucleic acid. *J Med Virol*. 2021;93: 448–455.
20. Yesudhas D, Srivastava A, Gromiha MM. COVID-19 outbreak: history, mechanism, transmission, structural studies and therapeutics. *Infection*. 2021;49:199–213.

21. Lu X, Zhang L, Du H, et al. SARS-CoV-2 infection in children. *N Engl J Med*. 2020;382: 1663–1665.
22. Wolff D, Nee S, Hickey NS, Marschollek M. Risk factors for Covid-19 severity and fatality: a structured literature review. *Infection*. 2021;49:15–28.
23. Boopathi S, Poma AB, Kolandaivel P. Novel 2019 coronavirus structure, mechanism of action, antiviral drug promises and rule out against its treatment. *J Biomol Struct Dyn*. 2021;39:3409–3418.
24. Mir T, Almas T, Kaur J, et al. Coronavirus disease 2019 (COVID-19): multisystem review of pathophysiology. *Ann Med Surg*. 2021;69:102745.
25. Alkhatib M, Svicher V, Salpini R, et al. SARS-CoV-2 variants and their relevant mutational profiles: update summer 2021. *Microbiol Spectr*. 2021:e0109621.
26. Zella D, Giovanetti M, Benedetti F, et al. The variants question: what is the problem? *J Med Virol*. 2021;93:6479–6485.
27. Galloway SE, Paul P, MacCannell DR, et al. Emergence of SARS-CoV-2 B.1.1.7 lineage—United States, December 29, 2020–January 12, 2021. *MMWR Morb Mortal Wkly Rep*. 2021;70:95–99.
28. Davies NG, Abbott S, Barnard RC, et al. Estimated transmissibility and impact of SARS-CoV-2 lineage B.1.1.7 in England. *Science*. 2021:372.
29. Tegally H, Wilkinson E, Giovanetti M, et al. Detection of a SARS-CoV-2 variant of concern in South Africa. *Nature*. 2021;592:438–443.
30. Weisblum Y, Schmidt F, Zhang F, et al. Escape from neutralizing antibodies by SARS-CoV-2 spike protein variants. *bioRxiv*. 2020. https://doi.org/10.1101/2020.07.21.214759.
31. Gutierrez B, Marquez S, Prado-Vivar B, et al. Genomic epidemiology of SARS-CoV-2 transmission lineages in Ecuador. *Virus Evol*. 2021;7(2):051.
32. Franco D, Gonzalez C, Abrego LE, et al. Early transmission dynamics, spread, and genomic characterization of SARS-CoV-2 in Panama. *Emerg Infect Dis*. 2021;27: 612–615.
33. Freitas ARR, Beckedorff OA, Cavalcanti LPG, et al. The emergence of novel SARS-CoV-2 variant P.1 in Amazonas (Brazil) was temporally associated with a change in the age and sex profile of COVID-19 mortality: a population based ecological study. *Lancet Reg Health Am*. 2021;1:100021.
34. Wang P, Casner RG, Nair MS, et al. Increased resistance of SARS-CoV-2 variant P.1 to antibody neutralization. *Cell Host Microbe*. e744. 2021;29:747–751.
35. Shiehzadegan S, Alaghemand N, Fox M, Venketaraman V. Analysis of the delta variant B.1.617.2 COVID-19. *Clin Pract*. 2021;11:778–784.
36. Jhun H, Park HY, Hisham Y, Song CS, Kim S. SARS-CoV-2 delta (B.1.617.2) variant: a unique T478K mutation in receptor binding motif (rbm) of spike gene. *Immune Netw*. 2021;21:e32.
37. Li Z, Nie K, Li K, et al. Genome characterization of the first outbreak of COVID-19 delta variant B.1.617.2—Guangzhou city, Guangdong province, China, May 2021. *China CDC Wkly*. 2021;3:587–589.
38. Ong SWX, Chiew CJ, Ang LW, et al. Clinical and virological features of SARS-CoV-2 variants of concern: a retrospective cohort study comparing B.1.1.7 (Alpha), B.1.315 (Beta), and B.1.617.2 (Delta). *Clin Infect Dis*. 2021;75:e1128–e1136. https://doi.org/10.1093/cid/ciab721.
39. Callaway E. *Heavily mutated Omicron variant puts scientists on alert*. *Nature*. 2021. Nov 25.
40. Baker N, Van Noorden R. Coronapod: everything we know about the new COVID variant. *Nature*. 2021. https://doi.org/10.1038/d41586-021-03562-8.

41. Gutierrez S, Svahn SL, Johansson ME. Effects of omega-3 fatty acids on immune cells. *Int J Mol Sci*. 2019;20.
42. Doaei S, Gholami S, Rastgoo S, et al. The effect of omega-3 fatty acid supplementation on clinical and biochemical parameters of critically ill patients with COVID-19: a randomized clinical trial. *J Transl Med*. 2021;19:128.
43. Akhtar S, Das JK, Ismail T, Wahid M, Saeed W, Bhutta ZA. Nutritional perspectives for the prevention and mitigation of COVID-19. *Nutr Rev*. 2021;79:289−300.
44. Parkin J, Cohen B. An overview of the immune system. *Lancet*. 2001;357:1777−1789.
45. Chaplin DD. Overview of the immune response. *J Allergy Clin Immunol*. 2010;125: S3−S23.
46. Hoebe K, Janssen E, Beutler B. The interface between innate and adaptive immunity. *Nat Immunol*. 2004;5:971−974.
47. Tomar N, De RK. A brief outline of the immune system. *Methods Mol Biol*. 2014;1184: 3−12.
48. Beutler B. Innate immunity: an overview. *Mol Immunol*. 2004;40:845−859.
49. Niyonsaba F, Kiatsurayanon C, Chieosilapatham P, Ogawa H. Friends or Foes? Host defense (antimicrobial) peptides and proteins in human skin diseases. *Exp Dermatol*. 2017; 26:989−998.
50. Idborg H, Oke V. Cytokines as Biomarkers in systemic Lupus Erythematosus: value for diagnosis and drug therapy. *Int J Mol Sci*. 2021;22.
51. Liew PX, Kubes P. The neutrophil's role during health and disease. *Physiol Rev*. 2019;99: 1223−1248.
52. Bonilla FA, Oettgen HC. Adaptive immunity. *J Allergy Clin Immunol*. 2010;125: S33−S40.
53. Dong C. Cytokine regulation and function in T cells. *Annu Rev Immunol*. 2021;39:51−76.
54. Seifert M, Kuppers R. Human memory B cells. *Leukemia*. 2016;30:2283−2292.
55. Hillion S, Arleevskaya MI, Blanco P, et al. The innate part of the adaptive immune system. *Clin Rev Allergy Immunol*. 2020;58:151−154.
56. Lian J, Yue Y, Yu W, Zhang Y. Immunosenescence: a key player in cancer development. *J Hematol Oncol*. 2020;13:151.
57. Galmes S, Serra F, Palou A. Current state of evidence: influence of nutritional and nutrigenetic factors on immunity in the COVID-19 pandemic framework. *Nutrients*. 2020; 12(9).
58. Dropulic LK, Lederman HM. Overview of infections in the immunocompromised host. *Microbiol Spectr*. 2016;4:1−50.
59. Dadson P, Tetteh CD, Rebelos E, Badeau RM, Moczulski D. Underlying kidney diseases and complications for COVID-19: a review. *Front Med*. 2020;7:600144.
60. Velavan TP, Pallerla SR, Ruter J, et al. Host genetic factors determining COVID-19 susceptibility and severity. *EBioMedicine*. 2021;72:103629.
61. Kompaniyets L, Agathis NT, Nelson JM, et al. Underlying medical conditions associated with severe COVID-19 illness among children. *JAMA Netw Open*. 2021;4:e2111182.
62. Chen Y, Klein SL, Garibaldi BT, et al. Aging in COVID-19: vulnerability, immunity and intervention. *Ageing Res Rev*. 2021;65:101205.
63. Ono K, Yamada M. Vitamin A and Alzheimer's disease. *Geriatr Gerontol Int*. 2012;12: 180−188.
64. Clegg A, Young J, Iliffe S, Rikkert MO, Rockwood K. Frailty in elderly people. *Lancet*. 2013;381:752−762.
65. Slominski A, Zbytek B, Nikolakis G, et al. Steroidogenesis in the skin: implications for local immune functions. *J Steroid Biochem Mol Biol*. 2013;137:107−123.

66. Manna PR, Stetson CL, Daugherty C, et al. Up-regulation of steroid biosynthesis by retinoid signaling: implications for aging. *Mech Ageing Dev*. 2015;150:74–82.
67. Beattie MC, Chen H, Fan J, Papadopoulos V, Miller P, Zirkin BR. Aging and luteinizing hormone effects on reactive oxygen species production and DNA damage in rat Leydig cells. *Biol Reprod*. 2013;88:100.
68. Vitale G, Salvioli S, Franceschi C. Oxidative stress and the ageing endocrine system. *Nat Rev Endocrinol*. 2013;9:228–240.
69. Kong Y, Trabucco SE, Zhang H. Oxidative stress, mitochondrial dysfunction and the mitochondria theory of aging. *Interdiscipl Top Gerontol*. 2014;39:86–107.
70. Babbar M, Basu S, Yang B, Croteau DL, Bohr VA. Mitophagy and DNA damage signaling in human aging. *Mech Ageing Dev*. 2020;186:111207.
71. Peckham H, de Gruijter NM, Raine C, et al. Male sex identified by global COVID-19 meta-analysis as a risk factor for death and ITU admission. *Nat Commun*. 2020;11:6317.
72. Lewis A, Duch R. Gender differences in perceived risk of COVID-19. *Soc Sci Q*. 2021; 102:2124–2133.
73. Rodrigues Siqueira I, Fochesatto C, da Silva Torres IL, Dalmaz C, Alexandre Netto C. Aging affects oxidative state in hippocampus, hypothalamus and adrenal glands of Wistar rats. *Life Sci*. 2005;78:271–278.
74. Manna PR, Sennoune SR, Martinez-Zaguilan R, Slominski AT, Pruitt K. Regulation of retinoid mediated cholesterol efflux involves liver X receptor activation in mouse macrophages. *Biochem Biophys Res Commun*. 2015;464:312–317.
75. Feldman EL, Savelieff MG, Hayek SS, Pennathur S, Kretzler M, Pop-Busui R. COVID-19 and diabetes: a collision and collusion of two diseases. *Diabetes*. 2020;69:2549–2565.
76. Muniyappa R, Wilkins KJ. Diabetes, obesity, and risk prediction of severe COVID-19. *J Clin Endocrinol Metab*. 2020:105.
77. Lim S, Bae JH, Kwon HS, Nauck MA. COVID-19 and diabetes mellitus: from pathophysiology to clinical management. *Nat Rev Endocrinol*. 2021;17:11–30.
78. Pal R, Bhansali A. COVID-19, diabetes mellitus and ACE2: the conundrum. *Diabetes Res Clin Pract*. 2020;162:108132.
79. Sabri S, Bourron O, Phan F, Nguyen LS. Interactions between diabetes and COVID-19: a narrative review. *World J Diabetes*. 2021;12:1674–1692.
80. Ganesan SK, Venkatratnam P, Mahendra J, Devarajan N. Increased mortality of COVID-19 infected diabetes patients: role of furin proteases. *Int J Obes*. 2020;44:2486–2488.
81. Belikina DV, Malysheva ES, Petrov AV, et al. COVID-19 in patients with diabetes: clinical course, metabolic status, inflammation, and coagulation disorder. *Sovrem Tekhnologii Med*. 2021;12:6–16.
82. Mahamat-Saleh Y, Fiolet T, Rebeaud ME, et al. Diabetes, hypertension, body mass index, smoking and COVID-19-related mortality: a systematic review and meta-analysis of observational studies. *BMJ Open*. 2021;11:e052777.
83. Hales CM, Carroll MD, Fryar CD, Ogden CL. *Prevalence of Obesity Among Adults and Youth: United States, 2015–2016*. NCHS Data Brief; 2017:1–8.
84. Sanchis-Gomar F, Lavie CJ, Mehra MR, Henry BM, Lippi G. Obesity and outcomes in COVID-19: when an epidemic and pandemic collide. *Mayo Clin Proc*. 2020;95: 1445–1453.
85. Srour B, Fezeu LK, Kesse-Guyot E, et al. Ultra-processed food intake and risk of cardiovascular disease: prospective cohort study (NutriNet-Sante). *BMJ*. 2019;365:l1451.
86. Martinez Steele E, Baraldi LG, Louzada ML, Moubarac JC, Mozaffarian D, Monteiro CA. Ultra-processed foods and added sugars in the US diet: evidence from a nationally representative cross-sectional study. *BMJ Open*. 2016;6:e009892.

87. Kane H, Lynch L. Innate immune control of adipose tissue homeostasis. *Trends Immunol.* 2019;40:857–872.
88. de Frel DL, Atsma DE, Pijl H, et al. The impact of obesity and lifestyle on the immune system and susceptibility to infections such as COVID-19. *Front Nutr.* 2020;7:597600.
89. Dixon AE, Peters U. The effect of obesity on lung function. *Expet Rev Respir Med.* 2018; 12:755–767.
90. Manna PR, Molehin D, Ahmed AU. Dysregulation of Aromatase in breast, Endometrial, and ovarian cancers: an overview of therapeutic strategies. *Prog Mol Biol Transl Sci.* 2016; 144:487–537.
91. Manna PR, Ahmed AU, Yang S, et al. Genomic profiling of the steroidogenic acute regulatory protein in breast cancer: in silico assessments and a mechanistic perspective. *Cancers.* 2019;11.
92. Liang W, Guan W, Chen R, et al. Cancer patients in SARS-CoV-2 infection: a nationwide analysis in China. *Lancet Oncol.* 2020;21:335–337.
93. Dai M, Liu D, Liu M, et al. Patients with cancer appear more vulnerable to SARS-CoV-2: a multicenter study during the COVID-19 outbreak. *Cancer Discov.* 2020;10:783–791.
94. Sarfati D, Koczwara B, Jackson C. The impact of comorbidity on cancer and its treatment. *CA Cancer J Clin.* 2016;66:337–350.
95. Gosain R, Abdou Y, Singh A, Rana N, Puzanov I, Ernstoff MS. COVID-19 and cancer: a comprehensive review. *Curr Oncol Rep.* 2020;22:53.
96. Han HJ, Nwagwu C, Anyim O, Ekweremadu C, Kim S. COVID-19 and cancer: from basic mechanisms to vaccine development using nanotechnology. *Int Immunopharm.* 2021;90: 107247.
97. du Plessis M, Fourie C, Riedemann J, de Villiers WJS, Engelbrecht AM. Cancer and COVID-19: collectively catastrophic. *Cytokine Growth Factor Rev.* 2022;63:78–89.
98. Gottschalk G, Knox K, Roy A. ACE2: at the crossroad of COVID-19 and lung cancer. *Gene Rep.* 2021;23:101077.
99. Stewart CA, Gay CM, Ramkumar K, et al. Lung cancer models reveal severe acute respiratory syndrome coronavirus 2-induced epithelial-to-mesenchymal transition contributes to coronavirus disease 2019 pathophysiology. *J Thorac Oncol.* 2021;16:1821–1839.
100. Pugazhenthi S, Qin L, Reddy PH. Common neurodegenerative pathways in obesity, diabetes, and Alzheimer's disease. *Biochim Biophys Acta Mol Basis Dis.* 2017;1863: 1037–1045.
101. Fotuhi M, Mian A, Meysami S, Raji CA. Neurobiology of COVID-19. *J Alzheimers Dis.* 2020;76:3–19.
102. Dewanjee S, Vallamkondu J, Kalra RS, Puvvada N, Kandimalla R, Reddy PH. Emerging COVID-19 neurological manifestations: present outlook and potential neurological challenges in COVID-19 pandemic. *Mol Neurobiol.* 2021;58:4694–4715.
103. Mandik F, Vos M. Neurodegenerative disorders: spotlight on sphingolipids. *Int J Mol Sci.* 2021;22.
104. Raz L, Knoefel J, Bhaskar K. The neuropathology and cerebrovascular mechanisms of dementia. *J Cerebr Blood Flow Metabol.* 2016;36(1):172–186.
105. Khan H, Rafiq A, Shabaneh O, Gittner LS, Reddy PH. Current issues in chronic diseases: a focus on dementia and hypertension in rural west Texans. *J Alzheimers Dis.* 2019;72(s1): S59–S69.
106. Reddy PH, Yin X, Manczak M, et al. Mutant APP and amyloid beta-induced defective autophagy, mitophagy, mitochondrial structural and functional changes and synaptic damage in hippocampal neurons from Alzheimer's disease. *Hum Mol Genet.* 2018;27: 2502–2516.

107. Amakiri N, Kubosumi A, Tran J, Reddy PH. Amyloid beta and MicroRNAs in Alzheimer's disease. *Front Neurosci*. 2019;13:430.
108. Lim KH, Yang S, Kim SH, Joo JY. Elevation of ACE2 as a SARS-CoV-2 entry receptor gene expression in Alzheimer's disease. *J Infect*. 2020;81:e33–e34.
109. Cojocaru IM, Cojocaru M, Miu G, Sapira V. Study of interleukin-6 production in Alzheimer's disease. *Rom J Intern Med*. 2011;49:55–58.
110. Chen X, Zhao B, Qu Y, et al. Detectable serum severe acute respiratory syndrome coronavirus 2 viral load (RNAemia) is closely correlated with drastically Elevated interleukin 6 level in critically ill patients with coronavirus disease 2019. *Clin Infect Dis*. 2020; 71:1937–1942.
111. Ijaz MK, Nims RW, de Szalay S, Rubino JR. Soap, water, and severe acute respiratory syndrome coronavirus 2 (SARS-CoV-2): an ancient handwashing strategy for preventing dissemination of a novel virus. *PeerJ*. 2021;9:e12041.
112. Freeman MC, Stocks ME, Cumming O, et al. Hygiene and health: systematic review of handwashing practices worldwide and update of health effects. *Trop Med Int Health*. 2014; 19:906–916.
113. Veys K, Dockx K, Van Remoortel H, Vandekerckhove P, De Buck E. The effect of hand hygiene promotion programs during epidemics and pandemics of respiratory droplet-transmissible infections on health outcomes: a rapid systematic review. *BMC Publ Health*. 2021;21:1745.
114. Wang Y, Tian H, Zhang L, et al. Reduction of secondary transmission of SARS-CoV-2 in households by face mask use, disinfection and social distancing: a cohort study in Beijing, China. *BMJ Glob Health*. 2020;5.
115. Chazelet S, Pacault S. Efficiency of community face coverings and surgical masks to limit the spread of aerosol. *Ann Work Expo Health*. 2021;66:495–509. https://doi.org/10.1093/annweh/wxab089.
116. Chu D, Wei L. Systematic analysis reveals cis and trans determinants affecting C-to-U RNA editing in *Arabidopsis thaliana*. *BMC Genet*. 2020;21:98.
117. Girum T, Lentiro K, Geremew M, Migora B, Shewamare S, Shimbre MS. Optimal strategies for COVID-19 prevention from global evidence achieved through social distancing, stay at home, travel restriction and lockdown: a systematic review. *Arch Publ Health*. 2021;79:150.
118. Rabie T, Curtis V. Handwashing and risk of respiratory infections: a quantitative systematic review. *Trop Med Int Health*. 2006;11:258–267.
119. Roncati L, Vadala M, Corazzari V, Palmieri B. COVID-19 vaccine and boosted immunity: nothing ad interim to do? *Vaccine*. 2020;38:7581–7584.
120. Scurr MJ, Zelek WM, Lippiatt G, et al. Whole blood-based measurement of SARS-CoV-2-specific T cells reveals asymptomatic infection and vaccine immunogenicity in healthy subjects and patients with solid organ cancers. *Immunology*. 2021;165:250–259.
121. Lee GY, Han SN. Direct-to-Consumer genetic testing in Korea: current status and significance in clinical nutrition. *Clin Nutr Res*. 2021;10:279–291.
122. Shakoor H, Feehan J, Al Dhaheri AS, et al. Immune-boosting role of vitamins D, C, E, zinc, selenium and omega-3 fatty acids: could they help against COVID-19? *Maturitas*. 2021;143:1–9.
123. Gao J, Zhang L, Liu X, et al. Repurposing low-molecular-weight drugs against the main protease of severe acute respiratory syndrome coronavirus 2. *J Phys Chem Lett*. 2020;11: 7267–7272.

124. BourBour F, Mirzaei Dahka S, Gholamalizadeh M, et al. Nutrients in prevention, treatment, and management of viral infections; special focus on Coronavirus. *Arch Physiol Biochem*. 2020:1–10.
125. Liu S, Zhu X, Qiu Y, et al. Effect of niacin on growth performance, intestinal morphology, mucosal immunity and microbiota composition in weaned piglets. *Animals*. 2021:11.
126. Tardy AL, Pouteau E, Marquez D, Yilmaz C, Scholey A. Vitamins and minerals for energy, fatigue and cognition: a narrative review of the biochemical and clinical evidence. *Nutrients*. 2020;12.
127. Tang J, Feng Y, Zhang B, et al. Severe pantothenic acid deficiency induces alterations in the intestinal mucosal proteome of starter Pekin ducks. *BMC Genom*. 2021;22(1):491.
128. Stach K, Stach W, Augoff K. Vitamin B6 in health and disease. *Nutrients*. 2021;13.
129. Shakeri H, Azimian A, Ghasemzadeh-Moghaddam H, et al. Evaluation of the relationship between serum levels of zinc, vitamin B12, vitamin D, and clinical outcomes in patients with COVID-19. *J Med Virol*. 2022;94:141–146.
130. Bae M, Kim H. Mini-Review on the roles of vitamin C, vitamin D, and selenium in the immune system against COVID-19. *Molecules*. 2020;25.
131. Rawat D, Roy A, Maitra S, Gulati A, Khanna P, Baidya DK. Vitamin C and COVID-19 treatment: a systematic review and meta-analysis of randomized controlled trials. *Diabetes Metabol Syndr*. 2021;15:102324.
132. Rawat D, Roy A, Maitra S, Shankar V, Khanna P, Baidya DK. Vitamin D supplementation and COVID-19 treatment: a systematic review and meta-analysis. *Diabetes Metabol Syndr*. 2021;15:102189.
133. Taha R, Abureesh S, Alghamdi S, et al. The relationship between vitamin D and infections including COVID-19: any hopes? *Int J Gen Med*. 2021;14:3849–3870.
134. Iddir M, Brito A, Dingeo G, et al. Strengthening the immune system and reducing inflammation and oxidative stress through diet and nutrition: considerations during the COVID-19 crisis. *Nutrients*. 2020:12.
135. Shioi A, Morioka T, Shoji T, Emoto M. The inhibitory roles of vitamin K in progression of Vascular calcification. *Nutrients*. 2020;12.
136. Anastasi E, Ialongo C, Labriola R, Ferraguti G, Lucarelli M, Angeloni A. Vitamin K deficiency and covid-19. *Scand J Clin Lab Invest*. 2020;80:525–527.
137. Vogrig A, Gigli GL, Bna C, Morassi M. Stroke in patients with COVID-19: clinical and neuroimaging characteristics. *Neurosci Lett*. 2021;743:135564.
138. Pal A, Squitti R, Picozza M, et al. Zinc and COVID-19: basis of current clinical trials. *Biol Trace Elem Res*. 2021;199:2882–2892.
139. Taneri PE, Gomez-Ochoa SA, Llanaj E, et al. Anemia and iron metabolism in COVID-19: a systematic review and meta-analysis. *Eur J Epidemiol*. 2020;35:763–773.
140. Steinbrenner H, Al-Quraishy S, Dkhil MA, Wunderlich F, Sies H. Dietary selenium in adjuvant therapy of viral and bacterial infections. *Adv Nutr*. 2015;6:73–82.
141. Avery JC, Hoffmann PR. Selenium, selenoproteins, and immunity. *Nutrients*. 2018;10(9).
142. Zhang R, Wang X, Ni L, et al. COVID-19: melatonin as a potential adjuvant treatment. *Life Sci*. 2020;250:117583.
143. Cooper ID, Crofts CAP, DiNicolantonio JJ, et al. Relationships between hyperinsulinaemia, magnesium, vitamin D, thrombosis and COVID-19: rationale for clinical management. *Open Heart*. 2020;7.
144. Mathur P, Jain N, Gupta A, Gunjiyal J, Nair S, Misra MC. Hand hygiene in developing nations: experience at a busy level-1 trauma center in India. *Am J Infect Control*. 2011;39: 705–706.

145. Ferreira AO, Polonini HC, Dijkers ECF. Postulated adjuvant therapeutic strategies for COVID-19. *J Personalized Med*. 2020:10.
146. Tukhvatulin AI, Dolzhikova IV, Shcheblyakov DV, et al. An open, non-randomised, phase 1/2 trial on the safety, tolerability, and immunogenicity of single-dose vaccine "Sputnik Light" for prevention of coronavirus infection in healthy adults. *Lancet Reg Health Eur*. 2021;11:100241.
147. Singh A, Khillan R, Mishra Y, Khurana S. The safety profile of COVID-19 vaccinations in the United States. *Am J Infect Control*. 2022;50:15−19.
148. Hu Y, Liu L, Lu X. Regulation of angiotensin-converting enzyme 2: a potential target to prevent COVID-19? *Front Endocrinol*. 2021;12:725967.
149. Leng Z, Zhu R, Hou W, et al. Transplantation of ACE2(-) mesenchymal stem cells improves the outcome of patients with COVID-19 pneumonia. *Aging Dis*. 2020;11: 216−228.
150. Elekhnawy E, Kamar AA, Sonbol F. Present and future treatment strategies for coronavirus disease 2019. *Futur J Pharm Sci*. 2021;7:84.
151. Asselah T, Durantel D, Pasmant E, Lau G, Schinazi RF. COVID-19: discovery, diagnostics and drug development. *J Hepatol*. 2021;74:168−184.
152. Jin Y, Yang H, Ji W, et al. Virology, epidemiology, pathogenesis, and control of COVID-19. *Viruses*. 2020;12.
153. Kumar S, Saurabh MK, Maharshi V, Saikia D. A narrative review of antiviral drugs used for COVID-19 pharmacotherapy. *J Pharm Bioall Sci*. 2021;13:163−171.
154. Wahid B, Amir A, Ameen A, Idrees M. Current status of therapeutic approaches and vaccines for SARS-CoV-2. *Future Microbiol*. 2021;16:1319−1326. https://doi.org/10.2217/fmb-2020-0147.
155. Chen XY, Yan BX, Man XY. TNFalpha inhibitor may be effective for severe COVID-19: learning from toxic epidermal necrolysis. *Ther Adv Respir Dis*. 2020;14, 1753466620926800.
156. Guo Y, Hu K, Li Y, et al. Targeting TNF-alpha for COVID-19: recent advanced and controversies. *Front Public Health*. 2022;10:833967.
157. Tardif JC, Bouabdallaoui N, L'Allier PL, et al. Colchicine for community-treated patients with COVID-19 (COLCORONA): a phase 3, randomised, double-blinded, adaptive, placebo-controlled, multicentre trial. *Lancet Respir Med*. 2021;9:924−932.
158. Facente SN, Reiersen AM, Lenze EJ, Boulware DR, Klausner JD. Fluvoxamine for the early treatment of SARS-CoV-2 infection: a review of current evidence. *Drugs*. 2021;81: 2081−2089.
159. Bonaventura A, Vecchie A, Dagna L, Tangianu F, Abbate A, Dentali F. Colchicine for COVID-19: targeting NLRP3 inflammasome to blunt hyperinflammation. *Inflamm Res*. 2022;71:293−307.
160. Kyriazopoulou E, Huet T, Cavalli G, et al. Effect of anakinra on mortality in patients with COVID-19: a systematic review and patient-level meta-analysis. *Lancet Rheumatol*. 2021; 3:e690−e697.
161. Davidson M, Menon S, Chaimani A, et al. Interleukin-1 blocking agents for treating COVID-19. *Cochrane Database Syst Rev*. 2022;1:CD015308.
162. Liu D, Zhang T, Wang Y, Xia L. Tocilizumab: the key to stop coronavirus disease 2019 (COVID-19)-Induced cytokine release syndrome (CRS)? *Front Med*. 2020;7:571597.
163. Peng J, Fu M, Mei H, et al. Efficacy and secondary infection risk of tocilizumab, sarilumab and anakinra in COVID-19 patients: a systematic review and meta-analysis. *Rev Med Virol*. 2021;32:e2295.

164. van de Veerdonk FL, Giamarellos-Bourboulis E, Pickkers P, et al. A guide to immunotherapy for COVID-19. *Nat Med*. 2022;28:39–50.
165. Cao Y, Wei J, Zou L, et al. Ruxolitinib in treatment of severe coronavirus disease 2019 (COVID-19): a multicenter, single-blind, randomized controlled trial. *J Allergy Clin Immunol*. 2020;146:137–146. e133.
166. Saber-Ayad M, Hammoudeh S, Abu-Gharbieh E, et al. Current status of Baricitinib as a repurposed therapy for COVID-19. *Pharmaceuticals*. 2021:14.
167. Guimaraes PO, Quirk D, Furtado RH, et al. Tofacitinib in patients hospitalized with covid-19 pneumonia. *N Engl J Med*. July 29, 2021;385:406–415.
168. Stauffer WM, Alpern JD, Walker PF. COVID-19 and dexamethasone: a potential strategy to avoid steroid-related strongyloides hyperinfection. *JAMA*. 2020;324:623–624.
169. Yu LM, Bafadhel M, Dorward J, et al. Inhaled budesonide for COVID-19 in people at high risk of complications in the community in the UK (PRINCIPLE): a randomised, controlled, open-label, adaptive platform trial. *Lancet*. 2021;398:843–855.
170. Ramakrishnan S, Nicolau Jr DV, Langford B, et al. Inhaled budesonide in the treatment of early COVID-19 (STOIC): a phase 2, open-label, randomised controlled trial. *Lancet Respir Med*. 2021;9:763–772.
171. Clemency BM, Varughese R, Gonzalez-Rojas Y, et al. Efficacy of inhaled ciclesonide for outpatient treatment of adolescents and adults with symptomatic COVID-19: a randomized clinical trial. *JAMA Intern Med*. 2022;182:42–49.
172. Xu Y, Li M, Zhou L, et al. Ribavirin treatment for critically ill COVID-19 patients: an observational study. *Infect Drug Resist*. 2021;14:5287–5291.
173. Magro P, Zanella I, Pescarolo M, Castelli F, Quiros-Roldan E. Lopinavir/ritonavir: repurposing an old drug for HIV infection in COVID-19 treatment. *Biomed J*. 2021;44: 43–53.
174. Saravolatz LD, Depcinski S, Sharma M. Molnupiravir and nirmatrelvir-ritonavir: oral COVID antiviral drugs. *Clin Infect Dis*. 2022. https://doi.org/10.1093/cid/ciac180.
175. McCreary EK, Angus DC. Efficacy of Remdesivir in COVID-19. *JAMA*. 2020;324: 1041–1042.
176. Al Bishawi A, Abdel Hadi H, Elmekaty E, et al. Remdesivir for COVID-19 pneumonia in patients with severe chronic kidney disease: a Case series and review of the literature. *Clin Case Rep*. 2022;10:e05467.
177. Hosogaya N, Miyazaki T, Fukushige Y, et al. Efficacy and safety of nelfinavir in asymptomatic and mild COVID-19 patients: a structured summary of a study protocol for a multicenter, randomized controlled trial. *Trials*. 2021;22:309.
178. Cao B, Wang Y, Wen D, et al. A trial of lopinavir-ritonavir in adults hospitalized with severe COVID-19. *N Engl J Med*. 2020;382:1787–1799.
179. Vargas M, Servillo G, Einav S. Lopinavir/ritonavir for the treatment of SARS, MERS and COVID-19: a systematic review. *Eur Rev Med Pharmacol Sci*. 2020;24:8592–8605.
180. Alavi Darazam I, Shokouhi S, Mardani M, et al. Umifenovir in hospitalized moderate to severe COVID-19 patients: a randomized clinical trial. *Int Immunopharm*. 2021;99: 107969.
181. Singh B, Ryan H, Kredo T, Chaplin M, Fletcher T. Chloroquine or hydroxychloroquine for prevention and treatment of COVID-19. *Cochrane Database Syst Rev*. 2021;2: CD013587.
182. Whitley R. Molnupiravir—a step toward orally bioavailable therapies for COVID-19. *N Engl J Med*. 2022;386:592–593. https://doi.org/10.1056/NEJMe2117814.
183. Pourkarim F, Pourtaghi-Anvarian S, Rezaee H. Molnupiravir: a new candidate for COVID-19 treatment. *Pharmacol Res Perspect*. 2022;10:e00909.

184. Mahase E. Covid-19: pfizer's paxlovid is 89% effective in patients at risk of serious illness, company reports. *BMJ*. 2021;375:n2713.
185. Wang Z, Yang L. In the age of Omicron variant: paxlovid raises new hopes of COVID-19 recovery. *J Med Virol*. 2022. https://doi.org/10.1002/jmv.27540.
186. O'Brien MP, Forleo-Neto E, Musser BJ, et al. Subcutaneous REGEN-COV antibody combination to prevent COVID-19. *N Engl J Med*. 2021;385:1184−1195.
187. Cohen MS, Nirula A, Mulligan MJ, et al. Effect of bamlanivimab vs placebo on incidence of COVID-19 among residents and staff of skilled nursing and assisted Living facilities: a randomized clinical trial. *JAMA*. 2021;326:46−55.
188. Bierle DM, Ganesh R, Razonable RR. Breakthrough COVID-19 and casirivimab-imdevimab treatment during a SARS-CoV-2 B1.617.2 (Delta) surge. *J Clin Virol*. 2021; 145:105026.
189. Wise J. Covid-19: Evusheld is approved in UK for prophylaxis in immunocompromised people. *BMJ*. 2022;376:o722.
190. Origuen J, Caro-Teller JM, Lopez-Medrano F. Early treatment with Sotrovimab for COVID-19. *N Engl J Med*. 2022. https://doi.org/10.1056/NEJMc2201606.
191. Palanques-Pastor T, Lopez-Briz E, Poveda Andres JL. Involvement of interleukin 6 in SARS-CoV-2 infection: siltuximab as a therapeutic option against COVID-19. *Eur J Hosp Pharm*. 2020;27(5):297−298.
192. Villaescusa L, Zaragoza F, Gayo-Abeleira I, Zaragoza C. A new approach to the management of COVID-19. Antagonists of IL-6: siltuximab. *Adv Ther*. 2022;39:1126−1148.
193. Tolksdorf B, Nie C, Niemeyer D, et al. Inhibition of SARS-CoV-2 replication by a small interfering RNA targeting the leader sequence. *Viruses*. 2021;13.
194. Ambike S, Cheng CC, Feuerherd M, et al. Targeting genomic SARS-CoV-2 RNA with siRNAs allows efficient inhibition of viral replication and spread. *Nucleic Acids Res*. 2022; 50:333−349.
195. Yamamoto N, Yang R, Yoshinaka Y, et al. HIV protease inhibitor nelfinavir inhibits replication of SARS-associated coronavirus. *Biochem Biophys Res Commun*. 2004;318: 719−725.
196. Kim UJ, Won EJ, Kee SJ, Jung SI, Jang HC. Combination therapy with lopinavir/ritonavir, ribavirin and interferon-alpha for Middle East respiratory syndrome. *Antivir Ther*. 2016; 21:455−459.
197. Wang M, Cao R, Zhang L, et al. Remdesivir and chloroquine effectively inhibit the recently emerged novel coronavirus (2019-nCoV) in vitro. *Cell Res*. 2020;30:269−271.
198. Bohn MK, Hall A, Sepiashvili L, Jung B, Steele S, Adeli K. Pathophysiology of COVID-19: mechanisms underlying disease severity and progression. *Physiology*. 2020; 35:288−301.

Role of oxidative stress in the severity of SARS-COV-2 infection

2

Sharda P. Singh, Sanjay Awasthi, Ashly Hindle and Chhanda Bose
Department of Internal Medicine, Division of Hematology and Oncology, Texas Tech University Health Sciences Center, Lubbock, TX, United States

Abstract

Since the World Health Organization declared SARS-CoV-2 (COVID-19) a pandemic in March 2020, serious efforts have been made to understand the epidemiology, molecular mechanisms, pathology, and clinical evolution of this disease. Oxidative stress (OX-S) has been implicated in the etiologies of many diseases, including SARS-CoV-2. Recent studies suggest that superoxide radicals and the products of lipid peroxidation, such as the electrophilic aldehyde, 4-hydroxynonenal (4-HNE), are important mediators of the pathological effects of oxidative stress during microbial and viral infections. Numerous studies have confirmed that viral infections induce inflammatory responses that generate excessive amounts of reactive oxygen species and 4-HNE protein adducts in plasma and in various tissues, including alveolar epithelium and endothelium. In this book chapter, we will highlight and discuss the apparent and plausible relationships between SARS-CoV-2 virulence and oxidative stress/lipid peroxidation, which affect cellular and DNA repair mechanisms and immune response.

Keywords: 4-HNE; COVID-19; Inflammation; Lipid peroxidation; Oxidative stress; ROS; SARS-CoV-2

1. Oxidative stress and lipid peroxidation

The oxygen molecule (O_2) and water, the fully reduced form of oxygen, are fairly nonreactive. However, partially reduced forms of O_2, i.e., the superoxide radical anion ($^{\bullet}O_2^-$), hydrogen peroxide (H_2O_2), and the hydroxyl radical ($^{\bullet}OH$), collectively known as ROS, are highly reactive and biologically damaging. In most cells, ROS are primarily generated as a by-product of the mitochondrial respiratory chain. In contrast, phagocytic cells of the innate immune system produce ROS using a specialized enzyme, the NADPH oxidase.[1] The resulting ROS, directly or via triggering the release of proteases,[2] kills invading pathogens.

Because of their high reactivity, ROS are relatively nonselective and can target all cellular constituents. However, the reaction of ROS with polyunsaturated fatty acids is unique in that it results in a chain reaction,[3] which amplifies the original event (Fig. 2.1). The products of the process, lipid hydroperoxides, change the properties of membranes, act as oxidants, or can be converted[4–6] to a variety of α, β-unsaturated

COVID-19 in Alzheimer's Disease and Dementia. https://doi.org/10.1016/B978-0-443-15256-6.00016-7

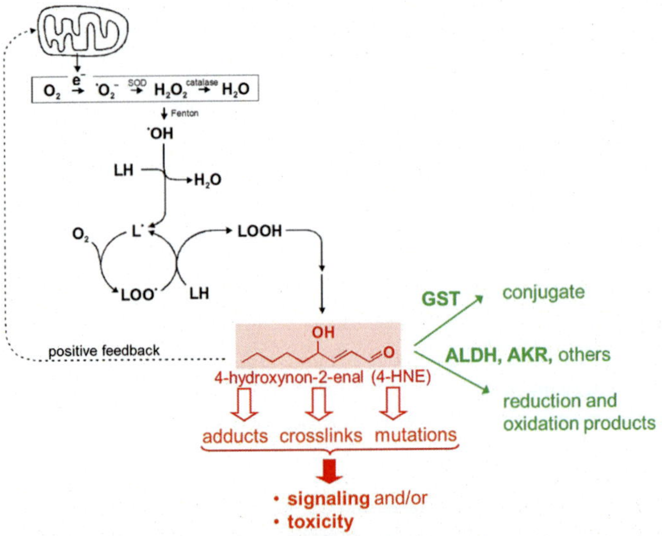

Figure 2.1 "Threshold theory" of electrophilic stress. Formation of 4-hydroxynonenal (HNE) requires reactive oxygen species (ROS) but is relatively insensitive to its level. The final concentration of electrophiles is primarily a function of the availability of suitably reactive PUFAs.

aldehydes of which 4-HNE is the most abundant and prototypical example. At physiological concentrations, 4-HNE is a signaling molecule, which modulates a variety of fundamental biological processes (reviewed in Refs. 7,8). The compound becomes toxic at higher levels due to its electrophilic character.[9]

1.1 Cellular origins and metabolism of ROS

The primary endogenous sources of OX-S are the mitochondria. Approximately 2% of oxygen consumed by mitochondria is converted to O_2^{-}•e.[10] Endogenous OX-S also arise from normal metabolism of lipids, carbohydrates, amino acids, and peptides. The lysosomes,[11–13] endoplasmic reticulum,[14,15] peroxisomes,[16,17] and plasma membranes[18,19] are also sites where normal oxidative metabolism generates oxygen- and nitrogen-free radicals. Numerous environmental exposures including viral and bacterial infections exacerbate endogenously generated OX-S. Though cigarette smoking represents the most common chronic environmental toxin exposure that causes continuing injury to pulmonary and vascular endothelium, a number of other environmental toxins are also implicated. These include heavy metal ions, polycyclic aromatic hydrocarbons (PAHs), oxidation of petroleum and other hydrocarbon fuels, polychlorinated biphenyls (PCBs) and other industrial chemicals, redox-cycling quinones, alkylating agents used in cancer chemotherapy and other medications, and cosmetic products. Although many of these poisons are not themselves electrophilic, their mono-oxygenation by the *P*450 enzymes results in formation of free-radicals, redox-cycling quinones, and electrophiles. Free radicals and electrophiles are also generated in large amounts upon exposure to high-energy radiant stress (heat, ultraviolet light, X-rays, γ-rays, or particle radiation).

2. Inflammatory stress

There is an inextricable relationship between OX-S and inflammation.[20–26] Inflammation occurs in response to a variety of physical, radiant, chemical, infectious, and immunological insults of varying severity. The first step in the inflammatory process is cellular injury caused by these insults that results in the release of a variety of intracellular constituents including proteins, lipoproteins, proteoglycans, peptides, phospholipids, glycolipids, nucleotides, complex carbohydrates, and metal ions. These intracellular constituents initiate and promote lipid peroxidation, a final common chemical manifestation of stress or injury. Lipid peroxidation yields a number of down-stream products (lipid hydroperoxy-radicals, reactive aldehydes, epoxides, leukotrienes, prostaglandins, prostacyclins, etc.) that exert potent local effects on leukocytes, blood vessels, and neurons in the vicinity. These lipid-peroxidation products are potent chemotactic simulators that recruit circulating leukocytes to the region of injury. This recruitment is aided by venoconstriction, arteriolar dilation, enhanced endothelial adhesiveness, and increased capillary permeability due to local effects of lipid-peroxidation products and other free-radicals. Stimulation of local leukocytes by these reactive chemical mediators releases potent cytokines and chemokines that exert local effects on capillary endothelium and systemic effects that promote leukocyte infiltration into the area of injury. The infiltrating leukocytes include granulocytes, lymphocytes, and monocytes in varying proportions depending on the etiology, nature, and severity of the injury. Release of lysosomal and peroxisomal contents at the site of injury exacerbates lipid peroxidation, causes endothelial injury and promotes further inflammation. Depending on the severity, extent or nature of the injury an acute or chronic inflammation can result. Injuries of greater severity usually initially cause acute inflammatory response, dominated by granulocytic infiltrates that can either resolve or transition to chronic inflammation dominated by lymphocytes, monocytes, and macrophages; both share the presence of varying degrees of OX-S. Insulin-resistance and its sequelae including diabetes mellitus, hypertension, dyslipidemia, and atherosclerosis are largely a result chronic inflammation.[27–31]

3. Quenching lipid peroxidation

Metal-binding proteins (ferritin, transferrin, etc.) act to prevent lipid peroxidation by reducing the ambient concentration of reduced iron and other heavy metals.[32] The chain reaction is terminated by free-radical scavengers and enzymatic antioxidants. Free radical-scavenging small molecules serve an important antioxidant function because they can accept unpaired electrons to become low-energy free-radicals due to delocalization of the unpaired electrons across conjugated double-bonds. Tocopherols (vitamins E) are the most important biological free-radical scavengers. They accept electrons from higher energy radicals and pass them to ascorbate (vitamin

C), which is then regenerated by reducing equivalents from GSH.[33,34] $OH^{\bullet}$ can be quenched by biological scavenging thiols that can accept the single electron; however, scavenger-derived radical species such as oxysulfur radicals may actually cause more damage.[35,36] Reactions of $OH^{\bullet}$ with other scavengers can also give rise to reactive free-radicals that can themselves cause damage.[37] Some free-radical scavengers such as mannitol or thiourea can also quench free-radical generation by complexing with metal ions.[32] Numerous naturally derived antioxidants containing conjugated double bonds are also excellent free-radical scavengers at low concentrations, but their ability to form free-radicals can render them damaging oxidants at high concentrations.[38,39]

Superoxide dismutase[40–42] and catalase[43] are enzymatic antioxidants that detoxify $O_2^{-}\bullet$ and H_2O_2, respectively. The glutathione-peroxidases (GPx) and glutathione S-transferases (GST) terminate lipid peroxidation by reducing the rate limiting metabolite, PUFA-OOH to the corresponding alcohol (PUFA-OH) coupled with GSH oxidation to GSH-disulfide (GSSG). GSSG is then recycled back to GSH by glutathione reductase (GR) using reducing equivalents that originate from the pentose phosphate shunt through glucose-6-phosphate dehydrogenase (G6PD) as well as from mitochondria through isocitrate dehydrogenase (IDH) activity.[39]

3.1 Oxidative stress and COVID-19

In the biological context, the broadest definition of stress is a potentially harmful challenge faced by a cell or organism that disturbs normal homeostasis and triggers a protective or adaptive response. The forms of stresses faced by humans are broadly characterized as psychological, infectious, mechanical, radiant, or chemical. Interestingly, all of these have been shown to generate OX-S, measurable as the peroxidation products of polyunsaturated fatty acids (PUFAs), the final common chemical manifestations of stress.[39,44–50]

COVID-19 primarily targets large areas of the lungs, which are the most oxygenated organ in the body. It is well documented that lung infections contribute to increased production of ROS due to oxidative stress and inflammation. COVID-19 infection triggers an inflammatory reaction, which releases proinflammatory cytokines, ROS, and lipid peroxidation end-products, causing metabolic and physiological alterations and various pathologies in the body; however, the majority of ROS formed in lung airways in COVID-19 infection appears to be due to the activity of NADPH oxidase of inappropriately recruited and activated immune cells. Oxidative stress due to the morbidities of aging,[23,51–55] smoking,[22,24,46,47,56–60] insulin-resistance,[55,61–68] and obesity[22,69–78] can further exacerbate the oxidative stress caused by SARS such as COVID-19.[79,80]

Increased biological markers of stress are closely correlated with biomarkers of inflammation.[20–26] The crucial role played by OX-S in the etiology of inflammation and the generation of oxidative stress during the inflammatory process indicate that these conditions propagate each other in a vicious cycle. Distinguishing their individual roles in promoting COVID-19 is particularly difficult because a crucial component of each is the chemical peroxidation of membrane PUFAs in a chain reaction called lipid peroxidation that is an integral part of oxidative as well as inflammatory stress.

A variety of cells relevant to COVID-19 can produce ROS. These are primarily inflammatory cells infiltrating the airways, such as neutrophils, eosinophils, and macrophages. In addition, ROS can be generated by airway structural cells such as epithelial, endothelial, and smooth muscle cells. Immunoglobulin E (IgE), which is typically elevated during COVID-19 infection, can bind to a receptor on monocytes and induce superoxide production.[81] One type of ROS, the superoxide radical anion, can react with nitric oxide (NO) and produce peroxynitrite, a compound thought to be more damaging than the parental ROS.[82] ROS and derived compounds, whether produced endogenously or inhaled, exacerbate COVID-19 pathology.

Multiple targets and affected processes have been identified (reviewed in Ref. 83). They include, among others, airway hyperresponsiveness, activation of inflammatory cells and cytokine release, goblet cell proliferation and mucus secretion, epithelial damage, and airway remodeling.[84,85] The molecular mechanisms by which ROS elicit these effects are only partially known but are thought to include protein oxidation (by ROS) and nitration (by peroxynitrite). In particular, modifications of histones and the histone acetylation/deacetylation system[86] have the capacity to alter gene expression patterns, which may lead to an increased production of proinflammatory cytokines. In addition, protein modification can directly affect function, e.g., peroxynitrite inhibits superoxide dismutase.[87] While the status of antioxidant defenses in COVID-19 is inconclusive, their possible impairment could additionally augment the pathology.[88] Conversely, the antioxidant supplements recommended in COVID-19 therapeutic strategies may ameliorate such effects.[89,90] While COVID-19 infection itself can generate ROS, the ROS produced by other causes can further exacerbate COVID-19 pathology. Therefore, there is a likely positive feedback, or vicious cycle, that may play a role in the self-sustaining characteristic of the disease.

4. 4-HNE in COVID-19

The pro-oxidant state prevalent in infected tissue would be expected to initiate lipid peroxidation and lead to the formation of 4-HNE and other lipid-derived α,β-unsaturated aldehydes (Fig. 2.2). Indeed, elevated levels of 4-HNE-adducts

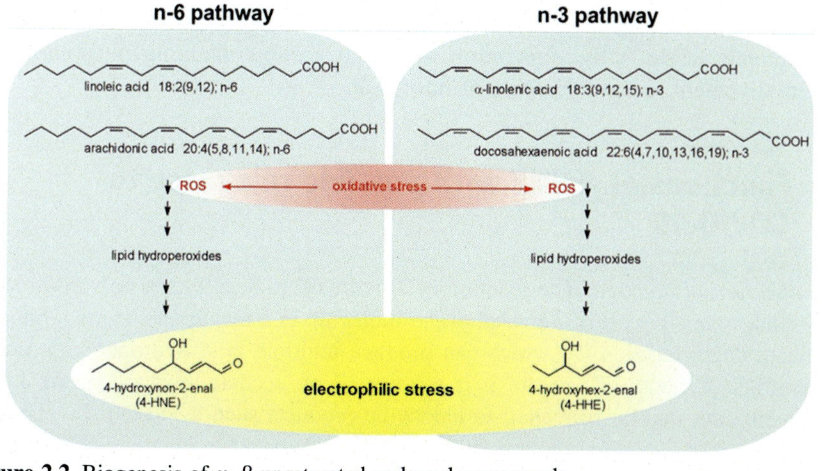

Figure 2.2 Biogenesis of α, β-unsaturated carbonyl compounds.

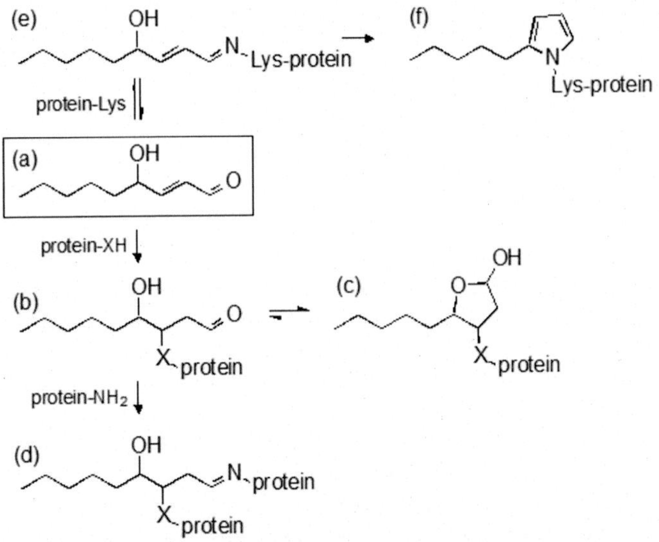

Figure 2.3 Structure of 4-hydroxynonenal (HNE) (A), and major types of protein adducts with 4-HNE: Michael adduct (B) in equilibrium with its hemiacetal (C); crosslink (D); schiff base (E) with resulting 2-pentylpyrrole adduct (F).

were observed in the blood of COVID-19 patients,[79] and protein carbonyl concentration correlated positively with the number of eosinophils, lymphocytes, and mast cells in bronchoalveolar lavage fluid from COVID-19 patients.[91] Increased levels of 4-HNE-modified proteins were found in the lungs of patients with another inflammatory lung disease, chronic obstructive pulmonary disease (COPD).[92] Both the modulation of protein function in 4-HNE-mediated signaling and the toxicity of the compound are thought to be largely mediated by the formation of Michael adducts with nucleophilic centers on proteins and by protein crosslinking[7] (Fig. 2.3). Modification of proteins with 4-HNE adds carbonyl groups (Fig. 2.3) and, along with directly oxidized proteins, can be detected as "protein carbonyls" a marker of oxidative stress. Other reactions of 4-HNE include adduct formation with nucleic acids,[7] the conjugation of amino group-containing phospholipids, and involvement in the formation of lipofuscin.[93,94]

5. Functions of 4-HNE with possible relevance to COVID-19

Amplification of damage: The reaction of a hydroxyl radical with a polyunsaturated fatty acid, typically part of a phospholipid molecule in a membrane, starts a radical-mediated chain reaction,[3,95] which can produce multiple lipid hydroperoxide molecules (Figs. 2.1 and 2.2) before it is terminated by a chain-breaking event. Lipid hydroperoxides can be converted to aldehydic products such as 4-HNE.[4–6] Thus, a single ROS molecule, which could otherwise modify a single site on a protein or nucleic acid, gives rise to multiple 4-HNE molecules, potentially amplifying the initial insult by two or three orders of magnitude.

Protein modification: As with most posttranslational modifications, adduct formation with 4-HNE modulates protein function. Depending on the protein, either a loss or a gain of function is observed. A large number of targets has been characterized (reviewed in Refs. 7,8,96–99). Examples include key regulatory proteins such as protein kinases, phospholipases, growth factor receptors, and ion transporters, among others. Modification of histones by 4-HNE[100] is of special interest as it could influence the expression of large sets of genes. In this way, 4-HNE could affect proteins with which it does not react directly. The effect of 4-HNE on cyclin expression[101] and on other signaling pathways[8,102,103] links the compound to the regulation of the cell cycle; in the human lung fibroblast cell line IMR-90, 4-HNE stimulates fibronectin production.[104] Both effects make 4-HNE directly relevant to COVID-19.

Cell proliferation: 4-HNE is usually considered to have cytostatic, proapoptotic, and/or prodifferentiating effects (e.g., Refs. 105,106). These conclusions are, however, based on studies of immortal transformed cell lines. In contrast, 4-HNE causes proliferation of aortic smooth muscle cells.[107,108] Given the parallels between arterial wall remodeling in atherosclerosis and airway remodeling in COVID-19, findings on the proproliferative action of 4-HNE on arterial smooth muscle cells may be relevant to the airway effects of COVID-19.

Chemotaxis: Various 4-hydroxyalkenals, including 4-HNE, are chemoattractants for neutrophils at micromolar and submicromolar concentrations (reviewed in Ref. 97); some are active at picomolar levels, significantly below the concentrations required for the compound's other functions. In vitro chemotaxis assays have been confirmed in vivo.[109] 4-HNE has also been shown to induce the secretion of MCP-1 by macrophages.[110] Monocyte chemoattractant protein-1, (MCP-1 Ref. 111) stimulates in target monocytes the synthesis of integrins, which are needed for the chemotactic mobility of the monocytes.[112,113] MCP-1 also acts as a chemoattractant for $CD4^+$ T-lymphocytes, which are involved in the pathogenesis of asthma.[114] Thus, 4-HNE appears to favor, directly or indirectly, the recruitment of several types of immune cells to the airways.

Effect on gene expression patterns. Examples of 4-HNE influencing the expression of specific genes are abundant. This function of 4-HNE has been shown by microarray analysis.[115] A comparison of gene expression patterns (obtained by microarray analysis, our unpublished results) in the lung of wild-type and *mGsta4* KO mice (which have an increased tissue level of 4-HNE) with a published microarray analysis of mRNAs modulated in COVID-19 experimental mouse models revealed significant similarities in the effects of the two interventions.

RALBP1 is a stress responsive enzyme of the mercapturic acid pathway that catalyzes the transmembrane efflux of glutathione (GSH)-electrophile thioether conjugates (GS-E) that are formed through the glutathione S-transferase (GST)-catalyzed conjugation of GSH with electrophilic toxins.[116–120] Among the efflux substrates of RALBP1 is the glutathione conjugate of 4-HNE (GS-HNE). While the MRP1 and MRP2 efflux pumps are also capable of transporting GS-HNE,[121] RALBP1 serves as the cell's low-affinity high-capacity efflux transporter.[122] RALBP1 may therefore be key in quickly clearing the large amounts of GS-HNE created during oxidative inflammatory conditions, and research into the role of RALBP1 in protecting against viral-induced lung damage is warranted.

References

1. Cross AR, Segal AW. The NADPH oxidase of professional phagocytes-prototype of the NOX electron transport chain systems. *Biochim Biophys Acta*. 2004;1657:1–22. PubMed PMID: 15238208.
2. Reeves EP, Lu H, Jacobs HL, et al. Killing activity of neutrophils is mediated through activation of proteases by K^+ flux. *Nature*. 2002;416:291–297. PubMed PMID: 11907569.
3. Gutteridge JMC, Halliwell B. The measurement and mechanism of lipid peroxidation in biological systems. *Trends Biochem Sci*. 1990;15:129–135.
4. Schneider C, Tallman KA, Porter NA, Brash AR. Two distinct pathways of formation of 4-hydroxynonenal: mechanisms of nonenzymatic transformation of the 9-and 13-hydroperoxides of linoleic acid to 4-hydroxyalkenals. *J Biol Chem*. 2001;276(24):20831–20838.
5. Schneider C, Porter NA, Brash AR. Autoxidative transformation of chiral w6 hydroxy linoleic and arachidonic acids to chiral 4-hydroxy-2*E*-nonenal. *Chem Res Toxicol*. 2004;17:937–941.
6. Sun M, Salomon RG. Oxidative fragmentation of hydroxy octadecadienoates generates biologically active g-hydroxyalkenals. *J Am Chem Soc*. 2004;126:5699–5708.
7. Petersen DR, Doorn JA. Reactions of 4-hydroxynonenal with proteins and cellular targets. *Free Radic Biol Med*. 2004;37(7):937–945.
8. Leonarduzzi G, Robbesyn F, Poli G. Signaling kinases modulated by 4-hydroxynonenal. *Free Radic Biol Med*. 2004;37(11):1694–1702. PubMed PMID: 15528028.
9. Esterbauer H, Schaur RJ, Zollner H. Chemistry and biochemistry of 4-hydroxynonenal, malonaldehyde and related aldehydes. *Free Radic Biol Med*. 1991;11:81–128.
10. Turrens JF. Mitochondrial formation of reactive oxygen species. *J Physiol*. 2003;552(Pt 2):335–344. https://doi.org/10.1113/jphysiol.2003.049478. PubMed PMID: PMC2343396.
11. Sun-Wada GH, Wada Y, Futai M. Lysosome and lysosome-related organelles responsible for specialized functions in higher organisms, with special emphasis on vacuolar-type proton ATPase. *Cell Struct Funct*. 2003;28(5):455–463. Epub 2004/01/28. PubMed PMID: 14745137.
12. Arai K, Kanaseki T, Ohkuma S. Isolation of highly purified lysosomes from rat liver: identification of electron carrier components on lysosomal membranes. *J Biochem*. 1991;110(4):541–547. Epub 1991/10/01. PubMed PMID: 1663946.
13. Gille L, Nohl H. The existence of a lysosomal redox chain and the role of ubiquinone. Epub 2000/03/04 *Arch Biochem Biophys*. 2000;375(2):347–354. https://doi.org/10.1006/abbi.1999.1649. PubMed PMID: 10700391.
14. Cheeseman KH, Slater TF. An introduction to free radical biochemistry. *Br Med Bull*. 1993;49(3):481–493. Epub 1993/07/01. PubMed PMID: 8221017.
15. Gross E, Sevier CS, Heldman N, et al. Generating disulfides enzymatically: reaction products and electron acceptors of the endoplasmic reticulum thiol oxidase Ero1p. Epub

2006/01/13 *Proc Natl Acad Sci U S A*. 2006;103(2):299–304. https://doi.org/10.1073/pnas.0506448103. PubMed PMID: 16407158; PMCID: PMC1326156.
16. Schrader M, Fahimi HD. Peroxisomes and oxidative stress. Epub 2006/10/13 *Biochim Biophys Acta*. 2006;1763(12):1755–1766. https://doi.org/10.1016/j.bbamcr.2006.09.006. PubMed PMID: 17034877.
17. De Duve C, Baudhuin P. Peroxisomes (microbodies and related particles). Epub 1966/04/01 *Physiol Rev*. 1966;46(2):323–357. https://doi.org/10.1152/physrev.1966.46.2.323. PubMed PMID: 5325972.
18. Cho K-J, Seo J-M, Kim J-H. Bioactive lipoxygenase metabolites stimulation of NADPH oxidases and reactive oxygen species. *Mol Cell*. 2011;32(1):1–5. https://doi.org/10.1007/s10059-011-1021-7.
19. Chanock SJ, el Benna J, Smith RM, Babior BM. The respiratory burst oxidase. *J Biol Chem*. 1994;269(40):24519–24522. Epub 1994/10/07. PubMed PMID: 7929117.
20. Barhoumi T, Briet M, Kasal DA, et al. Erythropoietin-induced hypertension and vascular injury in mice overexpressing human endothelin-1: exercise attenuated hypertension, oxidative stress, inflammation and immune response. Epub 2014/01/28 *J Hypertens*. 2014; 32(4):784–794. https://doi.org/10.1097/HJH.0000000000000101. PubMed PMID: 24463938.
21. Desideri G, Croce G, Tucci M, et al. Effects of bezafibrate and simvastatin on endothelial activation and lipid peroxidation in hypercholesterolemia: evidence of different vascular protection by different lipid-lowering treatments. Epub 2003/11/07 *J Clin Endocrinol Metab*. 2003;88(11):5341–5347. https://doi.org/10.1210/jc.2003-030724. PubMed PMID: 14602771.
22. Keaney Jr JF, Larson MG, Vasan RS, et al. Obesity and systemic oxidative stress: clinical correlates of oxidative stress in the Framingham Study. Epub 2003/03/05 *Arterioscler Thromb Vasc Biol*. 2003;23(3):434–439. https://doi.org/10.1161/01.ATV.0000058402.34138.11. PubMed PMID: 12615693.
23. Marchal J, Pifferi F, Aujard F. Resveratrol in mammals: effects on aging biomarkers, age-related diseases, and life span. Epub 2013/07/17 *Ann N Y Acad Sci*. 2013;1290:67–73. https://doi.org/10.1111/nyas.12214. PubMed PMID: 23855467.
24. Resch U, Tatzber F, Budinsky A, Sinzinger H. Reduction of oxidative stress and modulation of autoantibodies against modified low-density lipoprotein after rosuvastatin therapy. Epub 2006/02/21 *Br J Clin Pharmacol*. 2006;61(3):262–274. https://doi.org/10.1111/j.1365-2125.2005.02568.x. PubMed PMID: 16487219; PMCID: PMC1885020.
25. Shishehbor MH, Zhang R, Medina H, et al. Systemic elevations of free radical oxidation products of arachidonic acid are associated with angiographic evidence of coronary artery disease. *Free Radic Biol Med*. 2006;41(11):1678–1683. https://doi.org/10.1016/j.freeradbiomed.2006.09.001.
26. Singh U, Devaraj S, Jialal I, Siegel D. Comparison effect of atorvastatin (10 versus 80 mg) on biomarkers of inflammation and oxidative stress in subjects with metabolic syndrome. Epub 2008/07/22 *Am J Cardiol*. 2008;102(3):321–325. https://doi.org/10.1016/j.amjcard.2008.03.057. PubMed PMID: 18638594; PMCID: PMC2676172.
27. Schmidt MI, Duncan BB, Sharrett AR, et al. Markers of inflammation and prediction of diabetes mellitus in adults (Atherosclerosis Risk in Communities study): a cohort study. *Lancet*. 1999;353(9165):1649–1652. Epub 1999/05/21. PubMed PMID: 10335783.
28. Shoelson SE, Herrero L, Naaz A. Obesity, inflammation, and insulin resistance. *Gastroenterology*. 2007;132(6):2169–2180. https://doi.org/10.1053/j.gastro.2007.03.059.

29. Arkan MC, Hevener AL, Greten FR, et al. IKK-beta links inflammation to obesity-induced insulin resistance. Epub 2005/02/03 *Nat Med.* 2005;11(2):191–198. https://doi.org/10.1038/nm1185. PubMed PMID: 15685170.
30. Niskanen L, Laaksonen DE, Nyyssonen K, et al. Inflammation, abdominal obesity, and smoking as predictors of hypertension. Epub 2004/10/20 *Hypertension.* 2004;44(6): 859–865. https://doi.org/10.1161/01.HYP.0000146691.51307.84. PubMed PMID: 15492131.
31. Ross R. Atherosclerosis — an inflammatory disease. *N Engl J Med.* 1999;340(2): 115–126. https://doi.org/10.1056/nejm199901143400207. PubMed PMID: 9887164.
32. Halliwell B. Oxidants and human disease: some new concepts. *Faseb J.* 1987;1(5): 358–364. Epub 1987/11/01. PubMed PMID: 2824268.
33. Kittridge KJ, Willson RL. Uric acid substantially enhances the free radical-induced inactivation of alcohol dehydrogenase. *FEBS Lett.* 1984;170(1):162–164. Epub 1984/05/07. PubMed PMID: 6373370.
34. Packer JE, Slater TF, Willson RL. Direct observation of a free radical interaction between vitamin E and vitamin C. *Nature.* 1979;278(5706):737–738. Epub 1979/04/19. PubMed PMID: 431730.
35. Redpath JL, Willson RL. Reducing compounds in radioprotection and radiosensitization: model experiments using ascorbic acid. *Int J Radiat Biol Relat Stud Phys Chem Med.* 1973; 23(1):51–65. Epub 1973/01/01. PubMed PMID: 4567291.
36. Rowley DA, Halliwell B. Formation of hydroxyl radicals from hydrogen peroxide and iron salts by superoxide- and ascorbate-dependent mechanisms: relevance to the pathology of rheumatoid disease. *Clin Sci (Lond).* 1983;64(6):649–653. Epub 1983/06/01. PubMed PMID: 6301745.
37. Kalyanaraman B, Janzen EG, Mason RP. Spin trapping of the azidyl radical in azide/catalase/H2O2 and various azide/peroxidase/H2O2 peroxidizing systems. *J Biol Chem.* 1985;260(7):4003–4006. Epub 1985/04/10. PubMed PMID: 2984193.
38. Laughton MJ, Halliwell B, Evans PJ, Hoult JR. Antioxidant and pro-oxidant actions of the plant phenolics quercetin, gossypol and myricetin. Effects on lipid peroxidation, hydroxyl radical generation and bleomycin-dependent damage to DNA. *Biochem Pharmacol.* 1989; 38(17):2859–2865. Epub 1989/09/01. PubMed PMID: 2476132.
39. Awasthi YC, Singhal SS, Awasthi S. *Mechanisms of Anti-carcinogenic Effects of Antioxidant Nutrients. Nutrition and Cancer Prevention.* Boca Raton, FL: CRC Press; 1996: 139–172.
40. Farr SB, D'Ari R, Touati D. Oxygen-dependent mutagenesis in *Escherichia coli* lacking superoxide dismutase. *Proc Natl Acad Sci U S A.* 1986;83(21):8268–8272. Epub 1986/11/01. PubMed PMID: 3022287; PMCID: PMC386909.
41. Getzoff ED, Tainer JA, Weiner PK, Kollman PA, Richardson JS, Richardson DC. Electrostatic recognition between superoxide and copper, zinc superoxide dismutase. *Nature.* 1983;306(5940):287–290. Epub 1983/11/17. PubMed PMID: 6646211.
42. Willson RL. From nitric oxide to desferal: nitrogen free radicals and iron in oxidative injury. *Basic Life Sci.* 1988;49:87–99. Epub 1988/01/01. PubMed PMID: 3074798.
43. Gordon T. Purity of catalase preparations: contamination by endotoxin and its role in the inhibition of airway inflammation. *J Free Radic Biol Med.* 1986;2(5–6):373–375. Epub 1986/01/01. PubMed PMID: 3036930.
44. Aoi W, Naito Y, Yoshikawa T. Role of oxidative stress in impaired insulin signaling associated with exercise-induced muscle damage. *Free Radic Biol Med.* 2013;65: 1265–1272. https://doi.org/10.1016/j.freeradbiomed.2013.09.014.

45. Cheng JZ, Sharma R, Yang Y, et al. Accelerated metabolism and exclusion of 4-hydroxynonenal through induction of RLIP76 and hGST5.8 is an early adaptive response of cells to heat and oxidative stress. *J Biol Chem.* 2001;276(44):41213–41223. https://doi.org/10.1074/jbc.m106838200.
46. Liu J, Liang Q, Frost-Pineda K, et al. Relationship between biomarkers of cigarette smoke exposure and biomarkers of inflammation, oxidative stress, and platelet activation in adult cigarette smokers. Epub 2011/06/29 *Cancer Epidemiol Biomarkers Prev.* 2011;20(8): 1760–1769. https://doi.org/10.1158/1055-9965.EPI-10-0987. PubMed PMID: 21708936.
47. Miri R, Saadati H, Ardi P, Firuzi O. Alterations in oxidative stress biomarkers associated with mild hyperlipidemia and smoking. Epub 2012/01/10 *Food Chem Toxicol.* 2012; 50(3–4):920–926. https://doi.org/10.1016/j.fct.2011.12.031. PubMed PMID: 22227215.
48. Ruperez M, Lorenzo O, Blanco-Colio LM, Esteban V, Egido J, Ruiz-Ortega M. Connective tissue growth factor is a mediator of angiotensin II-induced fibrosis. Epub 2003/09/04 *Circulation.* 2003;108(12):1499–1505. https://doi.org/10.1161/01.CIR.0000089129.51288.BA. PubMed PMID: 12952842.
49. Tamashiro KL, Sakai RR, Shively CA, Karatsoreos IN, Reagan LP. Chronic stress, metabolism, and metabolic syndrome. *Stress.* 2011;14(5):468–474. https://doi.org/10.3109/10253890.2011.606341.
50. Miller E, Morel A, Saso L, Saluk J. Isoprostanes and neuroprostanes as biomarkers of oxidative stress in neurodegenerative diseases. Epub 2014/05/29 *Oxid Med Cell Longev.* 2014;2014:572491. https://doi.org/10.1155/2014/572491. PubMed PMID: 24868314; PMCID: PMC4020162.
51. Barja G. Updating the mitochondrial free radical theory of aging: an integrated view, key aspects, and confounding concepts. Epub 2013/05/07 *Antioxidants Redox Signal.* 2013; 19(12):1420–1445. https://doi.org/10.1089/ars.2012.5148. PubMed PMID: 23642158; PMCID: PMC3791058.
52. Beckman KB, Ames BN. The free radical theory of aging matures. Epub 1998/04/30 *Physiol Rev.* 1998;78(2):547–581. https://doi.org/10.1152/physrev.1998.78.2.547. PubMed PMID: 9562038.
53. Holvoet P, Harris TB, Tracy RP, et al. Association of high coronary heart disease risk status with circulating oxidized LDL in the well-functioning elderly: findings from the Health, Aging, and Body Composition study. Epub 2003/06/07 *Arterioscler Thromb Vasc Biol.* 2003;23(8):1444–1448. https://doi.org/10.1161/01.ATV.0000080379.05071.22. PubMed PMID: 12791672.
54. Hong SH, Kwak JH, Paik JK, Chae JS, Lee JH. Association of polymorphisms in FADS gene with age-related changes in serum phospholipid polyunsaturated fatty acids and oxidative stress markers in middle-aged nonobese men. Epub 2013/07/03 *Clin Interv Aging.* 2013;8:585–596. https://doi.org/10.2147/CIA.S42096. PubMed PMID: 2381 8766; PMCID: PMC3693593.
55. Nguyen D, Samson SL, Reddy VT, Gonzalez EV, Sekhar RV. Impaired mitochondrial fatty acid oxidation and insulin resistance in aging: novel protective role of glutathione. *Aging Cell.* 2013;12(3):415–425. https://doi.org/10.1111/acel.12073.
56. Agewall S, Hernberg A. Atorvastatin normalizes endothelial function in healthy smokers. Epub 2006/04/13 *Clin Sci (Lond).* 2006;111(1):87–91. https://doi.org/10.1042/CS20060033. PubMed PMID: 16608440.
57. Bielicki JK, Forte TM, McCall MR. Gas-phase cigarette smoke inhibits plasma lecithin-cholesterol acyltransferase activity by modification of the enzyme's free thiols. *Biochim Biophys Acta.* 1995;1258(1):35–40. Epub 1995/08/24. PubMed PMID: 7654778.

58. Chen C, Loo G. Inhibition of lecithin: cholesterol acyltransferase activity in human blood plasma by cigarette smoke extract and reactive aldehydes. *J Biochem Toxicol*. 1995;10(3): 121–128. Epub 1995/06/01. PubMed PMID: 7473602.
59. Levitzky YS, Guo CY, Rong J, et al. Relation of smoking status to a panel of inflammatory markers: the framingham offspring. Epub 2008/02/22 *Atherosclerosis*. 2008;201(1): 217–224. https://doi.org/10.1016/j.atherosclerosis.2007.12.058. PubMed PMID: 18289552; PMCID: PMC2783981.
60. Rutter MK, Meigs JB, Sullivan LM, D'Agostino RB, Wilson PW. C-reactive protein, the metabolic syndrome, and prediction of cardiovascular events in the Framingham Offspring Study. Epub 2004/07/21 *Circulation*. 2004;110(4):380–385. https://doi.org/10.1161/01.CIR.0000136581.59584.0E. PubMed PMID: 15262834.
61. Awasthi S, Singhal SS, Yadav S, et al. A central role of RLIP76 in regulation of glycemic control. Epub 2009/12/17 *Diabetes*. 2010;59(3):714–725. https://doi.org/10.2337/db09-0911. PubMed PMID: 20007934; PMCID: PMC2828645.
62. Faure P, Rossini E, Wiernsperger N, Richard MJ, Favier A, Halimi S. An insulin sensitizer improves the free radical defense system potential and insulin sensitivity in high fructose-fed rats. *Diabetes*. 1999;48(2):353–357. https://doi.org/10.2337/diabetes.48.2.353.
63. Fernandes AB, Guarino MP, Macedo MP. Understanding the in-vivo relevance ofS-nitrosothiols in insulin action. *Can J Physiol Pharmacol*. 2012;90(7):887–894. https://doi.org/10.1139/y2012-090.
64. Gage MC, Yuldasheva NY, Viswambharan H, et al. Endothelium-specific insulin resistance leads to accelerated atherosclerosis in areas with disturbed flow patterns: a role for reactive oxygen species. *Atherosclerosis*. 2013;230(1):131–139. https://doi.org/10.1016/j.atherosclerosis.2013.06.017.
65. Khamaisi M, Kavel O, Rosenstock M, et al. Effect of inhibition of glutathione synthesis on insulin action: in vivo and in vitro studies using buthionine sulfoximine. *Biochem J*. 2000; 349(2):579–586. https://doi.org/10.1042/bj3490579.
66. Shinozaki K, Hirayama A, Nishio Y, et al. Coronary endothelial dysfunction in the insulin-resistant state is linked to abnormal pteridine metabolism and vascular oxidative stress. *J Am Coll Cardiol*. 2001;38(7):1821–1828. https://doi.org/10.1016/s0735-1097(01)01659-x.
67. Yadav S, Zajac E, Singhal SS, Awasthi S. Linking stress-signaling, glutathione metabolism, signaling pathways and xenobiotic transporters. Epub 2007/01/30 *Cancer Metastasis Rev*. 2007;26(1):59–69. https://doi.org/10.1007/s10555-007-9043-5. PubMed PMID: 17260165.
68. Szypowska AA, Burgering BM. The peroxide dilemma: opposing and mediating insulin action. Epub 2011/01/18 *Antioxidants Redox Signal*. 2011;15(1):219–232. https://doi.org/10.1089/ars.2010.3794. PubMed PMID: 21235358.
69. Bastard J-P, Bruckert E, Porquet D, et al. Evidence for a relationship between plasminogen activator inhibitor-1 and gamma glutamyl transferase. *Thromb Res*. 1996;81(2):271–275. https://doi.org/10.1016/0049-3848(95)00244-8.
70. Bigornia SJ, Farb MG, Mott MM, et al. Relation of depot-specific adipose inflammation to insulin resistance in human obesity. Epub 2012/01/01 *Nutr Diabetes*. 2012;2:e30. https://doi.org/10.1038/nutd.2012.3. PubMed PMID: 23449529; PMCID: PMC3341707.
71. Chu NF, Spiegelman D, Rifai N, Hotamisligil GS, Rimm EB. Glycemic status and soluble tumor necrosis factor receptor levels in relation to plasma leptin concentrations among normal weight and overweight US men. *Int J Obes*. 2000;24(9):1085–1092. https://doi.org/10.1038/sj.ijo.0801361. PubMed PMID: WOS:000088984900001.
72. Gougeon R. Insulin resistance of protein metabolism in type 2 diabetes and impact on dietary needs: a review. Epub 2013/09/28 *Can J Diabetes*. 2013;37(2):115–120. https://doi.org/10.1016/j.jcjd.2013.01.007. PubMed PMID: 24070802.

73. Higdon JV, Frei B. Obesity and oxidative stress: a direct link to CVD?. Epub 2003/03/18 *Arterioscler Thromb Vasc Biol.* 2003;23(3):365–367. https://doi.org/10.1161/01.ATV.0000063608.43095.E2. PubMed PMID: 12639823.
74. Hotamisligil GS, Arner P, Caro JF, Atkinson RL, Spiegelman BM. Increased adipose tissue expression of tumor necrosis factor-alpha in human obesity and insulin resistance. *J Clin Invest.* 1995;95(5):2409–2415. https://doi.org/10.1172/jci117936.
75. Laight DW, Desai KM, Gopaul NK, Änggård EE, Carrier MJ. F2-isoprostane evidence of oxidant stress in the insulin resistant, obese Zucker rat: effects of vitamin E. *Eur J Pharmacol.* 1999;377(1):89–92. https://doi.org/10.1016/s0014-2999(99)00407-0.
76. Russo I, Del Mese P, Doronzo G, et al. Resistance to the nitric oxide/cyclic guanosine 5'-monophosphate/protein kinase G pathway in vascular smooth muscle cells from the obese Zucker rat, a classical animal model of insulin resistance: role of oxidative stress. Epub 2007/12/15 *Endocrinology.* 2008;149(4):1480–1489. https://doi.org/10.1210/en.2007-0920. PubMed PMID: 18079207.
77. Shimabukuro M, Zhou YT, Levi M, Unger RH. Fatty acid-induced cell apoptosis: a link between obesity and diabetes. *Proc Natl Acad Sci USA.* 1998;95(5):2498–2502. https://doi.org/10.1073/pnas.95.5.2498.
78. Singhal SS, Figarola J, Singhal J, et al. RLIP76 protein knockdown attenuates obesity due to a high-fat diet. *J Biol Chem.* 2013;288(32):23394–23406. https://doi.org/10.1074/jbc.m113.480194.
79. Zarkovic N, Orehovec B, Milkovic L, et al. Preliminary findings on the association of the lipid peroxidation product 4-hydroxynonenal with the lethal outcome of aggressive COVID-19. Epub 2021/09/29 *Antioxidants.* 2021;10(9). https://doi.org/10.3390/antiox10091341. PubMed PMID: 34572973; PMCID: PMC8472532.
80. Suhail S, Zajac J, Fossum C, et al. Role of oxidative stress on SARS-CoV (SARS) and SARS-CoV-2 (COVID-19) infection: a review. Epub 2020/10/28 *Protein J.* 2020;39(6):644–656. https://doi.org/10.1007/s10930-020-09935-8. PubMed PMID: 33106987; PMCID: PMC7587547.
81. Demoly P, Vachier I, Pene J, Michel FB, Godard P, Damon M. IgE produces monocyte superoxide anion release: correlation with CD23 expression. Comparison of patients with asthma, patients with rhinitis, and normal subjects. *J Allergy Clin Immunol.* 1994;93:108–116. PubMed PMID: 8308176.
82. Crow JP. Peroxynitrite scavenging by metalloporphyrins and thiolates. *Free Radic Biol Med.* 2000;28(10):1487–1494.
83. Henricks PA, Nijkamp FP. Reactive oxygen species as mediators in asthma. *Pulm Pharmacol Ther.* 2001;14:409–420. PubMed PMID: 11782121.
84. Delpino MV, Quarleri J. SARS-CoV-2 pathogenesis: imbalance in the Renin-angiotensin system favors lung fibrosis. *Front Cell Infect Microbiol.* 2020;10:340. https://doi.org/10.3389/fcimb.2020.00340. PubMed PMID: 32596170.
85. Robinot R, Hubert M, de Melo GD, et al. SARS-CoV-2 infection induces the dedifferentiation of multiciliated cells and impairs mucociliary clearance. *Nat Commun.* 2021;12(1):4354. https://doi.org/10.1038/s41467-021-24521-x.
86. Cosio BG, Mann B, Ito K, et al. Histone acetylase and deacetylase activity in alveolar macrophages and blood mononocytes in asthma. *Am J Respir Crit Care Med.* 2004;170:141–147. PubMed PMID: 15087294.
87. Radi R, Cassina A, Hodara R. Nitric oxide and peroxynitrite interactions with mitochondria. *Biol Chem.* 2002;383:401–409. PubMed PMID: 12033431.
88. Fujisawa T. Role of oxygen radicals on bronchial asthma. *Curr Drug Targets Inflamm Allergy.* 2005;4:505–509. PubMed PMID: 16101530.

89. Mrityunjaya M, Pavithra V, Neelam R, Janhavi P, Halami PM, Ravindra PV. Immune-boosting, antioxidant and anti-inflammatory food supplements targeting pathogenesis of COVID-19. *Front Immunol.* 2020;11:570122. https://doi.org/10.3389/fimmu.2020.570122. PubMed PMID: 33117359.
90. Sahebnasagh A, Saghafi F, Avan R, et al. The prophylaxis and treatment potential of supplements for COVID-19. Epub 2020/09/01 *Eur J Pharmacol.* 2020;887:173530. https://doi.org/10.1016/j.ejphar.2020.173530. PubMed PMID: 32882216.
91. Paul BD, Lemle MD, Komaroff AL, Snyder SH. Redox imbalance links COVID-19 and myalgic encephalomyelitis/chronic fatigue syndrome. *Proc Natl Acad Sci USA*. 2021; 118(34). https://doi.org/10.1073/pnas.2024358118.
92. Rahman I, van Schadewijk AA, Crowther AJ, et al. 4-Hydroxy-2-nonenal, a specific lipid peroxidation product, is elevated in lungs of patients with chronic obstructive pulmonary disease. *Am J Respir Crit Care Med.* 2002;166:490–495.
93. Xu G, Sayre LM. Structural characterization of a 4-hydroxy-2-alkenal-derived fluorophore that contributes to lipoperoxidation-dependent protein cross-linking in aging and degenerative disease. *Chem Res Toxicol.* 1998;11:247–251.
94. Xu GH, Liu YH, Sayre LM. Polyclonal antibodies to a fluorescent 4-hydroxy-2-nonenal (HNE)-derived lysine-lysine cross-link: characterization and application to HNE-treated protein and in vitro oxidized low-density lipoprotein. *Chem Res Toxicol.* 2000;13:406–413.
95. Niki E, Yoshida Y, Saito Y, Noguchi N. Lipid peroxidation: mechanisms, inhibition, and biological effects. *Biochem Biophys Res Commun.* 2005;338:668–676.
96. Dianzani MU. 4-Hydroxynonenal and cell signalling. *Free Radic Res.* 1998;28:553–560.
97. Dianzani MU, Barrera G, Parola M. 4-hydroxy-2,3-nonenal as a signal for cell function and differentiation. *Acta Biochim Pol.* 1999;46:61–75.
98. Yang Y, Sharma R, Sharma A, Awasthi S, Awasthi YC. Lipid peroxidation and cell cycle signaling: 4-hydroxynonenal, a key molecule in stress mediated signaling. *Acta Biochim Pol.* 2003;50:319–336.
99. Forman HJ, Dickinson DA, Iles KE. Hne - signaling pathways leading to its elimination. *Mol Aspect Med.* 2003;24:189–194.
100. Drake J, Petroze R, Castegna A, et al. 4-Hydroxynonenal oxidatively modifies histones: implications for Alzheimer's disease. *Neurosci Lett.* 2004;356:155–158.
101. Pizzimenti S, Barrera G, Dianzani MU, Brusselbach S. Inhibition of D1, D2, and A-cyclin expression in HL-60 cells by the lipid peroxydation product 4-hydroxynonenal. *Free Radic Biol Med.* 1999;26:1578–1586.
102. Barrera G, Pizzimenti S, Dianzani MU. 4-hydroxynonenal and regulation of cell cycle: effects on the pRb/E2F pathway. *Free Radic Biol Med.* 2004;37:597–606.
103. Laurora S, Tamagno E, Briatore F, et al. 4-Hydroxynonenal modulation of p53 family gene expression in the SK-N-BE neuroblastoma cell line. *Free Radic Biol Med.* 2005; 38(2):215–225. PubMed PMID: 15607904.
104. Tsukagoshi H, Kawata T, Shimizu Y, Ishizuka T, Dobashi K, Mori M. 4-Hydroxy-2-nonenal enhances fibronectin production by IMR-90 human lung fibroblasts partly via activation of epidermal growth factor receptor-linked extracellular signal-regulated kinase p44/42 pathway. *Toxicol Appl Pharmacol.* 2002;184:127–135. PubMed PMID: 12460740.
105. Fazio VM, Barrera G, Martinotti S, et al. 4-Hydroxynonenal, a product of cellular lipid peroxidation, which modulates c-myc and globin gene expression in K562 erythroleukemic cells. *Cancer Res.* 1992;52:4866–4871.
106. Cheng J-Z, Singhal SS, Saini M, et al. Effects of mGST A4 transfection on 4-hydroxynonenal-mediated apoptosis and differentiation of K562 human erythroleukemia cells. *Arch Biochem Biophys.* 1999;372:29–36.

107. Ruef J, Rao GN, Li F, et al. Induction of rat aortic smooth muscle cell growth by the lipid peroxidation product 4-hydroxy-2-nonenal. *Circulation*. 1998;97:1071–1078.
108. Lee TJ, Lee JT, Moon SK, Kim CH, Park JW, Kwon TK. Age-related differential growth rate and response to 4-hydroxynonenal in mouse aortic smooth muscle cells. *Int J Mol Med*. 2006;17:29–35.
109. Schaur RJ, Dussing G, Kink E, et al. The lipid peroxidation product 4-hydroxynonenal is formed by–and is able to attract–rat neutrophils in vivo. *Free Radic Res*. 1994;20: 365–373.
110. Nitti M, Domenicotti C, d'Abramo C, et al. Activation of PKC-beta isoforms mediates HNE-induced MCP-1 release by macrophages. *Biochem Biophys Res Commun*. 2002; 294(3):547–552.
111. Rollins BJ. Chemokines. *Blood*. 1997;90:909–928. PubMed PMID: 9242519.
112. Jiang Y, Beller DI, Frendl G, Graves DT. Monocyte chemoattractant protein-1 regulates adhesion molecule expression and cytokine production in human monocytes. *J Immunol*. 1992;148:2423–2428. PubMed PMID: 1348518.
113. Vaddi K, Newton RC. Regulation of monocyte integrin expression by β-family chemokines. *J Immunol*. 1994;153:4721–4732. PubMed PMID: 7525713.
114. Larche M, Robinson DS, Kay AB. The role of T lymphocytes in the pathogenesis of asthma. *J Allergy Clin Immunol*. 2003;111:450–463.
115. Patrick B, Li J, Jeyabal PVS, et al. Depletion of 4-hydroxynonenal in *hGSTA4*-transfected HLE B-3 cells results in profound changes in gene expression. *Biochem Biophys Res Commun*. 2005;334:425–432.
116. Awasthi S, Tompkins J, Singhal J, et al. Rlip depletion prevents spontaneous neoplasia in TP53 null mice. Epub 2018/03/25 *Proc Natl Acad Sci U S A*. 2018;115(15):3918–3923. https://doi.org/10.1073/pnas.1719586115. PubMed PMID: 29572430; PMCID: PMC5899455.
117. Jinesh GG, Kamat AM. RalBP1 and p19-VHL play an oncogenic role, and p30-VHL plays a tumor suppressor role during the blebbishield emergency program. Epub 2017/06/06 *Cell death discovery*. 2017;3:17023. https://doi.org/10.1038/cddiscovery.2017.23. PubMed PMID: 28580172; PMCID: PMC5447132.
118. Sahu M, Sharma R, Yadav S, et al. Lens specific RLIP76 transgenic mice show a phenotype similar to microphthalmia. Epub 2013/11/06 *Exp Eye Res*. 2014;118:125–134. https://doi.org/10.1016/j.exer.2013.10.018. PubMed PMID: 24188744.
119. Awasthi YC, Chaudhary P, Vatsyayan R, Sharma A, Awasthi S, Sharma R. Physiological and pharmacological significance of glutathione-conjugate transport. Epub 2010/02/26 *J Toxicol Environ Health B Crit Rev*. 2009;12(7):540–551. https://doi.org/10.1080/10937400903358975. PubMed PMID: 20183533.
120. Awasthi S, Singhal SS, Sharma R, Zimniak P, Awasthi YC. Transport of glutathione conjugates and chemotherapeutic drugs by RLIP76 (RALBP1): a novel link between G-protein and tyrosine kinase signaling and drug resistance. Epub 2003/07/17 *Int J Cancer*. 2003;106(5):635–646. https://doi.org/10.1002/ijc.11260. PubMed PMID: 12866021.
121. Renes J, de Vries EE, Hooiveld GJ, Krikken I, Jansen PL, Muller M. Multidrug resistance protein MRP1 protects against the toxicity of the major lipid peroxidation product 4-hydroxynonenal. *Biochem J*. 2000;350 Pt 2:555–561. Epub 2000/08/19. PubMed PMID: 10947971; PMCID: PMC1221284.
122. Awasthi S, Cheng J, Singhal SS, et al. Novel function of human RLIP76: ATP-dependent transport of glutathione conjugates and doxorubicin. Epub 2000/08/05 *Biochemistry*. 2000;39(31):9327–9334. https://doi.org/10.1021/bi992964c. PubMed PMID: 10924126.

Immune enhancers for COVID-19

3

Katherine G. Holder[1], *Bernardo Galvan*[1], *Pulak R. Manna*[2], *Zachery C. Gray*[2] and *P. Hemachandra Reddy*[3]
[1]School of Medicine, Texas Tech University Health Sciences Center, Lubbock, TX, United States; [2]Department of Internal Medicine, School of Medicine, Texas Tech University Health Sciences Center, Lubbock, TX, United States; [3]Department of Internal Medicine, Texas Tech University Health Sciences Center, Lubbock, TX, United States

Abstract

SARS-CoV-2, also known as COVID-19, is a novel coronavirus that began sweeping the globe at the end of 2019, causing mild illness in some patients while leading to devastating shock, immune dysregulation, multiorgan failure, and even death in others. Immune dysregulation may lead to increased susceptibility to severe disease from COVID-19. Immune enhancers could aid in immune regulation and protect against severe COVID-19 infection. Herbal supplements, spices, and lifestyle modifications have been shown to enhance immune responses to a number of pathogens, which may include COVID-19. These immune enhancers could be used adjunctively with vaccines, social distancing, and pharmacologic treatments to prevent life-threatening infection in susceptible patients.

Keywords: COVID-19; Immune dysregulation; Immune enhancement; Lifestyle modifications; Supplements; Vitamins

1. Introduction

At the beginning of 2020, as severe acute respiratory syndrome coronavirus 2 (SARS-CoV-2) spread, data from the outbreak clarified that the novel virus does not affect all patient populations equally, causing asymptomatic infection in some and critical illness with respiratory failure, shock, and multisystem dysfunction in others.[1–3] COVID-19 mortality rates increased with patient age and underlying comorbidities, making the ill and the elderly more susceptible to severe disease and associated death.[4] The greatest underlying health risks for severe COVID-19 infection included diabetes, obesity, cardiovascular disease, respiratory disease, cancer, and aging.[5] In fact, age and comorbidities were strong predictors of hospital admission, critical illness, and mortality in COVID-19 patients.[6] Increased inflammation, as measured with D-dimer and C reactive protein, also strongly correlated with mortality and critical illness among COVID-19 patients. Inflammation dampens patients' immune system and prevents optimized protection from pathogens like COVID-19.

The immune system is a network of intricately regulated biological processes and organs, and it prevents an organism from a variety of pathogens.[7,8] A healthy immune system is the first line of defense against viruses, bacteria, fungi, and chemicals/toxins.

COVID-19 in Alzheimer's Disease and Dementia. https://doi.org/10.1016/B978-0-443-15256-6.00014-3

As such, the immune system acts as a barrier between environmental invaders and the internal milieu, as the former has the ability to recognize diverse issues and factors by danger- and/or pathogen-associated molecular patterns.[9,10] The immune system is a complex and pervasive system that can be divided into two parts: the innate immune system and the adaptive immune system.

The innate immune system relies upon the physical features and gene expressions of an organism and is present at birth. Innate immune response provides the initial defense against invading pathogens.[7,8,10] This system involves skin, epithelial tissue linings, respiratory tract, and genitourinary tract, as well as the mucus layers that coat these tissues and includes neutrophils, monocytes, macrophages, cytokines, and specific proteins.[11–13] The skin is composed of three layers, that is, epidermis, dermis, and a subcutaneous fat layer, which provide a solid and sturdy barrier for the internal milieu from environmental factors.[14–16] Skin possesses various biomolecules that neutralize pathogenic organisms, including antimicrobial peptides that break down bacterial membranes.[12] The epithelial cells have direct contact with outside molecules such as the inner linings of airways, lungs, and digestive tracts, which are tightly packed and function under a consistently replenished mucus layer to remove foreign matter through literal movements of cilia and other mucus migrating systems.[8] Additionally, neutrophils are the body's first cellular line of defense for external pathogens that are ingested through phagocytosis and subsequently metabolized.[11,17] Macrophages, being part of innate response, are capable of engulfing and consuming foreign substances through toll-like receptor-mediated mechanisms. Cytokines provide a large influx of blood and immune cells to the points of infection in combating pathogens.[9,10] Several enzymes, including lactoferrin and lysozyme, are also involved under innate immune responses in protecting the body from a variety of invaders.[18]

The adaptive (also called acquired) immune system, in coordination with the innate system, is generally protects an organism exposed to microbes or toxins. This immune process produces/develops antibodies against the pathogen or an antigen to protect the body from a specific invader.[19,20] The adaptive immune response is reserved for complex vertebrates and has shown an evolutionary trend stemming off the innate response. Of note, T and B lymphocytes mainly comprise the adaptive immune response, and these immune cells are matured in the thymus and made in the bone marrow, respectively.[19] Mature T-cells influence immunity through cytokine production, antigen destruction, and maintenance of other immune cells as needed.[7,10,21] In addition, macrophages play a key role in recognizing pathogens, attaching to them, and carrying them to T- and B-lymphocytes for destruction.[11] B-lymphocytes are produced rapidly and serve as memory cells to specific antigens. Once B-cells recognize a pathogen, they work to prevents future infections by producing antibodie that target specific antigens.[21–23] This acquired memory is unique to each individual and provides a tailored immune response against the pathogens.[21–23]

A heathy immune system is not only the first line of defense but often the best line of defense against viruses, bacteria, and parasites. While balanced foods rich in vitamins and minerals help boost immunity in preventing invaders, nutritional deficiencies result in impaired immunity. Contextually, COVID-19 infections and outcomes have been shown to be disastrous among people with dampened immune health in contrast to those with healthy immunity.[24] Therefore, the immune system and its functional efficacies have been under intense investigation, which has led to the investigation of nutritional benefits, vaccines, and preventive drugs that ward off pathogens.[25] As mentioned above, vitamins, nutrients, and antioxidants are key factors in strengthening the immune system. Maintenance of healthy immunity with various micronutrients results in long-term health benefits. Known infectious diseases, such as COVID-19, have strong implications toward high mortality rates for those with immunocompromized conditions in which the immune system fails in fighting off pathogens, antigens, and cancers.[26,27] A variety of approaches have targeted immune boosting, especially with vitamins D, C, and E, for patients who are suffering from acute respiratory distress syndromes and other symptoms in conjunction with COVID-19.[28,29] Accordingly, the immune system is a major focus of modern medicine and its function as the center of prevention and treatment for various diseases.

Immune dysregulation may be innate or can be triggered by environmental factors, lifestyle habits, or pathogens. The relationship between chronic illness, aging, and severe COVID-19 infection may be attributed primarily to the immune dysregulation inflicted by chronic inflammation. Aging alone is associated with higher baseline inflammation, coining the term "inflammaging." Elevated baseline inflammation from a myriad of causes may result in intrinsic defects in B- and T-cells, the main champions of the human adaptive immune response. Existing dysregulation of these immune cells predisposes patients to severe COVID-19 infection by limiting the immune system's ability to fight off novel infection and to utilize immune memory for host defense. For example, when a healthy subject's lungs are infected with COVID-19, tissue destruction triggers a local immune response that releases cytokines, primes adaptive B- and T-cells, and effectively eradicates the virus. However, when a patient with existing inflammation attempts to mount an immune response to the COVID-19 infection, the adaptive immune system is prone to dysregulation. In this scenario, one of two things can happen: either the immune response is not able to develop a strong enough response to quickly eradicate the infection or the immune response over produces and releases cytokines, causing a cytokine storm. Either of these scenarios may be detrimental to the patient, possibly leading to severe organ damage and failure of the cardiovascular, hepatic, and/or renal systems.

By late 2020, the United States had experienced more COVID-19 deaths than any other country and had one of the highest cumulative per capita death rates.[30] While this may be partially explained by weak public health infrastructure and decentralized response to the pandemic, country wide health data could also delineate American predisposition for severe COVID-19 infections.[30] Higher rates of obesity, cardiovascular disease, and diabetes may also lead to increased inflammation and severe COVID-19 infection in the United States. Additionally, general "wellness" factors that decrease inflammation, such as good sleep hygiene and a diet consisting of low sugar and whole foods, are not regularly practiced in the United States. Other potential risk factors that predispose to severe COVID-19 infection include lack of exposure to sunlight, certain medications, disadvantaged social and economic status, and lifestyle factors such as smoking, drinking alcohol, and excessive consumption of salt or sugar.[31]

At the height of the COVID-19 pandemic, most public health experts advocated for enhanced immune protection measures such as frequent hand washing, staying home more often, avoiding crowds, and wearing masks.[32] These precautions mitigated viral spread and prevented many people from viral exposure during the first part of the pandemic. However, significant precautions, such as staying home and avoiding crowds, are only sustainable for a short period of time. As a result, researchers have begun investigating alternative approaches such as immune enhancement. In this chapter, the terms "immunity" and "immune enhancement" are used liberally to discuss therapies, which may enhance the bodies natural immune response and upregulate immune defense against all pathogens, including COVID-19. These terms should not be interpreted to upregulate immune defense against only one pathogen nor should they confer that these methods alone are sufficient in protection from COVID-19. This chapter discusses how to optimize one's own immune system. Immune enhancement techniques include lifestyle changes, dietary adjustments, and vaccinations. Lifestyle changes may include instituting positive behaviors such as increasing exercise, practicing good sleep hygiene, or maintaining appropriate weight. These changes may also include terminating negative immunologic behaviors such as consuming alcohol, smoking, and overeating inflammatory foods.

2. Immune enhancement—supplements

Vitamins and minerals, including zinc, vitamins C, A, and D, are essential in maintaining a well-functioning immune system.[33] Research also encourages consumption of antioxidant rich spices to boost immunity and to aid the body's natural ability to fight infection. Generally, nutritional supplements that act as immune enhancers can be broken down into three categories: vitamins and minerals, spices, and grouped supplements. These categories are explored in Table 3.1.

2.1 *Fat soluble vitamins*

Vitamins can be classified as fat soluble and non-fat-soluble compounds. Vitamin D, A, K, and E are the main fat-soluble vitamins used by the human body. The fat-soluble vitamins that are thought to enhance immunity to COVID-19 infection include vitamin D, vitamin A, and vitamin E.

Vitamin A, a retinoid, is recognized for its crucial role in maintaining epithelial cells. Retinoids are small, lipophilic, hormone-like molecules that predominantly act through ligand-activated nuclear receptors, the retinoic acid receptors (RARs), and retinoid X receptors (RXRs), each of which have three subtypes (α, β, and γ), with additional isoforms resulting from alternative splicing.[34] Vitamin A is a fat-soluble nutrient that is found in a variety of foods, including fruits, leafy vegetables, dairy products, and carrots. Vitamin A is crucial for the maintenance of the immune system and vision, and it is effective in preventing cataracts, retinitis (eye diseases), and premature skin aging (Fig. 3.1). Retinoids have antioxidant properties and control various

Table 3.1 Table delineating mechanism of action and literature summaries for efficacy of vitamins and minerals found in spices and supplements used to maximize immune system function.

Name of micronutrient	Mechanism of action	Proof of efficacy against COVID-19
Vitamin A	• Promote growth and differentiation of epithelial cells • Anti-inflammatory • Enhancing immune function	• Intact mucous membranes act as a barrier within the body's innate immune system keeping viral particles from ever gaining access into the body • Less inflammation preserves cell to cell adhesion countering hematogenous spread • B and T cells are necessary to combat and clear viral antigens
Vitamin D	• Lowers viral replication rate • Suppresses key proinflammatory pathways including NF-kB, IL-6, and TNF • Enhance neurotropic factor (NGF)	• Decreases viral load and severity of infection • Anti-inflammatory and immunomodulatory especially in decreasing cytokine release syndrome associated with severe infections • Minimally preserves loss of neuronal sensation like smell and taste
Vitamin E	• Antioxidant • Alter membrane integrity, signal transduction, cellular division, and cytokine signaling	• Prevent free radical damage to lungs and vasculature generated by virus • Modulate T cell function necessary for viral clearance
Vitamin K	• Maintain integrity of vascular system by modulating coagulation cascade	• Evidence is not sufficient for or against currently
Vitamin C	• Enhance and support epithelial cell function • Antioxidant from environmental oxidative stress • Cofactor for biosynthetic and gene regulatory enzymes • Accumulates in phagocytic cells to enhance chemotaxis, phagocytosis, and generation of reactive oxygen species, which eventually enables intracellular microbial killing • Enhance differentiation and proliferation of B and T cells in the adaptive immune response	• Very well accepted to boost the overall immune system. Beneficial in curtailing initial susceptibility to viral infection. Also beneficial in strengthening adaptive immune response once an active infection has taken hold

Continued

Table 3.1 Table delineating mechanism of action and literature summaries for efficacy of vitamins and minerals found in spices and supplements used to maximize immune system function.—cont'd

Name of micronutrient	Mechanism of action	Proof of efficacy against COVID-19
B6 (pyridoxine)	• Lymphocyte differentiation and maturation	• Antibody production
B9 (folate)	• Essential for DNA and protein synthesis and cellular proliferation • Necessary for cell mediated blastogenic response of T lymphocytes and other mitogens • Cofactor of enzymes in thymus for production of new immune cells • Cofactor for enzymes in bone marrow to produce B cells • Necessary for PMN cells phagocytic and bactericidal capacities	• Recognized as a crucial vitamin in the production of new cells in the body. Hematopoietic cells of the immune system have an especially high turnover rate, which makes B9 indispensable in keeping up immune capabilities to combat high viral loads
B12 (cobalamin)	• DNA and protein synthesis of lymphocytes and natural killer cells	• Maintain adequate immune cell availability especially CD4/CD8 ratio required for appropriate immune response
Iron (Fe^{2+})	• Lymphocyte proliferation and maturation	• Contraindicated since also necessary for use by appropriated cells to produce viral antigens
Zinc (Zn)	• Regulates gene transcription and is required for modulating DNA replication, RNA transcription, cell division, and cellular activation • Used by scavenger cells and neutrophils to maintain innate immunity	• Mechanisms necessary for proliferation of cells in adaptive immune system • A strong innate immune reaction limits overall viral load at onset of infection
Copper (Cu)	• Lymphoproliferative response to mitogens • Cofactor for several enzymatic reactions that limit oxidative stress	• Similar to many of the micronutrients previously discussed. Cu is needed for ensuing steps of increased immune cell production attempting to ward off infection
Selenium (Se)	• Component of glutathione reductase an antioxidant mechanism that diffuses circulating hydrogen peroxides and limits oxidative stress	• Similar properties of vitamins C and E

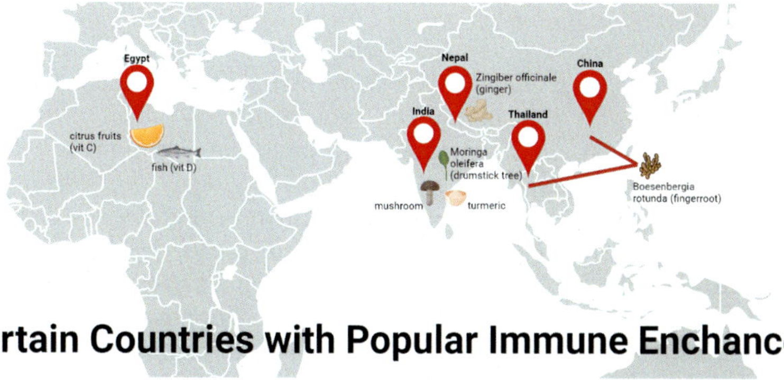

Figure 3.1 A map displaying common immune enhancers and regions that use them regularly.

genes involved in both innate and adaptive immune responses.[35,36] The recommended daily amount (RDA) of vitamin A is 700 and 900 μg for adult women and men, respectively. Deficiency of vitamin A is associated with decreases not only immune responses but also steroidogenic machineries influencing reproductive development and function.[14,37,38]

Epithelial cells make up many of our body "surfaces" including our skin, cornea, and mucous membranes. Vitamin A works on epithelial cells to enhance vision, promote growth, and most importantly, maintain the integrity of mucous membranes and skin barriers. Mucous membrane integrity is perhaps the body's most important innate defense against infection. Infectious agents may get stuck in the mucous membranes of the nares or oropharynx and never move past this barrier to infect the body's cells. Unfortunately, with viral particles as small and as resistant as COVID-19, mucous barriers are often not enough to prevent infection entirely. However, mucous barriers still play a role in containing the initial infection. Vitamin A is also an anti-inflammatory vitamin that plays a critical role in enhancing immune function, developing an adaptive immune system, and regulating cellular immune responses within the humoral immune system. Recent studies show that vitamin A reduces morbidity and mortality in viral diseases including measles, acute pneumonia, infant diarrhea, enteric infection, malaria, hand foot and mouth disease, and mumps. Vitamin A is regularly used as a mainstay of treatment for these severe viral infections. While the literature is still developing on vitamin A's use in response to the coronavirus family, prior viral research indicates that increasing vitamin A intake could reduce morbidity and mortality in the face of severe COVID-19 infection.[39] Nutritional status with suboptimal vitamin A has been shown to be correlated with COVID-19 incidence and mortality.[28] Interestingly, Yuan et al. 2019 have shown that RAR agonist acts on SARS and MERS through interruption of lipogenic pathways.[40] Retinoids are capable of increasing the actions of type 1 interferons (IFN−I), a cytokine released via innate immunity against viral infections, which could be used in combination with other antiviral drugs in the management of COVID-19 or other viral infections.[41] Vitamin A's role in barrier protection and immune support could make it a critical nutrient for enhancing immunity to COVID-19.

2.1.1 Vitamin D (Ergocalciferol)

Vitamin D may be the most well-researched micronutrient immune enhancer shown to reduce severe infection from COVID-19. Vitamin D is a group of fat-soluble secosteroids (especially D2, ergocalciferol; D3 cholecalciferol), which possess anti-inflammatory, antioxidant, and neuroprotective properties and primarily supports immune health, build, and maintain healthy bones.[42,43] Vitamin D is responsible for increasing intestinal absorption of calcium, magnesium, phosphate, and many other physiological activities. Vitamin D rich foods include meats, eggs, fish, and dairy products, and the RDA of vitamin D is 15 μg per day for both adult women and men. The biological action of vitamin D is mediated upon the binding of a ligand (calcitriol) to nuclear vitamin D receptor (VDR), which heterodimerizes with the RARs and regulates the transcription of genes in target cells.[44,45] VDR is ubiquitously expressed in various organs, including immune cells, skin keratinocytes, mammary epithelium, lungs, kidney, intestine, pituitary, bone and germ cells, suggesting that this micronutrient has diverse effects in various biological processes (Fig. 3.1). Vitamin D regulates a number of genetic pathways and influence several health conditions such as cancer, diabetes, respiratory tract infections, and autoimmune diseases.[46–49]

Over a century ago, during the influenza pandemic of 1918 (often referred to as the Spanish Flu), research suggested that vitamin D may play a nonclassic role in reducing lethal pneumonia and fatality from severe respiratory viruses. Since then, a robust body of literature, including clinical trials, has reported that vitamin D supplementation can reduce both the incidence of acute respiratory infection and the severity of respiratory tract diseases among all age groups.[50] More recently, studies have reviewed the correlation of vitamin D levels with severe COVID-19 cases and death. One study published in the final months of 2020 examined the mean vitamin D levels in 20 European countries and found that higher levels of vitamin D inversely correlated to the number of COVID-19 infections per 1 million people. Other retrospective studies demonstrated a similar inverse correlation between vitamin D status and COVID-19 severity and mortality. Several additional studies demonstrated the role of vitamin D in reducing the risk of acute viral respiratory tract infections and pneumonia.

Generally, the literature suggests that vitamin D may directly inhibit viral replication and enhance anti-inflammatory and immunomodulatory responses to infection. Vitamin D induces the production of antimicrobial peptides (e.g., cathelicidin and defensins) in immune cells. These peptides are a part of the innate immune system and neutralize various microbes through destruction of their envelope proteins, thus, providing a barrier for microbe entry to the host cells.[51,52] Vitamin D is known to critically influence both innate and adaptive immune responses; therefore, eating a diet rich in vitamin D might protect people from COVID-19 infections.[53–55] Vitamin D deficiency is linked with various complications, including skeletal deformities, atherosclerosis, and inflammatory bowel disease.[43,56] One contributor to severe outcomes in patients afflicted with COVID-19 is excessive and unregulated cytokine storms that inflame the body leading to a malfunctioning of organs and bioprocesses.[57–59] Vitamin D has been shown to reduce the proinflammatory cytokines such as IL-6 and IFN-γ that are known contributors of COVID-19 proinflammatory states, and their reduction is likely to prevent risk of viral infections.[42,60] Studies have shown that vitamin D supplementation not only decreases acute respiratory infections caused by COVID-19 but also enhances the immune system in lowering risk of infection, severity, and mortality.[61,62] Additionally, both vitamin D and calcitriol are capable of increasing blood oxygen and hemoglobin levels, which would be beneficial for either prevention and/or recovery of COVID-19-associated complications.[63]

In a recent meta-analysis, vitamin D supplementation was shown to be a safe and effective agent against acute respiratory infections. While immune support is essential for fighting viruses including COVID-19, the second most common cause of death in the face of COVID-19 is immune dysregulation leading to physiologic demise from processes such as the cytokine storm. Vitamin D might also act as a strong suppressor of cytokine release syndrome in response to COVID-19. Vitamin D has been shown to suppress key proinflammatory pathways including nuclear factor kappa B (NF-kB), interleukin-6 (IL-6), and tumor necrosis factor (TNF). Lastly, Vitamin D may enhance neutrophilic factors including nerve growth factor (NGF) to prevent loss of neural sensation in COVID-19. This may protect against symptoms of infection including loss of taste and smell.[64]

2.1.2 Vitamin E (Tocopherol)

Vitamin E is a potent antioxidant known as one of the most effective nutrients for modulating immune function. This lipid soluble compound is found in higher concentrations among immune cells compared to any other cell of the body. Vitamin E is a fat-soluble nutrient that includes four tocopherols and four tocotrienols, which play crucial roles in a wide variety of physiological processes, ranging from vision to skin health.[65,66] It should be noted that the most biologically active form of vitamin E is α-tocopherol, which can be obtained from various sources, including canola oil, olive oil, almonds, peanuts, cereals, meats, and dairy products. Vitamin E possesses antioxidant and anti-inflammatory properties and impacts many important biological processes, including regulation of enzymes involved in signal transduction pathways, the support of immune function, increased lymphocyte and IL-2 proliferation, and decreased IL-6 production.[67–69] As an antioxidant, vitamin E neutralizes reactive oxygen species (ROS), and, by doing so, it plays an important role in protection from heart diseases and certain cancers, in addition to block lung neutrophil inflammation. The RDA of vitamin E is 15 mg for both adult men and women.

Vitamin E deficiency has been shown to impair normal immune function in multiple human and animal studies. While vitamin E deficiency is rare, vitamin E supplementation has been shown to enhance immune integrity and reduce infection particularly in immunodeficient individuals such as the elderly. Vitamin E is also known to modulate T-cell function by altering membrane integrity, signal transduction, cellular division, and cytokine signaling. Respiratory infections and allergic diseases are commonly modified with vitamin E supplementation. While vitamin E reduces susceptibility to respiratory infections, its functional relevance in COVID-19 protection is lacking. Vitamin E has been shown to increase T-lymphocyte response through its assistance in the signal transduction pathways, which can assist patients with immunocompromized conditions against a COVID-19 infection.[70,71] Vitamin E deficiency is caused by a number of pathological conditions, including pancreatitis, cholestasis, cystic fibrosis, and Crohn's disease, and results are generally associated with decreased coordination, muscle dystrophy, and vision issues.[72,73] Importantly, deficiency of vitamin E is associated with a neurological disorder, called ataxia, in 5–15 year age groups and is characterized by clumsiness of the hands, loss of proprioception, and coordination.[74,75] Supplementation of α-tocopherol has been shown to reduce respiratory tract infections in aging populations. Considering this and the beneficial effect of vitamin E in immunity, elderly individuals are encouraged to intake this nutrient and other antioxidants for the prevention of COVID-19 infection. Knowing that the immunocompromized and the elderly are most susceptible to severe COVID-19 infection, vitamin E supplementation should be considered when evaluating immune enhancers against COVID-19.[76]

2.1.3 Vitamin K (Phylloquinone)

Vitamin K is a fat-soluble coenzyme that helps to generate proteins involved in blood coagulation, blood calcium regulation (prothrombin), bone mineralization, and bone health (osteocalcin).[77,78] This vitamin is naturally present in certain foods and is available in dietary supplements in two forms, phylloquinone (K1) and menaquinone (K2).[79] While phylloquinone is mostly found in leafy vegetables, artichokes, broccoli, cauliflower, spinach, kale, and collards, menaquinone is present in meats, seafoods, dairy products, and fermented foods. The RDA for vitamin K varies upon age and gender; however, the values are generally 90 and 120 μg for adult women and men, respectively (Table 3.1). Vitamin K decreases production of proinflammatory cytokines that can also help with anticoagulation to reduce the risk of strokes and embolisms.[80,81] Strokes and heart diseases are common in patients with COVID-19 infections, and vitamin K has been linked to a 50% reduction in atherosclerotic calcification and/or lesions.[82,83] This leads to a decrease in cardiovascular diseases occurring frequently with COVID-19 infections; thus, vitamin K may be influential in decreasing the severity of symptoms and fatal outcomes associated with COVID-19.[84]

Vitamin K deficiency can result in excessive bleeding/hemorrhage, osteoporosis, osteoarthritis, neurodegenerative, and cardiovascular diseases; and these occurrences can be treated with adequate vitamin K containing foods or drugs (e.g., phytonadione).[85–87] Vitamin K, by activating elastin, influences lung tissue flexibility for appropriate function. Severe vitamin K insufficiency has been reported in COVID-19 patients; thus, administration of this nutrient may be beneficial in the prevention of this devastating infectious disease.[88]

2.2 Water-soluble vitamins

2.2.1 Vitamin C (Ascorbic acid)

Water-soluble vitamins including vitamin C, folate, B6, and B12 are critical for proper immune function and may also enhance immunity to COVID-19. Vitamin C (also known as ascorbic acid or ascorbate) is a water-soluble nutrient with antiviral properties and has long been known to boost the immune system in the prevention of bacterial and viral infections.[89] It is predominantly found in citrus fruits, bell peppers, oranges, strawberries, kiwis, tomatoes, potatoes, broccoli, brussels sprouts, cabbage, kale, and cauliflower. As an antioxidant, vitamin C helps control harmful free radicals, the cytokine storm, and oxidative damages.[57,90,91] It acts as a cofactor in many enzymatic reactions that regulate important biological activities, including the formation of blood vessels, cartilage, muscle, and collagen biosynthesis (Fig. 3.1).[89] Vitamin C has been shown to protect against respiratory tract infections and assist in their healing and decrease and shorten common cold symptoms.[92] The RDA of Vitamin C is 75 and 90 mg for adult women and men, respectively (Table 3.1). It has been reported that vitamin C reduces not only infections with sepsis and ARDS but also clinical symptoms associated with COVID-19.[93,94]

This nutrient is most important to treat a vitamin C deficient disease, scurvy, which is characterized by fatigue, anemia, and skin hemorrhages. In addition, deficiency of vitamin C has been associated with a variety of complications, including fatigue, anemia, impaired wound healing, heart disease, dry skin, increased inflammation, enhanced oxidative stress, and poor immune function.[91,95] Vitamin C supplementation generally resolves many of these symptoms within a short period of time. Vitamin C has been demonstrated to effectively decrease pneumonia.[89] Of importance, vitamin C displays antioxidant, anti-inflammatory, and immuno-modulatory effects, which are relevant to many bacterial and/or viral infections; thus, this nutrient could be beneficial in the management of COVID-19.

Vitamin C is an electron donor and functions similarly to vitamin A by which it enhances and supports epithelial cell function. Vitamin C is a potent antioxidant and a cofactor for biosynthetic and gene regulatory enzymes. It supports various cellular functions of the innate and adaptive immune system by enhancing barrier function. Vitamin C protects against pathogens and promotes occident scavenging activity of the skin thereby aiding in protection from environmental oxidative stress, which may otherwise cause immune compromise. Perhaps most importantly, vitamin C accumulates in phagocytic cells such as neutrophils where it enhances chemotaxis, phagocytosis, and generation of reactive oxygen species, which eventually enable intracellular microbial killing. Vitamin C has also been shown to enhance differentiation and proliferation of B and T cells in the adaptive immune response likely due to its gene regulatory effects. The literature suggests that Vitamin C deficiency impairs immunity and causes higher susceptibility to infection. Additionally, severe infections may rapidly decrease stored levels of vitamin C due to increased metabolic requirements during inflammation. Some studies show that prophylactic prevention of infection can be achieved by increasing vitamin C consumption.[89] While the research is still unclear about vitamin C's role in COVID-19 infection, the known properties of vitamin C suggest that it may have the ability to enhance immunity and prevent severe infection from respiratory viruses like the novel coronavirus.

2.2.2 B vitamins

Vitamin B is essential for the appropriate functioning of the immune system and thus, many physiological processes (Fig. 3.1).[96,97] Notably, eight water-soluble vitamins in the B complex (not stored in the body to a major extent) are the following: B1, thiamine; B2, riboflavin; B3, niacin; B5, pantothenic acid; B6, pyridoxine; B7, biotin; B9, folic acid; and B12, cyanocobalamin.[98] All of these B vitamins are obtained from various food sources that people consume regularly and are important for a boosted immune health in protecting against various invaders.

2.2.3 Vitamin B1 (Thiamine)

Thiamine is an essential micronutrient that is primarily found in cereals, whole grains, meats, legumes, and nuts. Vitamin B1 is important in the breakdown of carbohydrates and, thus, energy metabolism, and it plays key roles in growth and development cells implicated in immune system functioning. The RDA of vitamin B1 is 1.1 and 1.2 mg for adult women and men, respectively (Table 3.1). Thiamine reduces high blood pressure and heart complications and influences kidney function in people with diabetes. As a coenzyme, thiamin becomes activated as thiamine pyrophosphate (TPP) and is used extensively in the mitochondria alongside other important enzymes. TPP functions specifically within immune cells as both energy production and synthesis of DNA and RNA that are necessary for a proper immune response.[99,100] An area of study coined immunometabolism works to demonstrate the connections between immune health and metabolic pathways.[101,102] Thiamine is known to influence the tricarboxylic acid (TCA) cycle, which is used for energy production in macrophages, naïve T cells, and T-regulatory cells.[97] The function of these immune cells is inhibited in the absence of thiamine, resulting in metabolic imbalance. Deficiency of thiamine is connected with various abnormalities, including Wernicke encephalopathy, beriberi, cardiac failure, and ataxia.[100,103] Thiamine has been shown to plays an important role in the protection of COVID-19 by building and boosting healthy immunity.[70]

2.2.4 Vitamin B2 (Riboflavin)

Riboflavin possesses antioxidant properties, and primarily influences energy production from foods, maintains heart and neurological health, and assists in immune system functioning.[104,105] Vitamin B2,riboflavin, is largely found in dairy products, meats (including fish), and dark green vegetables. Riboflavin presents a protection against oxidative stress, inflammatory responses, and vision loss, and it improves migraines and hair and skin complications.[97,106] It is also important for production of glutathione, which prevents oxidative damage by free radicals and influences aging processes. Like all B vitamins, riboflavin is necessary for normal development, physical performance, reproductive development, and function, and it preserves hormonal balance and healthy metabolism (Fig. 3.1). It is recommended that adult women and men should take 1.1 and 1.3 mg everyday, respectively. Vitamin B2 deficiency (though uncommon) affects endocrine abnormalities and other issues such as skin complications/lesions, hyperemia and edema, and folliculitis. An organism's ability to maintain healthy iron, especially in hemoglobin, is also influenced by adequate riboflavin supplementation.[107,108] As such, riboflavin may play an important role in efficient oxygen transportation, an aspect crucial for COVID-19 patients.

2.2.5 Vitamin B3 (Niacin)

Niacin is part of a B multivitamin, and it acts as an anti-inflammatory agent. It is mostly found in cereals, legumes, meats, and milk and helps maintain nervous, digestive, and integumentary systems. Niacin is frequently utilized to increase high-density lipoprotein (HDL, good cholesterol) and decrease low-density lipoprotein (LDL, bad cholesterol), a process that has been implicated in heart health by sustaining steroid hormone regulation.[109–111] In addition, niacin produces sex- and stress-related hormones and maintains cholesterol levels. Noteworthy, a niacin-deficient condition, called pellagra (includes the triad of dermatitis, dementia, and diarrhea), can be treated with niacinamide.[112] The RDA of daily niacin is 14 and 16 mg for adult women and men, respectively. Niacin plays a key role in macronutrient metabolism and is a precursor to NAD (nicotinamide adenine dinucleotide) and NAD phosphate, which reduce viral infections.[97] NAD is important for the TCA cycle and it could play a similar role as that of vitamin B1 in ensuring the function of immune cells.[113] Niacin deficiency has been linked to various symptoms, including depression, disorientation, apathy, and birth defects, and its supplementation generally reverses these issues. It provides some anti-inflammatory measures within the lungs and has been shown to decrease the effects of severe inflammatory responses on lung tissues.[114,115] Niacin decreases levels of interleukin 6 (IL-6), IL-1β, and tumor necrosis factor α in stimulated alveolar macrophages and has been reported to reduce inflammation in COVID-19 patients.[116,117]

2.2.6 Vitamin B5 (Pantothenic acid)

Pantothenic acid is naturally found in almost all foods, including meats, eggs, milk, legumes, nuts, and potatoes. It is an important B vitamin that synthesizes co-enzyme A and acyl carrier protein, both of which are essential cofactors in various biochemical reactions including the TCA cycle. Pantothenic acid, like other B vitamins, contributes to a heathy immune system.[97] For adult women and men, the RDA of vitamin B5 is 5 mg. Since vitamin B5 is commonly found in most plants and animal foods, its deficiency is rare. However, deficiency of vitamin B5 has been associated with multiple symptoms such as fatigue, depression, insomnia, stomach pains, glucose metabolism, nerve damage, and upper respiratory infections.[98,118] Metabolism of pantothenic acid to co-enzyme-A is affected by vitamin B5 deficiency and results in a genetic disease called pantothenate kinase-associated neurodegeneration.[119,120] Since pantothenic acid, like other B vitamins, contributes to a heathy immune system and, it helps protect the body from various pathogens such as COVID-19.

2.2.7 Vitamin B6 (Pyridoxine)

B6, also called pyridoxine, is also essential in maintaining both cellular and humoral immunity. Like folate, B6 deficiency has been shown to limit lymphocyte differentiation and maturation. Additionally, pyridoxine plays an important role in neurogenesis and helps develop normal functioning of the nervous and immune systems.[121,122] Vitamin B6 is found in many foods, including meats, poultry, nuts, beans, grains, fruits, cereals, and vegetables. The RDA of vitamin B6 is 1.7–1.3 mg for adult women and men, respectively. Vitamin B6, pyridoxine, plays key roles in interleukin-2 production and hemoglobin synthesis, in addition to its involvement in many enzymatic processes as a cofactor. Pyridoxine has been shown to influence innate and adaptive immunity by stimulating interferon-γ.[123] Vitamin B6 controls the level of blood homocysteine, whose upregulation has been linked to cardiovascular diseases, as well as Alzheimer's disease.[124] A derivative of vitamin B6, pyridoxal 5-phosphate, in low quantities, has been influential with chronic diseases such as atherosclerosis and cancers.[98,123] These diseases can create states of immunosuppression and allow for greater susceptibility to infections of bacteria and viruses, including COVID-19. As such, pyridoxine influences a person's ability to fend off infection. Nonetheless, it has innumerable health benefits, including improving cognitive function, healthier skin, eye health, rheumatoid arthritis, diabetes, and anemia. Deficiency of vitamin B6 has been associated with many signs and symptoms, which include skin rashes, cracked and sore lips, mood changes, seizures, and deteriorated immune function.[122] Vitamin B6/pyridoxine supplements can prevent and/or treat many of these issues effectively. Additionally, B6 deficiency reduces the efficacy of type IV hypersensitivity reactions and impairs antibody production. While adequate B6 intake is required to optimize immune response, mega dosing of the vitamin has not been shown to enhance immunity. Patients should focus on maintaining appropriate B6 levels rather than oversupplementing.[125]

2.2.8 Vitamin B7 (Biotin)

Biotin (also known as vitamin H) is an important component in the B complex, and it is known to influence a variety of metabolic processes.[97,126] Vitamin B7 is generally synthesized in the body, and it is mostly found in plant and animal foods, including meats, eggs, fish (especially tuna and salmon), cereals, starchy vegetables, oranges, cantaloupe, and potatoes. Biotin supplements are often used in treating a variety of nail, hair, and skin-associated conditions (Fig. 3.1).[127] Biotin is important for carbohydrate and fat metabolism, acts as a coenzyme in many catalytic reactions, and was recently identified as a regulator of glucose and lipid metabolism, suggesting its importance in the management of hyperglycemia.[126,128,129] There is no RDA for biotin, however, an amount of 20–30 mg per day for teenagers and adults is required for the prevention of biotin deficiency. Just as with B6, biotin aids in the prevention of certain conditions/diseases that led to immunosuppression.[130] Deficiency of biotin is rare, but its occurrence is associated with many complications, including brittle hair and nails with hair thinning, dry skin rashes, dry eyes and poor vision, and restless leg syndrome with movement issues. Alcoholism promotes biotin deficiency by blocking its absorption. Biotin supplementation improves several of these issues, including symptoms associated with an autoimmune disorder called multiple sclerosis.[131] The immune boosting properties of B vitamins, including biotin, may support effective protection from COVID-19 infection.

2.2.9 *Vitamin B9 (Folic acid, folate)*

Folic acid, commonly referred to as folate, is a water-soluble vitamin that plays a crucial role in DNA and protein synthesis and is essential for cellular proliferation. Vitamin B9, folate, is naturally obtained in foods, in which folic acid is the synthetic form of folate. Foods rich in vitamin B9 include dark leafy greens, beans, legumes, peanuts, asparagus, melons, oranges, berries, grains, meats, and eggs. Folic acid/folate is very important for growth and development of brain function. It controls blood homocysteine concentration that is associated with increased risk of cardiovascular diseases.[132] During pregnancy, women need an increased amount of folic acid for appropriate development of the fetus and the prevention of a number of birth defects associated with the brain and spine. During early 1998, the FDA-approved manufacturers to add folic acid to commonly eaten foods, including breads, cereals, and grains, in order to prevent neural tube defects.[133,134] Folate is an undoubtedly important molecule in humans as it promotes DNA and protein synthesis, as well as adaptive immunity.[135] This vitamin has been shown to not only protect against cancers (such as lung, colon, and cervical) but also cognitive decline during the process of aging.[135–137] The daily RDA value for folic acid is 400 μg for adult women and men. Deficiency of folic acid/folate is fairly common due to a variety of events, including alcoholism, pregnancy, digestive disorders, and certain prescription drugs. Symptoms connected this vitamin B9 deficiency have been linked to birth defects, depression, memory loss, digestive diseases (e.g., Crohn's disease and colitis) and atherosclerosis.[138,139]

Folate deficiency can contribute to a variety of hematologic impairments; it can also have devastating effects on the human immune system. Cell-mediated immunity is critically affected by folate deficiency through impairment of the blastogenic response of T-lymphocytes and other mitogens. Humoral immune response, which relies on B-cell proliferation, is also impaired in folate deficiency, and can inhibit phagocytic and bactericidal capacities of polymorphonuclear cells.[140] New retrospective cohort studies show that decreased serum folate levels are common among hospitalized patients with COVID-19, but the correlation of folate levels and clinical outcomes of infection is yet to be established.[141] Studies have shown that folic acid inhibits the enzyme furin (that facilitates entry of pathogens), which could be helpful in the prevention of COVID-19 infections. Folate supplementation should be considered when optimizing the immune system, especially when considering severe infections like COVID-19.

2.2.10 *Vitamin B12 (Cyanocobalamin, Cobalamin)*

Cyanocobalamin influences a variety of physiological functions, including the development of the nervous system, antibodies, cell metabolism, cytokine formation, and immune cell function. It plays an important role in regulating the nervous system, in which its dysregulation results in dementia, cognitive function, and neurological disorders.[142,143] This is an important nutrient for RBC production. Vitamin B12 is found in many foods, including meat (red meat, seafood, liver, poultry), eggs, dairy products, and cereals. Cyanocobalamin works in combination with vitamins B6 and B9 to aid in red blood cell genesis and nucleic acid synthesis.[144] The RDA of vitamin B12 is 2.4ug for both adult men and women. Deficiency of vitamin B12 is common in many developing counties such as India, Africa, and Mexico, in addition to vegan and aging populations.[124,145] Vitamin B12 deficiencies include weakness and dizziness, anemia, fatigue, intestinal problems, and neurovegetative diseases. Lower $CB8^+$ T-cells and natural killer T-cells suppression is related to B12 deficiency, demonstrating a correlation between B12 and immune responses. B12's role in DNA production may explain its complex implications in modulating the immune system. B12 deficient patients have been shown to have decreased lymphocyte levels, abnormally high CD4/CD8 ratios, and suppressed natural killer cell activity. Appropriate B12 levels are essential for optimizing and appropriately modulating immune response, but megadose supplementation has not been studied in relation to outcomes of COVID-19 infection.[146] This nutrient, in combination with vitamin B6 and B9, has been recently targeted as a potential therapy in the management of COVID-10.[147,148] More research is required to understand the effects of B12 on immune modulation when preventing or treating COVID-19.

2.3 Micronutrients

Other micronutrients including zinc, iron, copper, and selenium have also been deemed essential for normal immune function.[149] Zinc is arguably the most well-researched mineral immune supplement and has been used widely thus far in the COVID-19 pandemic. Zinc is an indispensable microelement required for a plethora of enzymatic physiological processes. It regulates gene transcription and is required for modulating DNA replication, RNA transcription, cell division, and cellular activation in many biological systems. Zinc is used by a variety of scavenger cells and neutrophils to maintain innate immunity. Its role in cell replication and division also makes it essential for optimal adaptive immune response.[150] The literature suggests that zinc deficiency predisposes patients to a variety of viral infections including herpes simplex, rhinovirus, hepatitis C, human immunodeficiency virus (HIV), and most pertinent, SARS-CoV-1. Zinc's confirmed role in optimizing immunity against SARS-CoV-1, which comes from the same viral family as COVID-19, makes it a promising micronutrient in enhancing immunity against COVID-19.[150] Zinc, along with zinc-ionophores, is classified as nutraceuticals, minerals that show promising antiviral activities and are taken medicinally around the globe. The US Federal Drug Administration classifies zinc as a Generally Recognized as Safe (GRAS) that can provide complementary treatment for COVID-19 through its immune enhancing and antiviral properties.[151] Zinc supplementation is one of the most well studied and promising nutritional supplements for enhancing immune response to COVID-19 infection.

Iron, copper, and selenium are also often recognized in the literature as essential immune enhancers. Iron is a fundamental element for immune system development. It is required for lymphocyte proliferation and maturation and helps modify the body's response to infection. Unlike other micronutrients, the body often sequesters iron and becomes functionally iron deficient in response to some infections or immunocompromized states. It does so to prevent proliferation of bacteria, parasites, and neoplastic cells as they need iron to thrive. While iron deficiency may dysregulate effective immune response, excess iron supplementation may be even more detrimental to immune function than deficiency. For this reason, excess supplementation is not recommended when considering immune enhancers for COVID-19.[152]

Unlike iron, copper, and selenium are preferred in excess for the body to mount an immune response. Although copper is rarely deficient in the modern diet, studies have linked copper deficiency to reduced lymphoproliferative responses to mitogens. Copper is also an essential cofactor for a variety of enzymatic reactions that limit oxidative stress, making it crucial for a functioning immune response. Selenium is also an essential component for antioxidant enzymes, particularly glutathione peroxidase, which plays an important role in diffusing hydrogen peroxides and preventing oxidative stress in humans. Selenium plays an important role in viral immunity against a variety of antigens including coxsackie virus and parainfluenza virus. While more research is needed to determine selenium's role in protecting against severe COVID-19 infection, adequate selenium intake is undoubtedly essential for optimizing immune response to similar viral infections.[53,153,154]

3. Immune enhancers—diet, herbs, and spices

Adjuvant nutritional treatment is commonly overlooked in treatment for viral infections such as COVID-19.[155] Optimal nutrition can decrease inflammation and improve immune response. This may help mitigate the risk of severe immune dysregulation, serious infection, and mortality in COVID-19-infected patients.[33] As a result, dietitians and other members of the healthcare team play an essential role in patient education over the basics of nutrition and may be used to help patients understand their optimal diet. For example, whole foods are usually nutrient dense, filling, and contain appropriate calories-per-gram. A diet that consists mostly of these foods decreases inflammation and boosts general immunity.

A variety of herbs and spices have also been implicated as immune boosters since the onset of the COVID-19 pandemic. Studies out of Thailand and Southeast Asia suggest that a variety of Thai plants may be effective in boosting immunity against COVID-19 infection as shown in Fig. 3.1. The most studied of these being *Boesenbergia rotunda,* which is a culinary herb generally grown in China and Southeast Asia. *B. rotunda* extract and its main phytochemical compound panduratin A have exhibited potent anti-COVID-19 activity in early studies by inhibiting entry and infectious phases of the virus. Treatment focused on supplementing patients with *B. rotunda* extract immediately after they contracted COVID-19 and found that supplementation after viral infection drastically suppressed COVID-19 infectivity in E6 immune cells. Supplementation has also been shown to suppress viral infectivity of human airway epithelial cells. Dietary intake of *B. rotunda* may be an efficient enhancer of immunity to COVID-19, particularly in developing communities where vaccination rates remain low.

In Nepal, healthcare remains rich in traditional medicinal herbs. Folk medicine, often called ethnomedicine, is especially reliant on locally grown and sourced herbal remedies. While more studies are needed to understand the efficacy of these ethnomedicines for treating COVID-19, early reports have demonstrated that increased use of herbs and spices may be responsible for placating Nepal's COVID-19 infection rates. The most cited species of plant used during the first year of the COVID-19 pandemic in Nepal comes from *Zingiber officinale*, an herb grown in family gardens and locally proposed to enhance viral immunity. Countless other plants and spices were utilized for ethnomedicine in the early months of the pandemic, and further investigations will be required to evaluate their efficacy.[156]

In Egypt, use of micronutrient supplementation including vitamin C and D is commonly used to boost immunity and prevent COVID-19 infection. A slew of immune boosting drinks, honey, and garlic products are also commonly used to enhance immunity. Some surveys showed that over 30% of Egyptian participants used these supplements at the onset of the COVID-19 pandemic.

In India, a country known for its rich spices and plethora of traditional medicine, interest has been rekindled in traditional medicine, which is thought to boast an abundance of herbal medicines and remedies. Among these are spices including root, rhizome, seed, fruit, leaf, bud, and flower of various plants that have long been used to flavor traditional food and are believed to also contain immense medicinal potential. Many clinical studies report the efficacy of these spices in their ability to promote healthy immune function and aid in treatment for a variety of ailments. Among these are garlic, turmeric, cinnamon, *Moringa oleifera*, honey, yogurt, nigella sativa, mushrooms, and spirulina. This list covers only the most researched of India's nutritional supplements, as a full investigation would surely require a book of its own. The immune enhancing benefits of these foods and spices have been outlined in the table below and should be considered when constructing a diet that could enhance immunity to COVID-19.

4. Conclusion

As evidenced by media reports and academic literature, the COVID-19 pandemic has left people searching for a simple dietary solution and has accelerated market demand for immune-boosting foods, supplements, and micronutrients.[157] From the literature gathered, it is difficult to recommend one supplement that would enhance immunity enough to protect fully against COVID-19. The human body's immune system is extremely complex and so is optimizing it. Obtaining natural immunity to severe infection with one dietary change alone is simply not possible currently.

Currently, the most effective prevention strategies against COVID-19 transmission and infection include diligent self-sanitation, social distancing protocols, and vaccination when available. Immunity against viruses like COVID-19 is also enhanced by consuming a balanced, nutrient rich diet full of antioxidants, vitamins, minerals, fatty acids, some nonnutrients (like polyphenols), and limited polysaccharides. Supplementation with the more heavily researched micronutrients like zinc and vitamin D may also be appropriate. A diet following these guidelines will reduce inflammation and optimize the immune system's response to infection.

Nutrients may also effectively protect the human body from viral infections via two primary mechanisms: directly by targeting the virus and interfering with its infectivity or indirectly by activating and optimizing the cells associated with the adaptive immune system. Since western medicine currently has no proper curative viral drug against COVID-19, enhancing the body's immunity through appropriate diet is perhaps the more appropriate treatment measure currently available. Additionally, maintaining an appropriate BMI, exercising regularly, and avoiding the consumption of alcohol, tobacco, salts, and processed sugars will decrease systemic inflammation and enhance the body's ability to mount an effective immune response to viruses including COVID-19.

It is important to continue highlighting evidence-based public health messages and prevent false and misleading claims about targeted foods or supplements. Clear communication of the benefits of micronutrients and COVID-19 infection is still being explored, and it is probable that no diet will prevent or cure severe COVID-19 infection. Rather, taking precautionary dietary and lifestyle measures to prime the human immune system for an efficient and effective response is the most beneficial strategy as we wait for vaccines to be globally available.

References

1. Beigel JH, Tomashek KM, Dodd LE, et al. Remdesivir for the treatment of covid-19—final report. *N Engl J Med*. 2020;383(19):1813−1826. https://doi.org/10.1056/NEJMoa2007764.
2. Sohrabi C, Alsafi Z, O'Neill N, et al. World Health Organization declares global emergency: a review of the 2019 novel coronavirus (COVID-19) [published correction appears in Int J Surg. 2020 May; 77:217]. *Int J Surg*. 2020;76:71−76. https://doi.org/10.1016/j.ijsu.2020.02.034.
3. Bulut C, Kato Y. Epidemiology of COVID-19. *Turk J Med Sci*. 2020;50(SI-1):563−570. https://doi.org/10.3906/sag-2004-172. Published 2020 Apr 21.
4. Li Q, Guan X, Wu P, et al. Early transmission dynamics in Wuhan, China, of novel coronavirus-infected pneumonia. *N Engl J Med*. 2020;382(13):1199−1207. https://doi.org/10.1056/NEJMoa2001316.
5. Wu Z, McGoogan JM. Characteristics of and important lessons from the coronavirus disease 2019 (COVID-19) outbreak in China: summary of a report of 72 314 cases from the Chinese center for disease control and prevention. *JAMA*. 2020;323(13):1239−1242. https://doi.org/10.1001/jama.2020.2648.
6. Petrilli CM, Jones SA, Yang J, et al. Factors associated with hospital admission and critical illness among 5279 people with coronavirus disease 2019 in New York City: prospective cohort study. *BMJ*. 2020;369:m1966. https://doi.org/10.1136/bmj.m1966. Published 2020 May 22.
7. Parkin J, Cohen B. An overview of the immune system. *Lancet*. 2001;357(9270):1777−1789. https://doi.org/10.1016/S0140-6736(00)04904-7.
8. Chaplin DD. Overview of the immune response. *J Allergy Clin Immunol*. 2010;125(2 Suppl 2):S3−S23. https://doi.org/10.1016/j.jaci.2009.12.980.
9. Hoebe K, Janssen E, Beutler B. The interface between innate and adaptive immunity. *Nat Immunol*. 2004;5(10):971−974. https://doi.org/10.1038/ni1004-971.

10. Tomar N, De RK. A brief outline of the immune system. *Methods Mol Biol.* 2014;1184: 3–12. https://doi.org/10.1007/978-1-4939-1115-8_1.
11. Beutler B. Innate immunity: an overview. *Mol Immunol.* 2004;40(12):845–859. https://doi.org/10.1016/j.molimm.2003.10.005.
12. Niyonsaba F, Kiatsurayanon C, Chieosilapatham P, Ogawa H. Friends or foes? Host defense (antimicrobial) peptides and proteins in human skin diseases. *Exp Dermatol.* 2017; 26(11):989–998. https://doi.org/10.1111/exd.13314.
13. Idborg H, Oke V. Cytokines as biomarkers in systemic lupus erythematosus: value for diagnosis and drug therapy. *Int J Mol Sci.* 2021;22(21):11327. https://doi.org/10.3390/ijms222111327. Published 2021 Oct 20.
14. Manna PR, Stetson CL, Daugherty C, et al. Up-regulation of steroid biosynthesis by retinoid signaling: implications for aging. *Mech Ageing Dev.* 2015;150:74–82. https://doi.org/10.1016/j.mad.2015.08.007.
15. Schmid-Wendtner MH, Korting HC. The pH of the skin surface and its impact on the barrier function. *Skin Pharmacol Physiol.* 2006;19(6):296–302. https://doi.org/10.1159/000094670.
16. Nguyen AV, Soulika AM. The dynamics of the skin's immune system. *Int J Mol Sci.* 2019; 20(8):1811. https://doi.org/10.3390/ijms20081811. Published 2019 Apr 12.
17. Liew PX, Kubes P. The neutrophil's role during health and disease. *Physiol Rev.* 2019; 99(2):1223–1248. https://doi.org/10.1152/physrev.00012.2018.
18. Singh PK, Parsek MR, Greenberg EP, Welsh MJ. A component of innate immunity prevents bacterial biofilm development. *Nature.* 2002;417(6888):552–555. https://doi.org/10.1038/417552a.
19. Bonilla FA, Oettgen HC. Adaptive immunity. *J Allergy Clin Immunol.* 2010;125(2 Suppl 2):S33–S40. https://doi.org/10.1016/j.jaci.2009.09.017\.
20. Cermakian N, Stegeman SK, Tekade K, Labrecque N. Circadian rhythms in adaptive immunity and vaccination. *Semin Immunopathol.* 2022;44(2):193–207. https://doi.org/10.1007/s00281-021-00903-7.
21. Dong C. Cytokine regulation and function in T cells. *Annu Rev Immunol.* 2021;39:51–76. https://doi.org/10.1146/annurev-immunol-061020-053702.
22. Seifert M, Küppers R. Human memory B cells. *Leukemia.* 2016;30(12):2283–2292. https://doi.org/10.1038/leu.2016.226.
23. Hillion S, Arleevskaya MI, Blanco P, et al. The innate part of the adaptive immune system. *Clin Rev Allergy Immunol.* 2020;58(2):151–154. https://doi.org/10.1007/s12016-019-08740-1.
24. Brodin P. Immune determinants of COVID-19 disease presentation and severity. *Nat Med.* 2021;27(1):28–33. https://doi.org/10.1038/s41591-020-01202-8.
25. Kaufmann SHE. Immunology's coming of age [published correction appears in Front Immunol. 2019 Jun 06; 10:1214]. *Front Immunol.* 2019;10:684. https://doi.org/10.3389/fimmu.2019.00684. Published 2019 Apr 3.
26. Dropulic LK, Lederman HM. Overview of infections in the immunocompromised host. *Microbiol Spectr.* 2016;4(4). https://doi.org/10.1128/microbiolspec.DMIH2-0026-2016, 10.1128/microbiolspec.DMIH2-0026-2016.
27. Zabetakis I, Lordan R, Norton C, Tsoupras A. COVID-19: the inflammation link and the role of nutrition in potential mitigation. *Nutrients.* 2020;12(5):1466. https://doi.org/10.3390/nu12051466. Published 2020 May 19.
28. Ahmed MH, Hassan A, Molnár J. The role of micronutrients to support immunity for COVID-19 prevention. *Rev Bras Farmacogn.* 2021;31(4):361–374. https://doi.org/10.1007/s43450-021-00179-w.

29. Grant WB, Lahore H, McDonnell SL, et al. Evidence that vitamin D supplementation could reduce risk of influenza and COVID-19 infections and deaths. *Nutrients*. 2020;12(4): 988. https://doi.org/10.3390/nu12040988. Published 2020 Apr 2.
30. Bilinski A, Emanuel EJ. COVID-19 and excess all-cause mortality in the US and 18 comparison countries. *JAMA*. 2020;324(20):2100–2102. https://doi.org/10.1001/jama.2020.20717.
31. Richardson DP, Lovegrove JA. Nutritional status of micronutrients as a possible and modifiable risk factor for COVID-19: a UK perspective. *Br J Nutr*. 2021;125(6):678–684. https://doi.org/10.1017/S000711452000330X.
32. Ahmed I, Hasan M, Akter R, et al. Behavioral preventive measures and the use of medicines and herbal products among the public in response to Covid-19 in Bangladesh: a cross-sectional study. *PLoS One*. 2020;15(12):e0243706. https://doi.org/10.1371/journal.pone.0243706. Published 2020 Dec 11.
33. Sińska B, Jaworski M, Panczyk M, Traczyk I, Kucharska A. The role of resilience and basic hope in the adherence to dietary recommendations in the polish population during the COVID-19 pandemic. *Nutrients*. 2021;13(6):2108. https://doi.org/10.3390/nu13062108. Published 2021 Jun 19.
34. Lefebvre P, Benomar Y, Staels B. Retinoid X receptors: common heterodimerization partners with distinct functions. *Trends Endocrinol Metabol*. 2010;21(11):676–683. https://doi.org/10.1016/j.tem.2010.06.009.
35. Manna PR, Slominski AT, King SR, Stetson CL, Stocco DM. Synergistic activation of steroidogenic acute regulatory protein expression and steroid biosynthesis by retinoids: involvement of cAMP/PKA signaling. *Endocrinology*. 2014;155(2):576–591. https://doi.org/10.1210/en.2013-1694.
36. Raverdeau M, Mills KH. Modulation of T cell and innate immune responses by retinoic Acid. *J Immunol*. 2014;192(7):2953–2958. https://doi.org/10.4049/jimmunol.1303245.
37. Clagett-Dame M, Knutson D. Vitamin A in reproduction and development. *Nutrients*. 2011;3(4):385–428. https://doi.org/10.3390/nu3040385.
38. Bikle DD. Vitamin D and the skin: physiology and pathophysiology. *Rev Endocr Metab Disord*. 2012;13(1):3–19. https://doi.org/10.1007/s11154-011-9194-0.
39. Huang Z, Liu Y, Qi G, Brand D, Zheng SG. Role of vitamin A in the immune system. *J Clin Med*. 2018;7(9):258. https://doi.org/10.3390/jcm7090258. Published 2018 Sep. 6.
40. Yuan S, Chu H, Chan JF, et al. SREBP-dependent lipidomic reprogramming as a broad-spectrum antiviral target. *Nat Commun*. 2019;10(1):120. https://doi.org/10.1038/s41467-018-08015-x. Published 2019 Jan 10.
41. Trasino SE. A role for retinoids in the treatment of COVID-19? *Clin Exp Pharmacol Physiol*. 2020;47(10):1765–1767. https://doi.org/10.1111/1440-1681.13354.
42. Zdrenghea MT, Makrinioti H, Bagacean C, Bush A, Johnston SL, Stanciu LA. Vitamin D modulation of innate immune responses to respiratory viral infections. *Rev Med Virol*. 2017;27(1). https://doi.org/10.1002/rmv.1909, 10.1002/rmv.1909.
43. Panfili FM, Roversi M, D'Argenio P, Rossi P, Cappa M, Fintini D. Possible role of vitamin D in Covid-19 infection in pediatric population. *J Endocrinol Invest*. 2021;44(1):27–35. https://doi.org/10.1007/s40618-020-01327-0.
44. Meza-Meza MR, Ruiz-Ballesteros AI, de la Cruz-Mosso U. Functional effects of vitamin D: from nutrient to immunomodulator [published online ahead of print, 2020 Dec 23]. *Crit Rev Food Sci Nutr*. 2020:1–21. https://doi.org/10.1080/10408398.2020.1862753.
45. Vergara D, Catherino WH, Trojano G, Tinelli A. Vitamin D: mechanism of action and biological effects in uterine fibroids. *Nutrients*. 2021;13(2):597. https://doi.org/10.3390/nu13020597. Published 2021 Feb 11.

46. Mitri J, Muraru MD, Pittas AG. Vitamin D and type 2 diabetes: a systematic review. *Eur J Clin Nutr*. 2011;65(9):1005–1015. https://doi.org/10.1038/ejcn.2011.118.
47. Manson JE, Cook NR, Lee IM, et al. Vitamin D supplements and prevention of cancer and cardiovascular disease. *N Engl J Med*. 2019;380(1):33–44. https://doi.org/10.1056/NEJMoa1809944.
48. Martineau AR, Jolliffe DA, Greenberg L, et al. Vitamin D supplementation to prevent acute respiratory infections: individual participant data meta-analysis. *Health Technol Assess*. 2019;23(2):1–44. https://doi.org/10.3310/hta23020.
49. Hayes CE, Ntambi JM. Multiple sclerosis: lipids, lymphocytes, and vitamin D. *Immunometabolism*. 2020;2(3):e200019. https://doi.org/10.20900/immunometab20200019.
50. Xu Y, Baylink DJ, Chen CS, et al. The importance of vitamin d metabolism as a potential prophylactic, immunoregulatory and neuroprotective treatment for COVID-19. *J Transl Med*. 2020;18(1):322. https://doi.org/10.1186/s12967-020-02488-5. Published 2020 Aug 26.
51. Beard JA, Bearden A, Striker R. Vitamin D and the anti-viral state. *J Clin Virol*. 2011; 50(3):194–200. https://doi.org/10.1016/j.jcv.2010.12.006.
52. Barlow PG, Findlay EG, Currie SM, Davidson DJ. Antiviral potential of cathelicidins. *Future Microbiol*. 2014;9(1):55–73. https://doi.org/10.2217/fmb.13.135.
53. Galmés S, Serra F, Palou A. Current state of evidence: influence of nutritional and nutrigenetic factors on immunity in the COVID-19 pandemic framework. *Nutrients*. 2020; 12(9):2738. https://doi.org/10.3390/nu12092738. Published 2020 Sep. 8.
54. Rawat D, Roy A, Maitra S, Shankar V, Khanna P, Baidya DK. Vitamin D supplementation and COVID-19 treatment: a systematic review and meta-analysis. *Diabetes Metabol Syndr*. 2021;15(4):102189. https://doi.org/10.1016/j.dsx.2021.102189.
55. Taha R, Abureesh S, Alghamdi S, et al. The relationship between vitamin D and infections including COVID-19: any hopes? *Int J Gen Med*. 2021;14:3849–3870. https://doi.org/10.2147/IJGM.S317421. Published 2021 Jul 24.
56. Yin K, Agrawal DK. Vitamin D and inflammatory diseases. *J Inflamm Res*. 2014;7:69–87. https://doi.org/10.2147/JIR.S63898. Published 2014 May 29.
57. Hu B, Huang S, Yin L. The cytokine storm and COVID-19. *J Med Virol*. 2021;93(1): 250–256. https://doi.org/10.1002/jmv.26232.
58. Ye Q, Wang B, Mao J. The pathogenesis and treatment of the 'Cytokine Storm' in COVID-19. *J Infect*. 2020;80(6):607–613. https://doi.org/10.1016/j.jinf.2020.03.037.
59. Pedersen SF, Ho YC. SARS-CoV-2: a storm is raging. *J Clin Invest*. 2020;130(5): 2202–2205. https://doi.org/10.1172/JCI137647.
60. Zhang Y, Leung DY, Richers BN, et al. Vitamin D inhibits monocyte/macrophage proinflammatory cytokine production by targeting MAPK phosphatase-1. *J Immunol*. 2012;188(5):2127–2135. https://doi.org/10.4049/jimmunol.1102412.
61. Bouillon R, Manousaki D, Rosen C, Trajanoska K, Rivadeneira F, Richards JB. The health effects of vitamin D supplementation: evidence from human studies. *Nat Rev Endocrinol*. 2022;18(2):96–110. https://doi.org/10.1038/s41574-021-00593-z.
62. Annweiler C, Cao Z, Sabatier JM. Point of view: should COVID-19 patients be supplemented with vitamin D? *Maturitas*. 2020;140:24–26. https://doi.org/10.1016/j.maturitas.2020.06.003.
63. Easty DJ, Farr CJ, Hennessy BT. New roles for vitamin D superagonists: from COVID to cancer. *Front Endocrinol*. 2021;12:644298. https://doi.org/10.3389/fendo.2021.644298. Published 2021 Mar 31.
64. Alexander J, Tinkov A, Strand TA, Alehagen U, Skalny A, Aaseth J. Early nutritional interventions with zinc, selenium and vitamin D for raising anti-viral resistance against

progressive COVID-19. *Nutrients*. 2020;12(8):2358. https://doi.org/10.3390/nu12082358. Published 2020 Aug 7.
65. Mohd Zaffarin AS, Ng SF, Ng MH, Hassan H, Alias E. Pharmacology and pharmacokinetics of vitamin E: nanoformulations to enhance bioavailability. *Int J Nanomed*. 2020;15: 9961–9974. https://doi.org/10.2147/IJN.S276355. Published 2020 Dec 8.
66. Fudalej E, Justyniarska M, Kasarełło K, Dziedziak J, Szaflik JP, Cudnoch-Jedrzejewska A. Neuroprotective factors of the retina and their role in promoting survival of retinal ganglion cells: a review. *Ophthalmic Res*. 2021;64(3):345–355. https://doi.org/10.1159/000514441.
67. Muller DP. Vitamin E and neurological function. *Mol Nutr Food Res*. 2010;54(5): 710–718. https://doi.org/10.1002/mnfr.200900460.
68. Galli F, Azzi A, Birringer M, et al. Vitamin E: emerging aspects and new directions. *Free Radic Biol Med*. 2017;102:16–36. https://doi.org/10.1016/j.freeradbiomed.2016.09.017.
69. Lee GY, Han SN. The role of vitamin E in immunity. *Nutrients*. 2018;10(11):1614. https://doi.org/10.3390/nu10111614. Published 2018 Nov 1.
70. Shakoor H, Feehan J, Al Dhaheri AS, et al. Immune-boosting role of vitamins D, C, E, zinc, selenium and omega-3 fatty acids: could they help against COVID-19? *Maturitas*. 2021;143:1–9. https://doi.org/10.1016/j.maturitas.2020.08.003.
71. Iddir M, Brito A, Dingeo G, et al. Strengthening the immune system and reducing inflammation and oxidative stress through diet and nutrition: considerations during the COVID-19 crisis. *Nutrients*. 2020;12(6):1562. https://doi.org/10.3390/nu12061562. Published 2020 May 27.
72. Walker M, Samii A. Delayed onset of ataxia in a patient with short bowel syndrome: a case of vitamin E deficiency. *Nutr Neurosci*. 2004;7(3):191–193. https://doi.org/10.1080/10284150400001995.
73. Traber MG. Vitamin E inadequacy in humans: causes and consequences. *Adv Nutr*. 2014; 5(5):503–514. https://doi.org/10.3945/an.114.006254.
74. Ulatowski LM, Manor D. Vitamin E and neurodegeneration. *Neurobiol Dis*. 2015;84: 78–83. https://doi.org/10.1016/j.nbd.2015.04.002.
75. Azzi A, Meydani SN, Meydani M, Zingg JM. The rise, the fall and the renaissance of vitamin E. *Arch Biochem Biophys*. 2016;595:100–108. https://doi.org/10.1016/j.abb.2015.11.010.
76. Lewis ED, Meydani SN, Wu D. Regulatory role of vitamin E in the immune system and inflammation. *IUBMB Life*. 2019;71(4):487–494. https://doi.org/10.1002/iub.1976.
77. Fusaro M, Cianciolo G, Brandi ML, et al. Vitamin K and osteoporosis. *Nutrients*. 2020; 12(12):3625. https://doi.org/10.3390/nu12123625. Published 2020 Nov 25.
78. Imbrescia K, Moszczynski Z. Vitamin K. In: *StatPearls*. Treasure Island (FL): StatPearls Publishing; July 13, 2021.
79. Akbulut AC, Pavlic A, Petsophonsakul P, et al. Vitamin K2 needs an RDI separate from vitamin K1. *Nutrients*. 2020;12(6):1852. https://doi.org/10.3390/nu12061852. Published 2020 Jun 21.
80. Suleiman L, Négrier C, Boukerche H. Protein S: a multifunctional anticoagulant vitamin K-dependent protein at the crossroads of coagulation, inflammation, angiogenesis, and cancer. *Crit Rev Oncol Hematol*. 2013;88(3):637–654. https://doi.org/10.1016/j.critrevonc.2013.07.004.
81. Simes DC, Viegas CSB, Araújo N, Marreiros C. Vitamin K as a diet supplement with impact in human health: current evidence in age-related diseases. *Nutrients*. 2020;12(1): 138. https://doi.org/10.3390/nu12010138. Published 2020 Jan 3.

82. Shioi A, Morioka T, Shoji T, Emoto M. The inhibitory roles of vitamin K in progression of vascular calcification. *Nutrients*. 2020;12(2):583. https://doi.org/10.3390/nu12020583. Published 2020 Feb 23.
83. Vogrig A, Gigli GL, Bnà C, Morassi M. Stroke in patients with COVID-19: clinical and neuroimaging characteristics. *Neurosci Lett*. 2021;743:135564. https://doi.org/10.1016/j.neulet.2020.135564.
84. Anastasi E, Ialongo C, Labriola R, Ferraguti G, Lucarelli M, Angeloni A. Vitamin K deficiency and covid-19. *Scand J Clin Lab Invest*. 2020;80(7):525–527. https://doi.org/10.1080/00365513.2020.1805122.
85. Sankar J, Lotha W, Ismail J, Anubhuti C, Meena RS, Sankar MJ. Vitamin D deficiency and length of pediatric intensive care unit stay: a prospective observational study. *Ann Intensive Care*. 2016;6(1):3. https://doi.org/10.1186/s13613-015-0102-8.
86. Levy DS, Grewal R, Le TH. Vitamin K deficiency: an emerging player in the pathogenesis of vascular calcification and an iatrogenic consequence of therapies in advanced renal disease. *Am J Physiol Ren Physiol*. 2020;319(4):F618–F623. https://doi.org/10.1152/ajprenal.00278.2020.
87. Eden RE, Coviello JM. Vitamin K deficiency. In: *StatPearls*. Treasure Island (FL): StatPearls Publishing; July 26, 2021.
88. Kumar P, Kumar M, Bedi O, et al. Role of vitamins and minerals as immunity boosters in COVID-19. *Inflammopharmacology*. 2021;29(4):1001–1016. https://doi.org/10.1007/s10787-021-00826-7.
89. Carr AC, Maggini S. Vitamin C and immune function. *Nutrients*. 2017;9(11):1211. https://doi.org/10.3390/nu9111211. Published 2017 Nov 3.
90. Hemilä H. Vitamin C in clinical therapeutics. *Clin Therapeut*. 2017;39(10):2110–2112. https://doi.org/10.1016/j.clinthera.2017.08.005.
91. Doseděl M, Jirkovský E, Macáková K, et al. Vitamin C-sources, physiological role, kinetics, deficiency, use, toxicity, and determination. *Nutrients*. 2021;13(2):615. https://doi.org/10.3390/nu13020615. Published 2021 Feb 13.
92. Hemilä H. Vitamin C and infections. *Nutrients*. 2017;9(4):339. https://doi.org/10.3390/nu9040339. Published 2017 Mar 29.
93. Bae M, Kim H. Mini-Review on the roles of vitamin C, vitamin D, and selenium in the immune system against COVID-19. *Molecules*. 2020;25(22):5346. https://doi.org/10.3390/molecules25225346. Published 2020 Nov 16.
94. Rawat D, Roy A, Maitra S, Gulati A, Khanna P, Baidya DK. Vitamin C and COVID-19 treatment: a systematic review and meta-analysis of randomized controlled trials. *Diabetes Metabol Syndr*. 2021;15(6):102324. https://doi.org/10.1016/j.dsx.2021.102324.
95. Ratajczak AE, Szymczak-Tomczak A, Skrzypczak-Zielińska M, et al. Vitamin C deficiency and the risk of osteoporosis in patients with an inflammatory bowel disease. *Nutrients*. 2020;12(8):2263. https://doi.org/10.3390/nu12082263. Published 2020 Jul 29.
96. Eckle SB, Corbett AJ, Keller AN, et al. Recognition of vitamin B precursors and byproducts by mucosal associated invariant T cells. *J Biol Chem*. 2015;290(51): 30204–30211. https://doi.org/10.1074/jbc.R115.685990.
97. Peterson CT, Rodionov DA, Osterman AL, Peterson SN. B vitamins and their role in immune regulation and cancer. *Nutrients*. 2020;12(11):3380. https://doi.org/10.3390/nu12113380. Published 2020 Nov 4.
98. Tardy AL, Pouteau E, Marquez D, Yilmaz C, Scholey A. Vitamins and minerals for energy, fatigue and cognition: a narrative review of the biochemical and clinical evidence. *Nutrients*. 2020;12(1):228. https://doi.org/10.3390/nu12010228. Published 2020 Jan 16.

99. Frank RA, Leeper FJ, Luisi BF. Structure, mechanism and catalytic duality of thiamine-dependent enzymes. *Cell Mol Life Sci.* 2007;64(7–8):892–905. https://doi.org/10.1007/s00018-007-6423-5.
100. Wiley KD, Gupta M. Vitamin B1 thiamine deficiency. In: *StatPearls*. Treasure Island (FL): StatPearls Publishing; June 21, 2021.
101. Mathis D, Shoelson SE. Immunometabolism: an emerging frontier. *Nat Rev Immunol.* 2011;11(2):81. https://doi.org/10.1038/nri2922.
102. Artyomov MN, Van den Bossche J. Immunometabolism in the single-cell era. *Cell Metabol.* 2020;32(5):710–725. https://doi.org/10.1016/j.cmet.2020.09.013.
103. Marrs C, Lonsdale D. Hiding in plain sight: modern thiamine deficiency. *Cells.* 2021;10(10):2595. https://doi.org/10.3390/cells10102595. Published 2021 Sep. 29.
104. Powers HJ. Riboflavin (vitamin B-2) and health. *Am J Clin Nutr.* 2003;77(6):1352–1360. https://doi.org/10.1093/ajcn/77.6.1352.
105. Thakur K, Tomar SK, Singh AK, Mandal S, Arora S. Riboflavin and health: a review of recent human research. *Crit Rev Food Sci Nutr.* 2017;57(17):3650–3660. https://doi.org/10.1080/10408398.2016.1145104.
106. Huang F, Zhang C, Liu Q, et al. Identification of amitriptyline HCl, flavin adenine dinucleotide, azacitidine and calcitriol as repurposing drugs for influenza A H5N1 virus-induced lung injury. *PLoS Pathog.* 2020;16(3):e1008341. https://doi.org/10.1371/journal.ppat.1008341. Published 2020 Mar 16.
107. Wojcieszyńska D, Hupert-Kocurek K, Guzik U. Flavin-dependent enzymes in cancer prevention. *Int J Mol Sci.* 2012;13(12):16751–16768. https://doi.org/10.3390/ijms131216751. Published 2012 Dec 7.
108. Saedisomeolia A, Ashoori M. Riboflavin in human health: a review of current evidences. *Adv Food Nutr Res.* 2018;83:57–81. https://doi.org/10.1016/bs.afnr.2017.11.002.
109. Handelsman Y, Shapiro MD. Triglycerides, atherosclerosis, and cardiovascular outcome studies: focus on omega-3 fatty acids. *Endocr Pract.* 2017;23(1):100–112. https://doi.org/10.4158/EP161445.RA.
110. Superko HR, Zhao XQ, Hodis HN, Guyton JR. Niacin and heart disease prevention: engraving its tombstone is a mistake. *J Clin Lipidol.* 2017;11(6):1309–1317. https://doi.org/10.1016/j.jacl.2017.08.005.
111. Cybulska B, Kłosiewicz-Latoszek L, Penson PE, Banach M. What do we know about the role of lipoprotein(a) in atherogenesis 57 years after its discovery? *Prog Cardiovasc Dis.* 2020;63(3):219–227. https://doi.org/10.1016/j.pcad.2020.04.004.
112. Fania L, Mazzanti C, Campione E, Candi E, Abeni D, Dellambra E. Role of nicotinamide in genomic stability and skin cancer chemoprevention. *Int J Mol Sci.* 2019;20(23):5946. https://doi.org/10.3390/ijms20235946. Published 2019 Nov 26.
113. Chu X, Raju RP. Regulation of NAD+ metabolism in aging and disease. *Metabolism.* 2022;126:154923. https://doi.org/10.1016/j.metabol.2021.154923.
114. Kwon WY, Suh GJ, Kim KS, Kwak YH. Niacin attenuates lung inflammation and improves survival during sepsis by downregulating the nuclear factor-κB pathway. *Crit Care Med.* 2011;39(2):328–334. https://doi.org/10.1097/CCM.0b013e3181feeae4.
115. Gasperi V, Sibilano M, Savini I, Catani MV. Niacin in the central nervous system: an update of biological aspects and clinical applications. *Int J Mol Sci.* 2019;20(4):974. https://doi.org/10.3390/ijms20040974. Published 2019 Feb 23.
116. Gao J, Zhang L, Liu X, et al. Repurposing low-molecular-weight drugs against the main protease of severe acute respiratory syndrome coronavirus 2. *J Phys Chem Lett.* 2020;11(17):7267–7272. https://doi.org/10.1021/acs.jpclett.0c01894.

117. Liu S, Zhu X, Qiu Y, et al. Effect of niacin on growth performance, intestinal morphology, mucosal immunity and microbiota composition in weaned piglets. *Animals*. 2021;11(8): 2186. https://doi.org/10.3390/ani11082186. Published 2021 Jul 23.
118. Tang J, Feng Y, Zhang B, et al. Severe pantothenic acid deficiency induces alterations in the intestinal mucosal proteome of starter Pekin ducks. *BMC Genom*. 2021;22(1):491. https://doi.org/10.1186/s12864-021-07820-x. Published 2021 Jun 30.
119. Gheita AA, Gheita TA, Kenawy SA. The potential role of B5: a stitch in time and switch in cytokine. *Phytother Res*. 2020;34(2):306–314. https://doi.org/10.1002/ptr.6537.
120. Kumar RR, Singh L, Thakur A, Singh S, Kumar B. Role of vitamins in neurodegenerative diseases: a review [published online ahead of print, 2021 Nov 19]. *CNS Neurol Disord: Drug Targets*. 2021. https://doi.org/10.2174/1871527320666211119122150, 10.2174/1871527320666211119122150.
121. Anogeianaki A, Castellani ML, Tripodi D, et al. Vitamins and mast cells. *Int J Immunopathol Pharmacol*. 2010;23(4):991–996. https://doi.org/10.1177/039463201002300403.
122. Stach K, Stach W, Augoff K. Vitamin B6 in health and disease. *Nutrients*. 2021;13(9): 3229. https://doi.org/10.3390/nu13093229. Published 2021 Sep. 17.
123. Ueland PM, McCann A, Ø M, Ulvik A. Inflammation, vitamin B6 and related pathways. *Mol Aspect Med*. 2017;53:10–27. https://doi.org/10.1016/j.mam.2016.08.001.
124. Zhang C, Luo J, Yuan C, Ding D. Vitamin B12, B6, or folate and cognitive function in community-dwelling older adults: a systematic review and meta-analysis. *J Alzheimers Dis*. 2020;77(2):781–794. https://doi.org/10.3233/JAD-200534.
125. Rall LC, Meydani SN. Vitamin B6 and immune competence. *Nutr Rev*. 1993;51(8): 217–225. https://doi.org/10.1111/j.1753-4887.1993.tb03109.x.
126. Fernandez-Mejia C. Pharmacological effects of biotin. *J Nutr Biochem*. 2005;16(7): 424–427. https://doi.org/10.1016/j.jnutbio.2005.03.018.
127. DiBaise M, Tarleton SM. Hair, nails, and skin: differentiating cutaneous manifestations of micronutrient deficiency. *Nutr Clin Pract*. 2019;34(4):490–503. https://doi.org/10.1002/ncp.10321.
128. Riveron-Negrete L, Fernandez-Mejia C. Pharmacological effects of biotin in animals. *Mini Rev Med Chem*. 2017;17(6):529–540. https://doi.org/10.2174/1389557516666160923132611.
129. León-Del-Río A. Biotin in metabolism, gene expression, and human disease. *J Inherit Metab Dis*. 2019;42(4):647–654. https://doi.org/10.1002/jimd.12073.
130. Mock DM. Biotin: from nutrition to therapeutics. *J Nutr*. 2017;147(8):1487–1492. https://doi.org/10.3945/jn.116.238956.
131. Piraccini BM, Berardesca E, Fabbrocini G, Micali G, Tosti A. Biotin: overview of the treatment of diseases of cutaneous appendages and of hyperseborrhea. *G Ital Dermatol Venereol*. 2019;154(5):557–566. https://doi.org/10.23736/S0392-0488.19.06434-4.
132. Shulpekova Y, Nechaev V, Kardasheva S, et al. The concept of folic acid in health and disease. *Molecules*. 2021;26(12):3731. https://doi.org/10.3390/molecules26123731. Published 2021 Jun 18.
133. Li B, Zhang X, Peng X, Zhang S, Wang X, Zhu C. Folic acid and risk of preterm birth: a meta-analysis. *Front Neurosci*. 2019;13:1284. https://doi.org/10.3389/fnins.2019.01284. Published 2019 Nov 28.
134. Watkins D, Rosenblatt DS. Immunodeficiency and inborn disorders of vitamin B12 and folate metabolism. *Curr Opin Clin Nutr Metab Care*. 2020;23(4):241–246. https://doi.org/10.1097/MCO.0000000000000668.
135. Craenen K, Verslegers M, Baatout S, Abderrafi Benotmane M. An appraisal of folates as key factors in cognition and ageing-related diseases. *Crit Rev Food Sci Nutr*. 2020;60(5): 722–739. https://doi.org/10.1080/10408398.2018.1549017.

136. Araújo JR, Martel F, Borges N, Araújo JM, Keating E. Folates and aging: role in mild cognitive impairment, dementia and depression. *Ageing Res Rev*. 2015;22:9–19. https://doi.org/10.1016/j.arr.2015.04.005.
137. Ebrahimnejad P, Sodagar Taleghani A, Asare-Addo K, Nokhodchi A. An updated review of folate-functionalized nanocarriers: a promising ligand in cancer. *Drug Discov Today*. 2022;27(2):471–489. https://doi.org/10.1016/j.drudis.2021.11.011.
138. Courtemanche C, Elson-Schwab I, Mashiyama ST, Kerry N, Ames BN. Folate deficiency inhibits the proliferation of primary human CD8+ T lymphocytes in vitro. *J Immunol*. 2004;173(5):3186–3192. https://doi.org/10.4049/jimmunol.173.5.3186.
139. Courtemanche C, Huang AC, Elson-Schwab I, Kerry N, Ng BY, Ames BN. Folate deficiency and ionizing radiation cause DNA breaks in primary human lymphocytes: a comparison. *FASEB J*. 2004;18(1):209–211. https://doi.org/10.1096/fj.03-0382fje.
140. Dhur A, Galan P, Hercberg S. Folate status and the immune system. *Prog Food Nutr Sci*. 1991;15(1–2):43–60.
141. Meisel E, Efros O, Bleier J, et al. Folate levels in patients hospitalized with coronavirus disease 2019. *Nutrients*. 2021;13(3):812. https://doi.org/10.3390/nu13030812. Published 2021 Mar 2.
142. Wiebe N, Field CJ, Tonelli M. A systematic review of the vitamin B12, folate and homocysteine triad across body mass index. *Obes Rev*. 2018;19(11):1608–1618. https://doi.org/10.1111/obr.12724.
143. Markun S, Gravestock I, Jäger L, Rosemann T, Pichierri G, Burgstaller JM. Effects of vitamin B12 supplementation on cognitive function, depressive symptoms, and fatigue: a systematic review, meta-analysis, and meta-regression. *Nutrients*. 2021;13(3):923. https://doi.org/10.3390/nu13030923. Published 2021 Mar 12.
144. Yoshii K, Hosomi K, Sawane K, Kunisawa J. Metabolism of dietary and microbial vitamin B family in the regulation of host immunity. *Front Nutr*. 2019;6:48. https://doi.org/10.3389/fnut.2019.00048. Published 2019 Apr 17.
145. Azzini E, Raguzzini A, Polito A. A brief review on vitamin B12 deficiency looking at some case study reports in adults. *Int J Mol Sci*. 2021;22(18):9694. https://doi.org/10.3390/ijms22189694. Published 2021 Sep. 7.
146. Tamura J, Kubota K, Murakami H, et al. Immunomodulation by vitamin B12: augmentation of CD8+ T lymphocytes and natural killer (NK) cell activity in vitamin B12-deficient patients by methyl-B12 treatment. *Clin Exp Immunol*. 1999;116(1):28–32. https://doi.org/10.1046/j.1365-2249.1999.00870.x.
147. BourBour F, Mirzaei Dahka S, Gholamalizadeh M, et al. Nutrients in prevention, treatment, and management of viral infections; special focus on coronavirus [published online ahead of print, 2020 Jul 9]. *Arch Physiol Biochem*. 2020:1–10. https://doi.org/10.1080/13813455.2020.1791188.
148. Shakeri H, Azimian A, Ghasemzadeh-Moghaddam H, et al. Evaluation of the relationship between serum levels of zinc, vitamin B12, vitamin D, and clinical outcomes in patients with COVID-19. *J Med Virol*. 2022;94(1):141–146. https://doi.org/10.1002/jmv.27277.
149. Subedi L, Tchen S, Gaire BP, Hu B, Hu K. Adjunctive nutraceutical therapies for COVID-19. *Int J Mol Sci*. 2021;22(4):1963. https://doi.org/10.3390/ijms22041963. Published 2021 Feb 16.
150. Samad N, Sodunke TE, Abubakar AR, et al. The implications of zinc therapy in combating the COVID-19 global pandemic. *J Inflamm Res*. 2021;14:527–550. https://doi.org/10.2147/JIR.S295377. Published 2021 Feb 26.

151. Celik C, Gencay A, Ocsoy I. Can food and food supplements be deployed in the fight against the COVID 19 pandemic? *Biochim Biophys Acta Gen Subj*. 2021;1865(2):129801. https://doi.org/10.1016/j.bbagen.2020.129801.
152. Soyano A, Gómez M. Participación del hierro en la inmunidad y su relación con las infecciones [Role of iron in immunity and its relation with infections]. *Arch Latinoam Nutr*. 1999;49(3 Suppl 2):40S–46S.
153. Institute of Medicine (US) Committee on Military Nutrition Research. *Military Strategies for Sustainment of Nutrition and Immune Function in the Field*. Washington (DC): National Academies Press (US); 1999.
154. Akhtar S, Das JK, Ismail T, Wahid M, Saeed W, Bhutta ZA. Nutritional perspectives for the prevention and mitigation of COVID-19. *Nutr Rev*. 2021;79(3):289–300. https://doi.org/10.1093/nutrit/nuaa063.
155. El Sayed SM, Aboonq MS, El Rashedy AG, et al. Promising preventive and therapeutic effects of TaibUVID nutritional supplements for COVID-19 pandemic: towards better public prophylaxis and treatment (A retrospective study). *Am J Blood Res*. 2020;10(5): 266–282. Published 2020 Oct 15.
156. Khadka D, Dhamala MK, Li F, et al. The use of medicinal plants to prevent COVID-19 in Nepal. *J Ethnobiol Ethnomed*. 2021;17(1):26. https://doi.org/10.1186/s13002-021-00449-w. Published 2021 Apr 8.
157. Galanakis CM, Aldawoud TMS, Rizou M, Rowan NJ, Ibrahim SA. Food ingredients and active compounds against the coronavirus disease (COVID-19) pandemic: a comprehensive review. *Foods*. 2020;9(11):1701. https://doi.org/10.3390/foods9111701. Published 2020 Nov 20.

Diabetes mellitus in relation to COVID-19

4

Bhagavathi Ramasubramanian [1,2], *Jonathan Kopel* [3], *Madison Hanson* [1] and *Cameron Griffith* [4]
[1]Department of Internal Medicine, Texas Tech University Health Sciences Center, Lubbock, TX, United States; [2]Department of Neurology, UT Southwestern Medical Center, Dallas, TX, United States; [3]School of Medicine, Texas Tech University Health Sciences Center, Lubbock, TX, United States; [4]Department of Anthropology, Texas Tech University, Lubbock, TX, United States

Abstract

The recent ongoing COVID-19 pandemic caused by the SARS-CoV-2 virus saw many hospitalizations and deaths among elderly patients. It has been reported that the most common underlying conditions in these patients were obesity and diabetes. While both type 1 and type 2 diabetes pose a higher risk of severe or fatal COVID-19 infections, patients with type 2 diabetes required ICU treatment at a greater frequency than those with type 1 diabetes. However, whether diabetes affects susceptibility for COVID-19 has yet to be explored. This chapter focuses on both type 1 and type 2 diabetes, with the main goal of understanding this chronic condition during the pandemic, based on currently available case studies.

Keywords: Biomarkers; COVID-19; Diabetes; Hyperglycemia; Lifestyle

1. Introduction

Diabetes is a chronic metabolic disease characterized by hyperglycemia. The prevalence of diabetes is rapidly increasing worldwide and is becoming a major health issue causing significant personal, social, and economic burdens. It has been reported recently by the International Diabetes Federation (IDF) that 463 million adults are currently living with diabetes, a statistic on course to exceed 578 million by 2030 and 700 million by 2045. The annual global health expenditure on diabetes has been estimated at USD 760 billion. Therefore, estimations of cost of managing diabetes are set to exceed USD 825 billion by 2030 and USD 845 billion by 2045.[1]

During the recent coronavirus pandemic, diabetes was identified as a common comorbidity in COVID-19-infected adults.[2,3] A study conducted on 5700 patients in New York state demonstrated that diabetes was the third most common comorbidity with COVID-19, following hypertension and obesity.[4] Interestingly, the grave prognosis

COVID-19 in Alzheimer's Disease and Dementia. https://doi.org/10.1016/B978-0-443-15256-6.00003-9

and the risk factors for patients with diabetes are well linked with hypertension and obesity. Patients with COVID-19 and diabetes are at greater risk for more severe infections, worse prognosis, and a much higher mortality rate compared to those without diabetes. In a recent study, 318 fatalities were seen in a sample population of 799 patients with COVID-19 with the highest odds of fatality risk associated with cardiovascular disease (CVD) or diabetes. The risk increased exponentially when CVD and diabetes coexisted in patients.[5] Furthermore, blood glucose management is challenging in COVID-19 patients because it requires a more detailed strategy for medical team integration.[6] In this chapter, we explore the risk factors of diabetes and obesity in relation to COVID-19.

2. Type 1 diabetes and type 2 diabetes

There are three main types of diabetes, type 1 diabetes (T1D), type 2 diabetes (T2D), and gestational diabetes. In this chapter, we focus only on T1D and T2D. T1D involves an autoimmune process whereby the body destroys the pancreatic cells responsible for producing insulin. Hence, the body does not produce insulin at all. In T2D, the body is unable to properly utilize insulin, resulting in a condition called "insulin resistance." T2D is more common and accounts for more than 90% of all diabetes cases in the world.

Patients with T1D or T2D have a risk of a higher mortality rate than those without diabetes. In a nationwide study conducted in Sweden in 2021, T2D was independently associated with increased risk of hospitalization, admission to intensive care, and death with concomitant COVID-19. There were fewer admissions into intensive care and deaths in T1D patients. Although hazards were significantly raised for all three outcomes, there was no independent risk persisting after adjustment for confounding factors.[7] In the UK, patients with both COVID-19 and T1D often presented with a higher BMI, worse renal function, and had evidence of microvascular damage that increased their risk of death and/or admission to adult intensive care unit. However, the risk of severe COVID-19 is reassuringly low in patients with T1D and who are under 55 years of age without microvascular or macrovascular disease.[8]

A 2021 study conducted in the UK on 88 patients with diabetic ketoacidosis (DKA), an acute complication of diabetes, noted that patients with T1D and COVID-19 presented with more severe hyperglycemia while patients with T2D presented with DKA. As a result, patients with T2D had a higher mortality rate and more complicated hospital course.[9]

3. Biomarkers and risk factors in COVID-19-infected patients

3.1 Inflammatory and blood biomarkers

Hyperglycemia is a marker of stress and inflammation that influences metabolic responses to infection. Hyperglycemia has also been reported previously in both diabetic and nondiabetic patients with conditions such as acute myocardial infarction, stroke, burns, and trauma.[10] Based on a targeted survey of extant studies, Roganović proposed a hypothesis that downregulation of circulating microRNA miR-146a is one of the factors underlying severe COVID-19. This downregulation was found in patients with diabetes, obesity, and hypertension.[11] Blood biomarkers including serum ferritin, D-dimer, alanine aminotransferase (ALT), troponin I, and hemoglobin A1c (Hb A1c) are significantly higher in COVID-19 patients. Inflammatory biomarkers such as C-reactive protein (CRP), neutrophils, IL-6, lactate dehydrogenase (LDH), and erythrocyte sedimentation rate (ESR) are increased in patients with diabetes compared to those without diabetes.[12,13] Ferritin and HbA1c levels were significantly higher in diabetic than nondiabetic COVID-19 patients.[14] Additionally, serum levels of intracellular adhesion molecule-1 (ICAM-1) are significantly elevated in patients with COVID-19 with diabetic foot ulcers (DFUs) compared with patients without COVID-19.[15]

3.2 Obesity as an independent risk factor in COVID-19 infection

Both obesity and diabetes are established comorbidities for COVID-19 infection. Adipose tissue highly expresses ACE2, a receptor used by SARS-CoV-2 to enter the host.[16,17] Studies in two cities in China in 2020 reported a higher incidence of pneumonia in obese patients infected with COVID-19 than in nonobese patients.[18,19] Another 2020 study conducted by Kass et al. in the US reported that in a subset of young patients infected with COVID-19, extreme obesity was an associated comorbidity.[20] Cao et al. conducted a study in 2021 with a large sample of 1637 adult inpatients in the first epicenter in Wuhan, China, found that obesity was significantly associated with the risk of severe pneumonia and in-hospital oxygen therapy. This study also found obesity to be an independent risk factor for COVID-19 severity in the elderly population and patients with diabetes.[16] These biomarkers and risk factors are outlined in Table 4.1.

4. Entry of SARS-CoV-2 into the host

Recently, it has been confirmed that the SARS-CoV-2 virus uses angiotensin-converting enzyme 2 (ACE2). ACE2 receptor is an inhibitor of the renin-angiotensin-aldosterone system (RAAS) to gain entry into the host along with the transmembrane protease serine protease-2 TMPRSS2 for S protein priming.[21] ACE2 is expressed in the respiratory system, the intestines, kidneys, myocardium, vasculature, and pancreatic islets. During the pandemic, it was observed that initial diagnosis of diabetes mellitus (DM) was common in COVID-19 patients who did not have any prior history of DM nor did they use glucocorticoids. This has been termed as new-onset hyperglycemia. It has been attributed to the binding of SARS-CoV-2 to ACE2 receptor in pancreatic islets.[22] Studies conducted in animal models have demonstrated that ACE2-knockout mice are more vulnerable to β cell dysfunction, while in diabetic mouse model, db/db provided with ACE2 therapy

Table 4.1 Biomarkers and risk factors significant in diabetes/obesity and COVID-19-infected patients.

Inflammatory biomarkers:	Regulation	Condition/s	References
C-reactive proteins (CRP)	Increased	Diabetes	Moghaddam et al., 2021[12]; Guo et al.[11]
Neutrophils	Increased	Diabetes	Moghaddam et al.[12]; Guo et al.[11]
IL-6	Increased	Diabetes	Moghaddam et al.[12] Guo et al.[11]
Lactate dehydrogenase (LDH)	Increased	Diabetes	Moghaddam et al.[12]; Guo et al.[11]
Erythrocyte sedimentation rate (ESR)	Increased	Diabetes	Moghaddam et al.[12]; Guo et al.[11]
ICAM-1	Increased	Diabetic foot ulcers	Oley et al.[12]
Hyperglycemia		Diabetes and nondiabetes	Mazori et al.[14]
Blood biomarkers:			
ICAM-1	Increased	Diabetic foot ulcers	Oley et al.[15]
Ferritin	Increased	Diabetes	Malik et al.[14]
D-dimer	Increased	Diabetes	Malik et al.[14]
Alanine aminotransferase (ALT)	Increased	Diabetes	Malik et al.[14]
Troponin-1	Increased	Diabetes	Malik et al.[14]
Hemoglobin A1c (HbA1c)	Increased	Diabetes	Malik et al.[14]
miR-146a	Downregulated	Diabetes, obesity, and hypertension	Roganović[11]
Stress markers:			
Hyperglycemia	Increased	Diabetes and nondiabetes	Mazori et al.[14]
Other risk factors:			
Obesity			Cao et al.[16]

directly in the pancreas, improved fasting glucose and glucose tolerance.[23] These studies indicate the possibility of why infection with SARS-CoV-2 causes hyperglycemia in humans without preexisting DM. D'Onofrio et al.[24] reported upregulation of ACE2 expression in DM cardiomyocytes along with nonenzymatic glycation, as a possibility to increase the susceptibility to COVID-19 infection in DM patients by favoring the cellular entry of SARS-CoV-2.[24]

5. Complications of diabetes during COVID-19 infection

5.1 Diabetic ketoacidosis

Diabetic ketoacidosis is an acute complication of diabetes characterized by hyperglycemia, excessive ketone body formation (ketosis), and acidosis. As noted in Section 2 of this chapter, T2D patients infected with COVID-19 presented with DKA, and greater ICU needs than T1D patients infected with COVID-19. During the pandemic, in COVID-19 infection, it has been observed that patients with preexisting diabetes were at increased risk of DKA.[25] Pasquel et al.[26] conducted a study in a large cohort of DKA cases during the COVID-19 pandemic. They found that patients with COVID-19 had a higher body mass index in all age strata, higher insulin requirements, prolonged time to resolution of DKA, and a much higher rate of mortality compared with patients without COVID-19.[26] DKA has been reported in new-onset diabetes during COVID-19 infection.[27]

5.2 Mucormycosis

COVID-19 pandemic also led to a wide range of opportunistic bacterial and fungal infections, particularly in patients with DM. Several cases of mucormycosis caused by mold fungi of the genus *Rhizopus, Mucor, Rhizomucor, Cunninghamella,* and *Absidia* of Order Mucorales, have been reported in India.[28] Symptoms in diabetes patients such as hypoxia or low oxygen, hyperglycemia, acidic environment due to ketoacidosis, increased ferritins leading to high levels of iron, decreased phagocytic activity of white blood cells due to immunosuppression appear to provide favorable conditions for the spores of Mucorales to germinate. Other factors also include hospitalization and use of mechanical ventilators and corticosteroids to manage COVID-19 infection. DM has been reported as the most common risk factor for mucormycosis in India.[28] While treatment involves complicated procedures such as a combination of surgical debridement and antifungal therapy, hyperglycemia control has been important for the management and prevention of mucormycosis.[29]

6. Treatment and management of diabetes during COVID-19 infection

Insulin treatment is a general therapy for both T1D and T2D and is recommended in patients with diabetes and COVID-19 infection. However, the severity of COVID-19 infection needs to be considered for insulin therapy.[6,30] A 2020 study at the Tongji Hospital in Wuhan, China, on 120 patients with both diabetes and COVID-19 found no significant difference among patients on insulin therapy compared to patients without insulin therapy with regard to the severity of illness upon admission.[30] Baseline glycemic control and access to care are important modifiable risk factors to optimize care of people with T1D during the worldwide COVID-19 pandemic.[31] Metformin is an FDA-approved drug commonly used in the treatment of T2D. Metformin is thought to act by exerting anti-inflammatory action and reducing circulating inflammation biomarkers in people with T2D.[32,33] Metformin produces antihyperglycemic action by suppressing hepatocyte gluconeogenesis. This causes inhibition of mitochondrial enzymes, including complex I in the electron transport chain. Alternatively, it has been suggested that metformin inhibits mitochondrial glycerophosphate dehydrogenase. The effect of the drug on hepatic glucose production is mediated by activation of AMP-activated protein kinase (AMPK) due to mitochondrial activation.[33]

However, there are limited data available on the effect of metformin in T2D in COVID-19-infected patients, although a significant reduction in the mortality rate has been reported in COVID-19 patients taking metformin compared to the patients not taking metformin.[32] Metformin and acarbose, when used together or separately, had a positive impact on T2D patients' mortality rates when compared to patients that used neither drug.[34] Itolizumab was found to help COVID-19 patients with the most prevalent comorbidities such as diabetes and hypertension, recovering by controlling immune hyperactivation. There was a reduction in inflammation markers. The patients were able to come off of oxygen support, and their hospital stays were shortened.[35] Other drugs that have a potential consideration in patients with both diabetes and COVID-19 infection are glucagon-like peptide-1 (GLP-1), dipeptidyl peptidase-4 (DPP4), sodium-glucose cotransporter-2 inhibitors (SGLT2) inhibitors, ACE inhibitors/ARBs, and statins.[32,36] Fig. 4.1 illustrates the available and potential treatments for diabetes during COVID-19 infection.

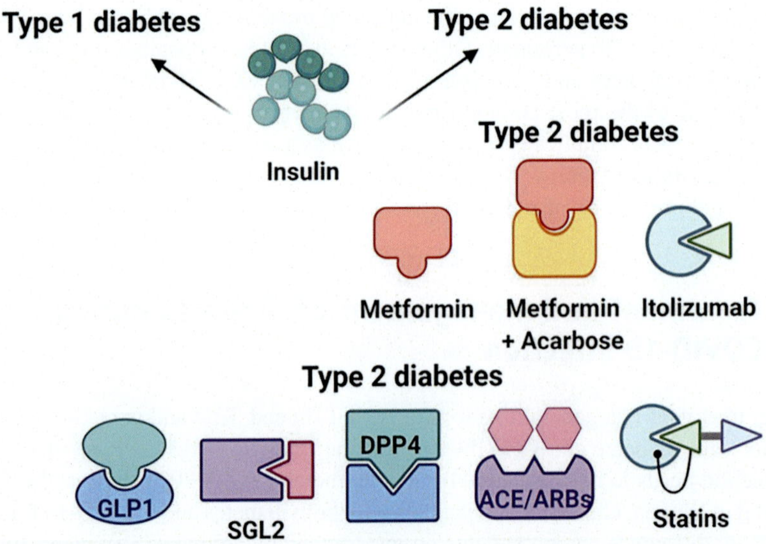

Figure 4.1 A schematic representation of the treatment and management of diabetes during COVID-19 infection. It is important to note that only insulin therapy is available for type 1 diabetes (T1D), whereas type 2 diabetes (T2D) has other treatment options besides insulin.

7. The effect of lockdowns on diabetes and obesity

Many countries worldwide introduced lockdowns to control the spread of COVID-19 infection. The lockdown due to the COVID-19 pandemic had a negative impact on body weight and glucose control in T2DM patients, in particular in those on insulin treatment.[37] Early findings and research showed that in countries that introduced lockdowns, patients with diabetes mellitus were subject to treatment delays, and discontinuation of care, service, and medicine supply. In fact, according to a report in the World Health Organization, diabetes treatment was either partially or completely disrupted in 49% of 155 countries surveyed in May 2020.[38] However, a study conducted in Japan by Tanji et al. in 2021, in an area where a lockdown was not introduced, revealed that glycemic control deteriorated in patients with type 2 diabetes during the pandemic even in the absence of a lockdown.[39] Fernández et al.[40] reported improved glycemic control in T1D patients after 8 weeks of lockdown in Spain compared to prior lockdown. This has been attributed to more time available for self-management and administration of multiple daily insulin injections or insulin pump therapy, using continuous glucose monitoring (CGM).[40] Similar findings in T1D patients using CGM have been reported from other countries where lockdowns were introduced.[39] Conversely, Verma et al. found that glycemic control worsened in patients not using CGM due to restricted availability of insulin or glucostrips in the UK during a lockdown in 2020.[41] Khare and Jindal also reported in 2020 that glycemic control worsened in patients with T2D after 3 weeks of lockdown in India.[42] They surmised that this was probably due to stress, difficulty in getting treatment, and the impact of lifestyle changes due to lockdown. There have also been additional studies from other countries that reported glycemic control either improved or remained unchanged during the lockdown.[39]

8. Lifestyle and diet during COVID-19 pandemic

Hyperglycemia has been indicated as a significant predictor of some viral infections. Wang et al.[43] demonstrated that increased secretion of glucocorticoids and catecholamines, insulin resistance, and increased blood glucose levels in diabetic patients with COVID-19 infection aggravated diabetic complications.[43] Therefore, it is important to pay attention to the glycemic control of viral-infected patients especially those with DM. Glycemic control and a balanced diet of both macronutrients and micronutrients are essential for DM patients with mild to moderate symptoms of COVID-19. Frequent snack consumption and a rise in the intake of carbohydrates with a high glycemic index, stress, and depression during quarantine have also led to unhealthy diets and reduced physical activity in some individuals during the pandemic.[44,45] The consequences of sedentary lifestyle due to lockdown during the pandemic resulted in increases in insulin resistance, total body fat, abdominal fat, and inflammatory cytokines.[46] During the lockdown period in the COVID-19 pandemic, increases in body weight, mainly caused by decreased exercise, solitude, anxiety/depression, and increased consumption of snacks, unhealthy foods, and sweets have been reported in obese patients. Patients with T2D reported high consumption of sugary food and snacks as well as increased stress and reduced exercise. Furthermore, associations were found between food cravings and snack consumption, and/or decreased exercise levels and body weight gain; increased total diet intake, and raised HbA1c concentrations. A poor glycemic control was found in T2D patients with an unhealthy diet and low physical activity. T2D patients with mental stress also had unhealthy dietary habits. The desire to consume excessive food, one of the eating disorders, is more common in people with obesity due to stress.[44–46]

8.1 Macronutrients- carbohydrates, proteins, and fatty acids

Mahluji et al. outlined a detailed review on the significance of diet in diabetes patients infected with COVID-19.[47] Complex carbohydrates, low simple sugars, and adequate intake of fibers have been recommended. Hospitalized diabetic patients with severe COVID-19 symptoms needed supportive nutrition. Enteral nutrition (EN) was usually well tolerated by the patients and was preferred over parenteral nutrition (PN). The amount of carbohydrates in enteral feeding formula depended on the patient's condition (e.g., glycemic control, ventilator dependency) and was usually 30%–50% of nonprotein calories, but the significant restriction on carbohydrates was not recommended as it can exacerbate viral outcomes.[47] Some evidence supported the beneficial effects of high protein diets on glycemic control, insulin resistance, and maintenance of lean body mass in diabetic people. The amount of protein intake also directly impacted the immune response. The American Society for Parenteral and Enteral Nutrition (ASPEN) recommended the protein supply of about 1.2–2.0 g/kg/day for nonobese patients (based on actual body weight) and 2–2.5 g/kg/day for class I– III obese patients (based on ideal body weight). Patients with DM were more exposed to oxidative stress.[47] Therefore, it was important to consider the quantity as well as the quality of dietary fatty acids especially during infections and inflammatory conditions. Since COVID-19 can cause severe lung injury and ARDS, increasing dietary fat intake may help to decrease CO_2 production as well as ventilator dependency.[47]

8.2 Micronutrients—vitamins, selenium, and zinc

Low serum levels of folate and B12 in type 2 diabetic patients are correlated with lower circulating level of glutathione and total antioxidant status. Diabetic patients are more prone to folate and B12 deficiency because of the established drug-nutrient interactions, especially those treated with metformin. There is insufficient evidence about the effectiveness of B vitamins in the treatment of COVID-19, and it has not been recommended by WHO Interim Guidance for COVID-19 management.[47] In the case of the current COVID-19 pandemic, it was suggested that vitamin D-deficient subjects may be more vulnerable to the infection. There is some evidence that vitamin C deficiency is more prevalent in patients with diabetes mellitus compared to nondiabetic individuals.[47] DM is usually accompanied by increased production of ROS and decreased levels of antioxidant enzymes like superoxide dismutase (SOD). Numerous studies reported the protective effects of vitamin C against oxidative damage of diabetes.[47]

Selenium (Se) plays a functional role in redox hemostasis, hormone metabolism, and protection against oxidative stress and inflammation.[47] It appeared that Se inhibited the entrance of different viruses into the cells and prevented their infectivity. Therefore, it may be useful against the pandemic of COVID-19. Since diabetic patients are susceptible to low antioxidative levels, Se supplementation might be a helpful choice for the prevention or treatment of diabetic patients infected by novel coronavirus.[47]

Zinc (Zn) has myriad functionality in the human body as it is a cofactor for more than 300 enzymes. In addition to the antioxidant and anti-inflammatory effects of Zn, this element has vital roles in both innate and adaptive immune cells and also is essential for activation and secretion of insulin.[47] There is evidence of its protective effects against DM. Since zinc depletion deteriorates antiviral immunity, it is hypothesized that Zn deficiency may increase host susceptibility to COVID−19.[47]

8.3 Mediterranean diet

Mediterranean lifestyle habits include diet, physical activity, and social interaction, and some Mediterranean pyramids include emotional balance or psychological wellness.[46,47] MD as well as low-carbohydrate, low-glycemic index, and high-protein diets decreased HbA1c compared with their control diets, but the greater effect was observed with the MD as it also reduced body mass significantly. Data from a meta-analysis in individuals with T2D and/or prediabetes showed that MD reduced fasting glucose and body mass compared to other diets, including the diet suggested by the American Diabetes Association and the European Association for the Study of Diabetes (50%−55% carbohydrates with mixed glycemic index, 30% fat, and 15%−20% protein).[46,47]

A study conducted by Grabia et al.[48] during the second wave of the COVID-19 pandemic in Poland among women with T1DM assessed the importance of health and nutritional behaviors, with particular reference to MD. The patients included in the study were overweight (32%) or obese (13%). Factors such as increased body weight, low physical activity, long screen time, and exposure to stressful situations often led to diabetic complications. Adherence to the MD had a crucial role in reducing the risk of health consequences.[48] Fig. 4.2 illustrates the recommended diet to manage diabetes during COVID-19 infection.

9. Conclusion

In conclusion, the COVID-19 pandemic caused by SARS-CoV-2 resulted in millions of hospitalizations and death across the globe. Of these cases, diabetes and obesity have been major risk factors for both hospitalization and mortality. A detailed review on diabetes during the COVID-19 pandemic by Corrao et al.[32] revealed a higher prevalence of diabetes in men than women and also a higher prevalence among the elderly population infected with COVID-19.[32] A definite mechanism behind severe COVID-19 infection is still unknown, but patients with metabolic disorders such as diabetes and obesity have a higher risk of developing severe cytokine storm and coagulopathy,

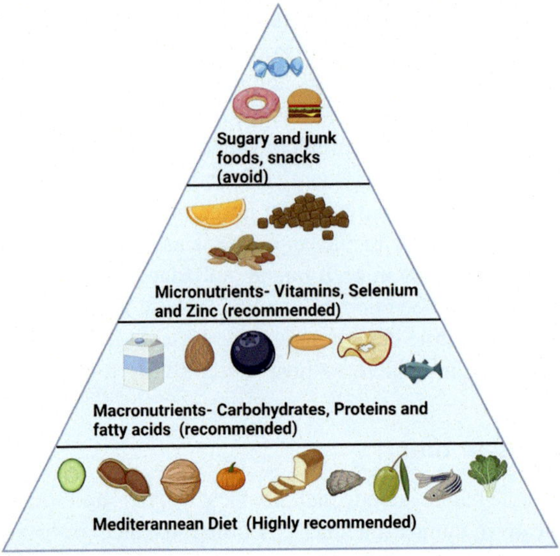

Figure 4.2 An illustration of recommended diet in diabetic patients during COVID-19 pandemic.

which eventually lead to critical clinical conditions.[49] This has been hypothesized to be one possible reason why the hospitalizations and mortality during the pandemic were significantly higher in diabetic patients than patients without diabetes.[49–51] Diabetic patients also have symptoms like anorexia and fatigue and have poor nutritional status. It is interesting to note that diabetes has also been implicated in infectious diseases and has been reported as a risk factor for morbidity and mortality in other viral infections such as 2009 influenza A (H1N1), MERS-CoV, and SARS-CoV-2.[52–55] However, a direct relationship between diabetes and infectious diseases has not yet been established. This chapter has discussed various factors such as biomarkers, risk factors, complications, treatment and management, lifestyle, and dietary influences from the perspective of diabetes during the current COVID-19 pandemic. Since the pandemic is still ongoing at the time of this writing, we expect to learn more from new research developments.

References

1. International Diabetes Federation. *Diabetes Atlas*. 9th ed.; 2019. https://www.idf.org/e-library/epidemiology-research/diabetes-atlas.html.
2. Fang L, Karakiulakis G, Roth M. Are patients with hypertension and diabetes mellitus at increased risk for COVID-19 infection? *Lancet Respir Med*. 2020;S2213–2600(20):30116–30118.
3. Gupta R, Ghosh A, Singh AK, Misra A. Clinical considerations for patients with diabetes in times of COVID-19 epidemic. *Diabetes Metabol Syndr*. 2020;14:211–212.

4. Richardson S, Hirsch JS, Narasimhan M, et al. Presenting characteristics, comorbidities, and outcomes among 5700 patients hospitalized with COVID-19 in the New York city area. *JAMA*. 2020;323(20):2052–2059.
5. Sharif N, Ahmed SN, Opu RR, et al. Prevalence and impact of diabetes and cardiovascular disease on clinical outcome among patients with COVID-19 in Bangladesh. *Diabetes Metabol Syndr*. 2021;15(3):1009–1016.
6. Jin S, Hu W. Severity of COVID-19 and treatment strategy for patient with diabetes. *Front Endocrinol*. 2021;12(469).
7. Rawshani A, Kjölhede EA, Rawshani A, et al. Severe COVID-19 in people with type 1 and type 2 diabetes in Sweden: a nationwide retrospective cohort study. *The Lancet Reg Health – Europe*. 2021:4.
8. Ruan Y, Ryder REJ, De P, et al. A UK nationwide study of people with type 1 diabetes admitted to hospital with COVID-19 infection. *Diabetologia*. 2021;64(8):1717–1724.
9. Kempegowda P, Melson E, Johnson A, et al. Effect of COVID-19 on the clinical course of diabetic ketoacidosis (DKA) in people with type 1 and type 2 diabetes. *Endocr Connect*. April 2021;10(4):371–377.
10. Mazori AY, Bass IR, Chan L, et al. Hyperglycemia is associated with increased mortality in critically Ill patients with COVID-19. *Endocr Pract*. February 2021;27(2):95–100.
11. Roganović J. Downregulation of microRNA-146a in diabetes, obesity and hypertension may contribute to severe COVID-19. *Med Hypotheses*. January 2021;146:110448.
12. Moghaddam Tabrizi F, Rasmi Y, Hosseinzadeh E, et al. Diabetes is associated with higher mortality and severity in hospitalized patients with COVID-19. *EXCLI J*. February 22, 2021;20:444–453.
13. Guo W, Li M, Dong Y, et al. Diabetes is a risk factor for the progression and prognosis of COVID-19. *Diabetes Metabol Res Rev*. 2020;36(7).
14. Malik SUF, Chowdhury PA, Hakim A, Islam MS, Alam MJ, Azad AK. Blood biochemical parameters for assessment of COVID-19 in diabetic and non-diabetic subjects: a cross-sectional study. *Int J Environ Health Res*. January 27, 2021:1–14.
15. Oley MH, Oley MC, Kepel BJ, et al. ICAM-1 levels in patients with covid-19 with diabetic foot ulcers: a prospective study in southeast asia. *Ann Med Surg (Lond)*. March 2021;63: 102171.
16. Cao P, Song Y, Zhuang Z, et al. Obesity and COVID-19 in adult patients with diabetes. *Diabetes*. May 2021;70(5):1061–1069.
17. Kruglikov IL, Shah M, Scherer PE. Obesity and diabetes as comorbidities for COVID-19: underlying mechanisms and the role of viral-bacterial interactions. *Elife*. September 15, 2020;9:e61330.
18. Cai Q, Chen F, Wang T, et al. Obesity and COVID-19 severity in a designated hospital in Shenzhen, China. *Diabetes Care*. July 2020;43(7):1392–1398.
19. Gao F, Zheng KI, Wang XB, et al. Obesity is a risk factor for greater COVID-19 severity. *Diabetes Care*. July 2020;43(7):e72–e74.
20. Kass DA, Duggal P, Cingolani O. Obesity could shift severe COVID-19 disease to younger ages. *Lancet*. May 16, 2020;395(10236):1544–1545.
21. Hoffmann M, Kleine-Weber H, Schroeder S, et al. SARS-CoV-2 cell entry depends on ACE2 and TMPRSS2 and is blocked by a clinically proven protease inhibitor. *Cell*. April 16, 2020;181(2):271–280. e8.
22. Michalakis K, Ilias I. COVID-19 and hyperglycemia/diabetes. *World J Diab*. May 15, 2021; 12(5):642–650.
23. Underwood PC, Adler GK. The renin angiotensin aldosterone system and insulin resistance in humans. *Curr Hypertens Rep*. February 2013;15(1):59–70.

24. D'Onofrio N, Scisciola L, Sardu C, et al. Glycated ACE2 receptor in diabetes: open door for SARS-CoV-2 entry in cardiomyocyte. *Cardiovasc Diabetol*. May 7, 2021;20(1):99.
25. Singh B, Patel P, Kaur P, Majachani N, Maroules M. COVID-19 and diabetic ketoacidosis: report of eight cases. *Cureus*. March 31, 2021;13(3):e14223.
26. Pasquel FJ, Messler J, Booth R, et al. Characteristics of and mortality associated with diabetic ketoacidosis among US patients hospitalized with or without COVID-19. *JAMA Netw Open*. March 1, 2021;4(3):e211091.
27. Eskandarani RM, Sawan S. Diabetic ketoacidosis on hospitalization with COVID-19 in a previously nondiabetic patient: a review of pathophysiology. *Clin Med Insight Endocrinol Diab*. December 24, 2020;13, 1179551420984125.
28. Singh AK, Singh R, Joshi SR, Misra A. Mucormycosis in COVID-19: a systematic review of cases reported worldwide and in India. *Diab Metabol Syndr*. 2021 Jul-Aug;15(4): 102146.
29. John TM, Jacob CN, Kontoyiannis DP. When uncontrolled diabetes mellitus and severe COVID-19 converge: the perfect storm for mucormycosis. *J Fungi (Basel)*. April 15, 2021; 7(4):298.
30. Chen Y, Yang D, Cheng B, et al. Clinical characteristics and outcomes of patients with diabetes and COVID-19 in association with glucose-lowering medication. *Diab Care*. July 2020;43(7):1399−1407.
31. O'Malley G, Ebekozien O, Desimone M, et al. COVID-19 hospitalization in adults with type 1 diabetes: results from the T1D exchange multicenter surveillance study. *J Clin Endocrinol Metab*. January 23, 2021;106(2):e936−e942.
32. Corrao S, Pinelli K, Vacca M, Raspanti M, Argano C. Type 2 diabetes mellitus and COVID-19: a narrative review. *Front Endocrinol*. March 31, 2021;12:609470.
33. Cameron AR, Morrison VL, Levin D, et al. Anti-inflammatory effects of metformin Irrespective of diabetes status. *Circ Res*. August 19, 2016;119(5):652−665.
34. Li J, Wei Q, McCowen KC, et al. Inpatient use of metformin and acarbose is associated with reduced mortality of COVID-19 patients with type 2 diabetes mellitus. *Endocrinol Diab Metab*. September 29, 2021:e00301.
35. Gore V, Kshirsagar DP, Bhat SM, Khatib KI, Mansukhani B. Itolizumab treatment for cytokine release syndrome in moderate to severe acute respiratory distress syndrome due to COVID-19: clinical outcomes, A retrospective study. *J Assoc Phys India*. February 2021; 69(2):13−18, 32.
36. Bornstein SR, Rubino F, Khunti K, et al. Practical recommendations for the management of diabetes in patients with COVID-19. *Lancet Diab Endocrinol*. June 2020;8(6):546−550.
37. Biamonte E, Pegoraro F, Carrone F, et al. Weight change and glycemic control in type 2 diabetes patients during COVID-19 pandemic: the lockdown effect. *Endocrine*. June 2021; 72(3):604−610.
38. Dyer O. Covid-19: pandemic is having "severe" impact on non-communicable disease care, WHO survey finds. *BMJ*. June 3, 2020;369:m2210.
39. Tanji Y, Sawada S, Watanabe T, et al. Impact of COVID-19 pandemic on glycemic control among outpatients with type 2 diabetes in Japan: a hospital-based survey from a country without lockdown. *Diabetes Res Clin Pract*. June 2021;176:108840.
40. Fernández E, Cortazar A, Bellido V. Impact of COVID-19 lockdown on glycemic control in patients with type 1 diabetes. *Diabetes Res Clin Pract*. August 2020;166:108348.
41. Verma A, Rajput R, Verma S, Balania VKB, Jangra B. Impact of lockdown in COVID 19 on glycemic control in patients with type 1 Diabetes Mellitus. *Diabetes Metabol Syndr*. 2020 Sep-Oct;14(5):1213−1216.

42. Khare J, Jindal S. Observational study on effect of lock down due to COVID 19 on glycemic control in patients with diabetes: experience from Central India. *Diabetes Metabol Syndr.* 2020 Nov-Dec;14(6):1571–1574.
43. Wang A, Zhao W, Xu Z, Gu J. Timely blood glucose management for the outbreak of 2019 novel coronavirus disease (COVID-19) is urgently needed. *Diab Res Clin Pract.* April 2020;162:108118.
44. Zupo R, Castellana F, Sardone R, et al. Preliminary trajectories in dietary behaviors during the COVID-19 pandemic: a public health call to action to face obesity. *Int J Environ Res Publ Health.* September 27, 2020;17(19):7073.
45. Bayram Deger V. Eating behavior changes of people with obesity during the COVID-19 pandemic. *Diab Metab Syndr Obes.* May 3, 2021;14:1987–1997.
46. Fedullo AL, Schiattarella A, Morlando M, et al. Mediterranean diet for the prevention of gestational diabetes in the covid-19 era: Implications of Il-6 in diabesity. *Int J Mol Sci.* January 26, 2021;22(3):1213.
47. Mahluji S, Jalili M, Ostadrahimi A, Hallajzadeh J, Ebrahimzadeh-Attari V, Saghafi-Asl M. Nutritional management of diabetes mellitus during the pandemic of COVID-19: a comprehensive narrative review. *J Diabetes Metab Disord.* April 5, 2021;20(1):1–10.
48. Grabia M, Puścion-Jakubik A, Markiewicz-Żukowska R, et al. Adherence to mediterranean diet and selected lifestyle elements among young women with type 1 diabetes mellitus from northeast Poland: a case-control COVID-19 survey. *Nutrients.* April 2, 2021;13(4):1173.
49. Roberts J, Pritchard AL, Treweeke AT, et al. Why is COVID-19 more severe in patients with diabetes? The role of angiotensin-converting enzyme 2, endothelial dysfunction and the Immunoinflammatory system. *Front Cardiovasc Med.* February 3, 2021;7:629933.
50. Tomar B, Anders HJ, Desai J, Mulay SR. Neutrophils and neutrophil extracellular traps drive necroinflammation in COVID-19. *Cells.* June 2, 2020;9(6):1383.
51. Webb BJ, Peltan ID, Jensen P, et al. Clinical criteria for COVID-19-associated hyperinflammatory syndrome: a cohort study. *Lancet Rheumatol.* December 2020;2(12): e754–e763.
52. Shang J, Wang Q, Zhang H, et al. The relationship between diabetes mellitus and COVID-19 prognosis: a retrospective cohort study in Wuhan, China. *Am J Med.* January 2021; 134(1):e6–e14.
53. Schoen K, Horvat N, Guerreiro NFC, de Castro I, de Giassi KS. Spectrum of clinical and radiographic findings in patients with diagnosis of H1N1 and correlation with clinical severity. *BMC Infect Dis.* November 12, 2019;19(1):964.
54. Yang JK, Feng Y, Yuan MY, et al. Plasma glucose levels and diabetes are independent predictors for mortality and morbidity in patients with SARS. *Diabet Med.* June 2006;23(6): 623–628.
55. Banik GR, Alqahtani AS, Booy R, Rashid H. Risk factors for severity and mortality in patients with MERS-CoV: analysis of publicly available data from Saudi Arabia. *Virol Sin.* February 2016;31(1):81–84.

Food bioactive compounds, sources, and their effectiveness during COVID-19

5

Giridhar Goudar[1], *Munikumar Manne*[2], *Jangampalli Adi Pradeepkiran*[3] *and Subodh Kumar*[4]

[1]Food Quality Analysis and Biochemistry Division, Biochem Research and Testing Laboratory, Dharwad, Karnataka, India; [2]Clinical Division, ICMR-National Institute of Nutrition, Hyderabad, Telangana, India; [3]Department of Internal Medicine, Texas Tech University Health Sciences Center, Lubbock, TX, United States; [4]Center of Emphasis in Neuroscience, Department of Molecular and Translational Medicine, Paul L. Foster School of Medicine, Texas Tech University Health Sciences Center El Paso, El Paso, TX, United States

Abstract

Since the occurrence of COVID-19 pandemic in the late 2019, it has impacted to almost all the countries throughout the world by its severity of infection. The vaccines or drugs to combat the SARS-CoV-2 infection were not available till recent time, and the people looked out for healthy foods, which help to increase the immune response. Even after the vaccines have been developed for the COVID-19, the importance of food-based diet cannot be neglected due to their health beneficial properties, which are required post-COVID-19 infection or during the incubation duration of infection. Several food groups and their products have been known to provide vital nutrients, which are required for proper functioning of human body. Along with nutrients, these foods are known to have various nonnutrients components, which help in reduction or management of several physiological disorders. These compounds are collectively known as bioactive compounds, which are naturally synthesized. These compounds are widely spread and abundantly found in plant-based foods. The major groups of bioactive compounds present in foods include polyphenols, carotenoids, bioactive carbohydrates, and phytosterols. These compounds are known to have mainly the properties of antioxidant, anti-inflammatory, anticancerous, and antimicrobial, which helps in keeping the human body healthy. In this chapter, the main bioactive compounds, their classification, and food sources have been highlighted, which will give an insight of their beneficial effects. The bioactive compounds will have a major role in reduction of inflammation and infection through several complex physiological mechanism, thereby involved in increasing the immunity. Hence, these compounds play a vital role in management or reduction of COVID-19 infection through the effectiveness of bioactive compounds.

Keywords: Carotenoids; Cereals; COVID-19; Fruits; Polyphenols; Vegetables.

COVID-19 in Alzheimer's Disease and Dementia. https://doi.org/10.1016/B978-0-443-15256-6.00008-8

1. Introduction

The coronavirus disease (COVID-19) pandemic has widely affected the food systems around the globe since its outbreak and spread during the early 2020. However, no evidence is available that the severe acute respiratory syndrome-coronavirus-2 (SARS-CoV-2) will cause the contamination or has spread through food systems, whereas it has impacted the routine daily life through food safety aspects, which include eating habits and implementation of safety measures throughout the entire food supply chain from farm to plate. As a result of multiple instabilities in the food economic system during the COVID-19 pandemic, nourishing food was scarce across several nations.

Healthy lifestyle habits, exercise, and well-balanced nutritious diets are considered to be essential factors in reducing the risk of COVID-19 infections. On the other hand, unhealthy and unbalanced diets along with reduced or no physical activity might cause degradation of the immune system. This is evidenced in the increased vulnerability of obese and overweight populations for developing pulmonary diseases. Since it has become a vital aspect and prime importance to maintain a healthy nutritional diet during the COVID-19 infection duration.

To combat the COVID-19 infection, the vaccines are available now; however, the importance of nutrition and dietary supplements provided through foods in the management of the viral infection cannot be neglected, which has important role in development of immune system. Foods contain several macro- and micro-nutrients along with secondary metabolites known as bioactive compounds, which have been considered as non-nutrients; however, they are helpful in less quantity that aids in the appropriate functioning of the human body. These bioactive compounds are well known for their physiological properties such as antioxidant, anti-inflammatory, antimicrobial, anticancerous, and antiviral activities. The foods which are known to provide health beneficial effects apart from providing nutrients to human well-being are known as functional foods.[1,2] Different studies have shown for the effectiveness of well-balanced diets containing abundant bioactive compounds and their role in managing the noncommunicable diseases such as type 2 diabetes mellitus, cardio-vascular diseases, hypertension, obesity. These compounds also have been effective against certain metabolic syndromes such as dementia and Alzheimer's disease.[3–6]

Several food groups or commodities are good sources of many bioactive compounds, which include cereals grains, fruits, vegetables, spices, etc. However, the content of various active compounds varies depending on the type of food, variety, regional variation, genetic factor, and environment.

2. COVID-19 and food safety

One of the aspects of food preparation during COVID-19 that was affected was food safety.[7] During all the stages of food processing and production, precautions and safety measures were put in place to avoid contamination with microorganisms. Around the world, food was studied as a possible COVID-19 vector.[8–12] Till date, there has been no scientific evidence for direct transmission or contamination of SARS-CoV-2 virus through foods.[13] However, the transmission may be possible through contamination of packing material or mishandling of food commodities by an infected person (sneezing, saliva droplet). Therefore, it is imperative that proper sanitation measures at food production and processing sites, good manufacturing, and stringent hygienic practices should be maintained by food handlers at every stage of food processing and transportation. COVID-19 is highly resistant to cold temperatures, stable up to −4°C, and capable of staying infectious for more than 2 years at a temperature of −20°C.[12] As such, there is a chance for transmission of viral infections through raw foods, which are shipped for longer duration at refrigerated conditions.[7] However, SARS-CoV-2 can be inactivated at 70°C,[14] suggesting that proper cooking of foods might be helpful in inactivating of virus.

3. Bioactive compounds

Bioactive compounds are the secondary metabolites synthesized by the plants and the main categories include polyphenols (phenolic acids, flavonoids, anthocyanins, etc.), carotenoids (lutein, lycopene, β-carotene), phytosterols (stigmasterol, β-sitosterol), bioactive carbohydrates (β-glucan, arabinoxylan), etc. Polyphenols comprise of a larger and most diverse group of bioactive compounds around the globe, which are dominantly found in plant kingdom; however, several phytochemical compounds are also found in microbial sources and animal-based foods. Phytochemicals are naturally produced from the phenylpropanoid pathway with the two precursor molecules of phenylalanine (amino acid from shikimate pathway) and malonyl-CoA (acetate pathway metabolite).[15]

3.1 *Polyphenols*

Polyphenols are group of bioactive compounds, which are having one or more than one aromatic ring to which hydroxyl functional groups are attached. Based on the number of benzene rings and the type and number of attached functional groups, they are classified into groups of phenolic acids, flavonoids, anthocyanins, etc.[16] These have been extensively studied for their antioxidant activity; however, consumption of foods rich in polyphenols has been inversely showed role in prevention or reduction of several metabolic disorders such as hypertension, cardiovascular diseases and cancer.[17,18] The mechanism of their antioxidant property is due to their free radical scavenging activity and inhibition of oxidative enzymes to produce free radicals.[19,20] The total types of polyphenols, which have been reviewed among several food commodities account for 501, among them different types comprise of flavonoids (279), phenolic acids (108), stilbenes (10), lignans (29), and others (80)[21] (Table 5.1).

Table 5.1 Bioactive compounds and their classified groups.

Group	Types	Compounds	References
Phenolic acids	Hydroxybenzoic acids	Gallic acid, protocatechuic acid, vanillic acid, syringic acid, p-hydroxybenzoic acid	Zhang et al.[22]
	Hydroxycinnamic acids	Ferulic acid, caffeic acid, p-coumaric acid, sinapic acid	Zhang et al.[22]
Flavonoids	Flavonoids	Flavanones, flavones, flavonols, dihydroflavonols, flavan-3-ols, flavan-4-ols, flavan-3,4-diols, anthocyanidins	Santana-Gálvez and Jacobo-Velázquez[23]
	Isoflavonoids	Isoflavans, isoflavones, isoflavanones, isoflav-3-enes, isoflavanols, rotenoids, coumestanes, 3-arylcoumarins, coumaronochromenes, coumaronochromones, pterocarpans	Santana-Gálvez and Jacobo-Velázquez[23]
	Neoflavonoids	4-Arylcoumarins, 3,4-dihydro-4-arylcoumarins, neoflavenes	Santana-Gálvez and Jacobo-Velázquez[23]
Anthocyanins	Anthocyanidins	Cyanidin, delphinidin, pelargonidin, peonidin, petunidin, malvidin	Khoo et al.[24]
Tannins	Hydrolysable	Tannic acid	Santana-Gálvez and Jacobo-Velázquez[23]
	Nonhydrolysable	Procyanidin C1	Santana-Gálvez and Jacobo-Velázquez[23]
Carotenoids	Xanthophylls	Lutein, zeaxanthin	Langi et al.[25]
	Carotenes	α-Carotene, β-carotene, lycopene	Langi et al.[25]
Bioactive carbohydrates	Dietary fiber	Arabinoxylan, resistant starch	Bartłomiej et al.[26]
	β-D-glucans	β-Glucan, α-glucan	Goudar et al.[27]
Phytosterols	Stanols	Campesterol, sitosterol, brassicasterol, stigmasterol	Bartłomiej et al.[26]

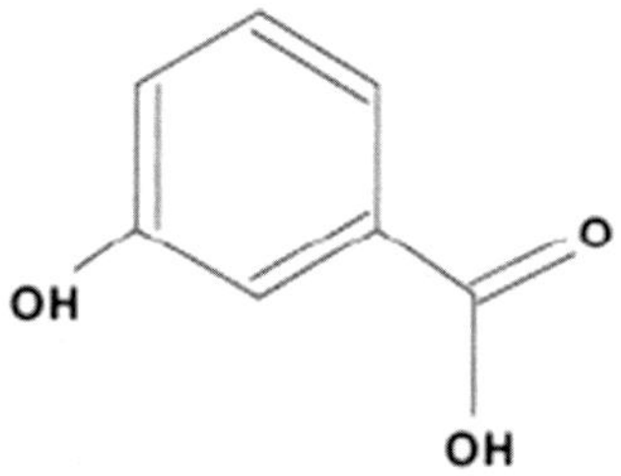

Figure 5.1 Basic structure of phenolic acid.

3.1.1 Phenolic acids

Phenolic acids comprise the simplest to complex group of polyphenols, which are dominantly found in the cereal grains. They are subclassified into two groups of hydroxybenzoic and hydroxycinnamic acids based on their chemical nature and structural arrangement (Fig. 5.1).

They are found in free (soluble) and bound (insoluble) form within the plant depending on the location and interaction with cell wall macromolecules or attached to them. In cereal grains, ~62% of phenolic compounds are in the bound form, while fruits and vegetables have ~24% of its phenolic compounds in its bound form.[28] The phenolic acids in the bound form are attached to the cell wall through ester or ether linkage with the components of proteins, cellulose, lignin, or arabinoxylans.[29]

3.1.2 Flavonoids

Flavonoids are ubiquitous in nature and the most found phenolic compounds among all the food groups. After chlorophyll and carotenoids, they are the most detected pigment molecules found in nature.[30] Structurally, the flavonoids have C6–C3–C6 carbon skeleton, which specifically is an aromatic ring attached to a benzopyran ring[31] (Fig. 5.2).

Based on the position of the aromatic ring attached to the benzopyran ring, they are classified into three major class of flavonoids, isoflavonoids, and neoflavonoids. Further, based on the oxidation and saturation status of the heterocyclic group of

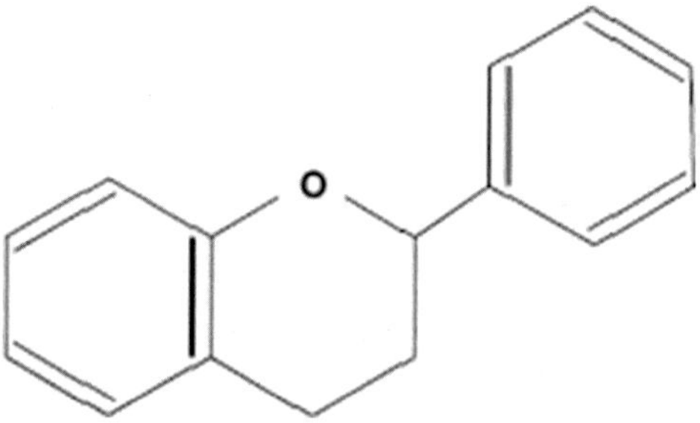

Figure 5.2 Basic structure of flavonoid.

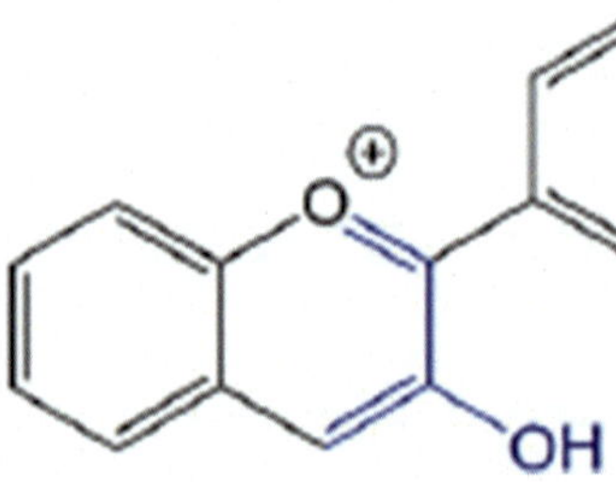

Figure 5.3 Basic structure of anthocyanin.

the benzopyran ring, these three groups of flavonoids, isoflavonoids, and neoflavonoids are further classified into 8, 11, and 3 groups, respectively.

3.1.3 Anthocyanins

Anthocyanins are the compounds, which exert red, purple, or blue color pigmentation especially in fruits, vegetables, flowers, and tubers. Anthocyanins are basically a type of flavonoid group (flavanols), which have the flavylium ion as their core molecular structure lacking in the ketonic oxygen at the para-position (Fig. 5.3).

The anthocyanin exists as blue color under alkaline condition, whereas it gives red color in acidic medium. The anthocyanin stability varies due to changes in temperature, lightness, and value of pH. Anthocyanins are chemically in glycosidic form, whereas the anthocyanidins are in aglycone forms. Anthocyanins are usually found as acylated or glycosides, while anthocyanidins are classified into 3-hydroxyanthocyanidins, 3-deoxyanthocyanidins, and methylated anthocyanidins.[24] The types of different anthocyanidins, which are commonly found in fruits and vegetables are cyanidin (50%), delphinidin (12%), peonidin (12%), petunidin (12%), pelargonidin (7%), and malvidin (7%).[32]

3.1.4 Tannins

Tannins are usually high-molecular-weight compounds, which are found in plant-based foods ranging between 500 and 20,000 Da. They are categorized into hydrolysable and nonhydrolysable tannins. Hydrolysable tannins are having the phenolic compounds esterified to the centered polyhydric alcohol or to a glucose moiety with some exceptions.[33] Nonhydrolysable tannins are in the form of condensed tannins and are also known as proanthocyanidins (Fig. 5.4).

They are mostly the polymeric forms of flavonoids such as (+)-catechin or (−)-epicatechin or leucoanthocyanidins or sometimes combinations of flavonoid compounds polymers. As the name suggests the hydrolysable tannins are easily hydrolyzed with treatment of acidic, basic or enzymatic, which yields the polyphenols and carbohydrate moieties.[33,23] Proanthocyanidins are wide spread among plant-based foods in comparison to hydrolysable tannins. The astringency taste of the foods is usually due to the tannin content of the particular food.[23]

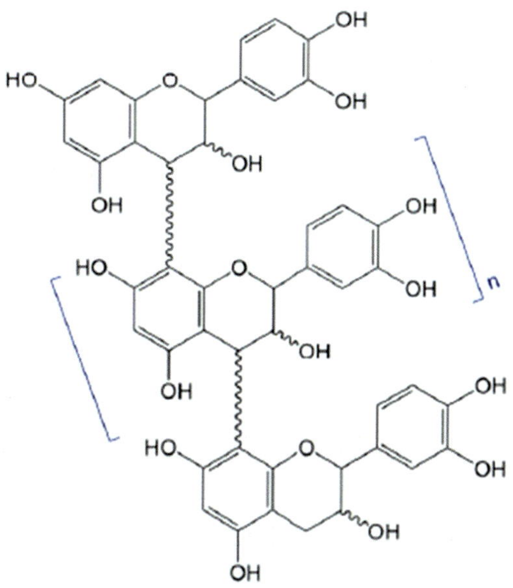

Figure 5.4 Tannin (proanthocyanidin).

3.2 Carotenoids

Carotenoids are the major pigment compounds found naturally in plants after chlorophyll; they are referred to as pigment molecules due to the different colors they impart on the food commodity such as red, yellow, or orange (Fig. 5.5).

Carotenoids are basically the hydrophobic compounds, which are lipophilic in nature and insoluble in water. They consist of isoprenoid units with eight numbers, which are linked together with specificity, which makes the isoprenoid units to converge in the center. They may occur either in cis or trans configuration due to the C—C bond isomerism in nature; hence, they differ in their solubility, stability, and physical characteristics.[25]

3.3 Bioactive carbohydrates

These are the carbohydrate components, which have several health beneficial properties. Some of the bioactive carbohydrate's components include dietary fiber, arabinoxylans, β-glucan, resistant starch, etc. Arabinoxylans are the pentosan polymeric

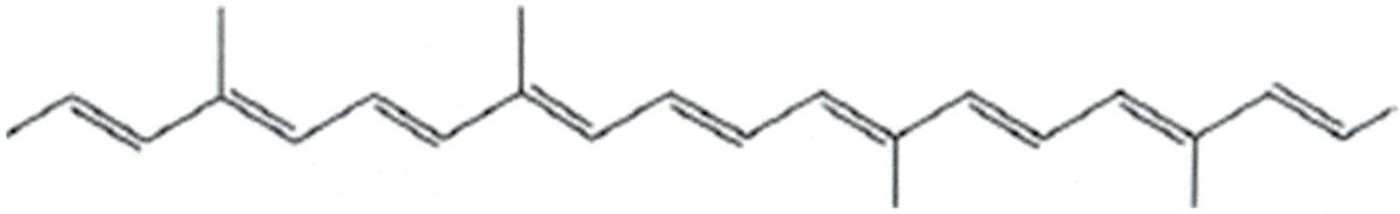

Figure 5.5 Carotenoid backbone.

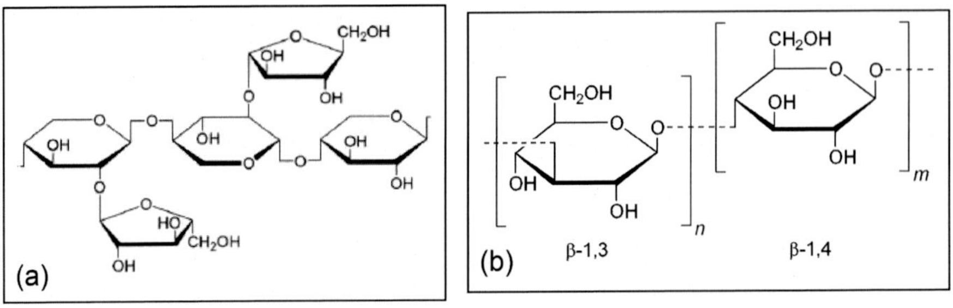

Figure 5.6 (A) Arabinoxylan. (B) β-Glucan.

structures containing the ferulic acid attached to it; hence, exerts the antioxidant activity. They are linear polymers, which are having D-xylopyranose attached with 1,4-β-linkages along with attachment of α-L-arabinofuranose residues. The degree of arabinoxylan branching depends on the composition of arabinose and xylose (A/X) ratio, which ranges between 0.3 and 1.1[26] (Fig. 5.6).

β-Glucans are the D-glucopyranose residues linked with β-(1 → 4) glycosidic linkages, which for every 2 to 3 β-(1 → 4)-linked residues are having one β-(1 → 3)-linked residue. The physiological properties of β-glucans such as its solubility, gel formation capacity, and viscosity depend on the structural arrangement of β-glucan and the ratio of linkages between β-(1 → 4) and β-(1 → 3).[27] These β-glucans are involved in the promotion of selective colonic microflora beneficial to health, short chain fatty acid formation, removal of harmful carcinogenic substances, management of type 2 diabetes mellitus and obesity.[34]

3.4 Phytosterols

Phytosterols are the plant-based sterols, which are secondary metabolites produced by cereal grains and oil seeds. Chemically, the sterols are the four-ring fused of 1,2-cyclopentanoperhydrophenantrene with the attachment of OH group at the carbon-3 position, with an unsaturated double bond probably between C5 and C6 or C7 and C8 position (Fig. 5.7).

Based on the position of the double bond and the type of functional group attached to the fused ring, there have been approximately 200 different identified phytosterols.

Figure 5.7 Phytosterols (stanols).

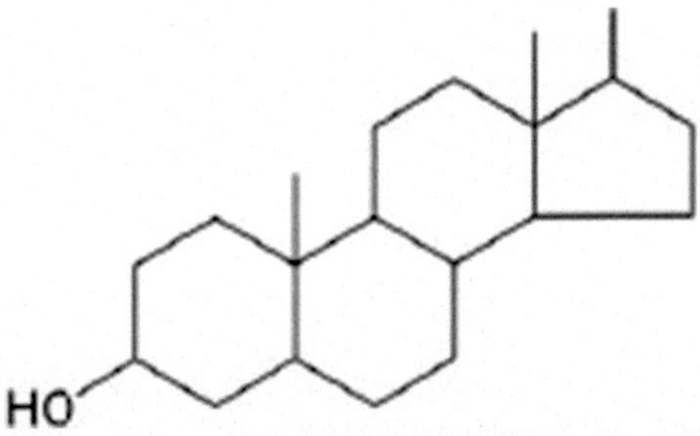

Usually in cereals, the phytosterols have been conjugated with the hydroxyl group attached at the carbon C3 of fused ring that are conjugated to the phenolic acids, carbohydrates, or fatty acids.[26]

4. Foods containing bioactive compounds helpful during COVID-19 infection

Most of the commonly consumed foods such as cereal grains, vegetables, and fruits are good sources of several bioactive compounds. Along with providing the necessary nutrients, they contain the specific bioactive compounds. The bioactive compounds among various food groups and their health beneficial properties have been highlighted in Table 5.2.

4.1 *Cereals and legumes*

Many cereal grains such as barley, millets, maize, quinoa, and sorghum comprise of several polyphenols such as phenolic acids, flavonoids, and anthocyanin compounds. Barley has been into use since olden times; however, in recent years, most of the barley is cultivated for the purpose of brewery. This grain is an excellent source of a bioactive carbohydrate component known as β-glucan,[27] it has been studied for its trained immunity (TRIM)-induced defense mechanism for COVID-19.[40,41] Barley is abundant in anthocyanin compounds such as cyanidin-3-glucoside, pelargonidin-3-glucoside, procyanidin B3, procyanidin C2, prodelphinidin B3, and prodelphinidin C2, which are having multiple health beneficial properties.[36] Pigmented grains of maize, sorghum, quinoa, and wheat are good sources of several nutrients and phenolic compounds such as ferulic acid, protocatechuic acid, vanillic acid, p-coumaric acid, quercetin, cyanidin-3-rutinoside, peonidin-3-glucoside, malvidin-3-glucoside, etc., which are helpful in the ailment of several disorders.[42–44] The grains that are good source of wide range of phenolic compounds include millets such as finger millet, foxtail millet, pearl millet, kodo millet, and barnyard millet, which might serve as an alternative for the normal rice and other types of commonly consumed cereals. These grains are also gluten-free and have antidiabetic properties.[35,45–48]

Cereals and legumes also contain the bioactive component of carotenoids, among which the major carotenoids found in cereals are lutein, α-carotene, β-carotene, β-cryptoxanthin, and zeaxanthin. Maize grain is one of the common sources of carotenoids among different cereal grains, the major carotenoids found in maize are lutein and zeaxanthin; however, the contents of different carotenoids vary depending on the pigmentation of the grain kernels and varieties.[42,49]

4.2 *Fruits and vegetables*

Fruits and vegetables also comprise a wide range of polyphenolic compounds. Resveratrol found in blueberries and grapes, hesperidin found in citrus fruits, and several

Table 5.2 Food groups, bioactive compounds and health beneficial effects.

Food group	Type	Bioactive compounds	Health benefits	References
Cereals and legumes	Rice, wheat, barley, sorghum, millet, and maize	Phenolic acids, flavonoids	Antioxidant property, antiproliferative activity on hepatic cancer cells, apoptotic activity (HCT-116 cells), inhibition of NF-κB pathway	Goudar et al.[35]; Sharma et al.[36]
	Rye, wheat, barley, oat	Phytosterols	Decrease the absorption of cholesterol, anti-inflammatory activity, retardation of oxidative degradation of lipids	Goudar et al.[27]; Bartłomiej et al.[26]
Fruits and vegetables	Raspberry, blackberry, grapefruit, cranberries, apricots, peaches, strawberries, pomegranate	Phenolic acids, flavonoids, tannins	Inhibition of DNA breakage and LDL cholesterol oxidation, cellular antioxidant property, inhibition of angiotensin converting enzyme (ACE)	Santana-Gálvez and Jacobo-Velázquez[23]
	Tomatoes, watermelon, red grapefruit	Carotenoids	Inhibition of proinflammatory/prothrombotic factors, hypolipemic	Langi et al.[25]
Spices	Ginger, cinnamon, oregano, clove	Phenolic acids, flavonoids	Inhibition of lipid peroxidation	Santana-Gálvez and Jacobo-Velázquez[23]
Beverages	Tea, coffee, wine	Phenolic acids, flavonoids	Antiproliferative activity of colon cancer HT-29 cells	Santana-Gálvez and Jacobo-Velázquez[23]
Herbs	Mint, basil, *giloy*	Phenolic acids, flavonoids, tannins	Antioxidative, anti-inflammatory, antimicrobial, antidiabetic, antihypertensive	Sharma et al.[37]; Singh and Chaudhuri[38]; Vastrad et al.[39]

other compounds such as naringenin, vitexin, rutin, quercetin, and catechin found in several fruits and vegetables may have antioxidative, antimicrobial, and antiviral activity.[50–52] Pomegranate (*Punica granatum* L.) peel is another rich source of polyphenols, which include phenolic acids, flavonoids, and anthocyanins and has been proved effective with antioxidative, antimicrobial, hypoglycemic, and antihypertensive. The polyphenols in the pomegranate extracts have been studied and proven for their inhibitory effects against viral infections such as herpes virus, human immunodeficiency virus, and influenza virus.[53]

Carotenoids are one of the important bioactive compounds abundant in commonly consumed fruits, which include mango, papaya, peach, apricot, and vegetables such as carrot, tomato, green leafy vegetables, red pepper, etc.[54,55] β-carotene is one of the major and most widely studied carotenoid present in fruits and vegetables. Pumpkin, mango, and apricot among fruits and carrot, spinach, lettuce, kale, green leafy vegetables, and pigmented vegetables (orange, yellow, red) among various vegetables are some of the major sources of β-carotene. Lycopene is predominantly found in tomato and its products such as ketchup and juices; however, it is also found in grape fruit (pink) and watermelon.[56] Lutein is another major carotenoid compound, which is found majorly in green leafy vegetables such as broccoli, spinach, lettuce, parsley, etc.[57]

4.3 Spices

Turmeric (*Curcuma longa* L.) was one of the important foods ingredients, which have been studied for its efficacy against viral infection. Some studies have shown its effective inhibition property against the COVID-19 infection through molecular docking of active ingredients (curcumin and its derivatives) by binding with spike protein of SARS-CoV-2.[58,59] Ginger (*Zingiber officinale* Roscoe) is a rhizome, which has been commonly used in the preparations of several food recipes, it is used as dried or in fresh form. Some phytochemicals of its rhizome include (−)-zingiberene, gingerenone A, 5-hydroxy-1-(4-hydroxy-3-methoxy phenyl) decan-3-one, (+)-curcumene, (−)-β-sesquiphellandrene, alloaromadendrene, etc., have shown antiviral properties against the virus such as chikungunya, human respiratory syncytial virus, and influenza virus.[60,61]

Garlic (*Allium sativum* L.) has been into use since ancient times as a spice or seasoning herb in several cuisines. It has been considered a functional food against various ailments, and its antiviral properties have been extensively reviewed.[62] The pungent odor in garlic is due to its organosulfur compounds (OSCs), which mainly belong to L-cysteine sulfoxides and γ-glutamyl-L-cysteine peptides and are considered as important bioactive compounds. Among almost 30 OSCs, alliin is most abundant in garlic, which is converted into allicin during processing such as chopping, crushing, mincing, etc. Allicin is one of the important bioactive compounds, which are having antioxidant, anti-inflammatory, immunomodulatory, and antiviral activity.[62,63] In vitro and in vivo studies have shown the effectiveness of garlic extract against viral infections such as flu, immunosuppression, respiratory problems, and genital herpes.[64,65] The synergistic effects of other phytochemicals such as flavonoids, saponins, bioactive carbohydrates, and lectins present in garlic other than OSC's play a major role for their effective mechanism against viral infections. *Nigella sativa* also known as black cumin is an oil seed used in food preparations. It has also been used as treatment for upper respiratory problems and stomach infections, and the extract has potential antimicrobial, antiviral, and antitumor activity. The main bioactive compounds present in *Nigella sativa* are thymoquinone, ρ-simen, longifolene, and karvakrol. The effect of *Nigella sativa* has been studied for its efficacy against replication of coronavirus.[66,67]

4.4 Beverages

Worldwide the most commonly consumed beverage is tea after water, and there are basically only three different types of beverages, which are consumed commonly across the globe and are nonalcoholic, which include tea, coffee, and cocoa. The major polyphenols among tea and cocoa are (+)-catechin, (−)-epicatechin, procyanidin C1, and B2. The green tea and oolong tea contain (−)-epigallocatechin gallate as a major polyphenol; however, the oxidized flavan-3-ols known as theaflavins are abundant in black tea. The major phenolic compounds in coffee include hydroxycinnamic acids such as geruloylquinic acid, dicaffeoylquinic acid, and mono-caffeoylquinic acid.[22] The bioactive compounds in tea (*Camellia sinensis* [L.] Kuntze) such as theaflavins, epigallocatechin gallate, epicatechingallate, and gallocatechin-3-gallate, have notable health beneficial properties including antiviral activity.[68,69] On the verge of COVID-19, the oxidized epigallocatechin gallate is studied for its effectiveness on inhibiting the M^{pro} of SARS-CoV-2.[70]

4.5 Herbs

Several plant-based products which have been commonly consumed in the form of decoctions and other consumable types during the COVID-19 infection are giloy (*Tinospora cordifolia*), tulsi (*Ocimum sanctum*), ashwagandha (*Withania somnifera*), amla (*Emblica officinalis*), liquorice (*Glycyrrhiza glabra* L.), aloevera, etc. These plants are known for their rich source of phenolic compounds such as phenolic acids, flavonoids, alkaloids, terpenoids, and anthocyanins with health beneficial properties of antioxidative, anti-inflammatory, antimicrobial, antidiabetic, and antihypertensive.[37–39] Some of the active principal compounds obtained from these plants have been proved for their antiviral properties toward COVID-19 infection, which include diferuloyl methane, withaferin A/D, glycyrrhizic acid, dehydroglyasperin C, liquiritin, glyasperin A, cordifolioside, berberine, catechin, quercetin, astragalin, pectolinarin, astragalin, ellagic acid, etc.[71–74]

5. Mechanistic activity of bioactive compounds helpful for COVID-19

Numerous food commodities have been proven for their immunomodulatory and antiviral efficacy. During the onset of COVID-19 pandemic, most of the natural products from medicinal plants as herbal source were studied through in vitro, in vivo, and in silico studies for their efficacy to combat COVID-19 infection.[75,76] These studies provided greater insights of these natural compounds' interaction, mechanism of action, their targets of virus inhibition, and specific bonding with active proteins of SARS-CoV-2 virus. The proteins with which these secondary metabolites might interact with SARS-CoV-2 virus are spike protein, M^{pro}, PL^{pro}, and RdRp.

The majority of bioactive compounds in plant functional foods are mainly the polyphenols compounds which are thought to provide antiviral efficacy. Also, the regulation of immune cells, production of proinflammatory cytokines, and suppression of proinflammatory gene expression are modulated by polyphenols.[77] These bioactive compounds inhibit the enzymes such as the serine/threonine specific protein kinase, cellular receptor kinases (MAPKs), and phosphatidylinositol-4,5-biphosphate-3-kinase (PI3-K), which thereby are involved in hindering the cellular cascades and signaling pathways to suppress the viral infection mechanism.[78]

A wide range of flavonoid compounds (hesperetin, catechin, naringin, quercetin) in plants have been proved for their efficacy against several viral infectious diseases as Rous sarcoma, pseudorabies, sindbis, parainfluenza virus type 3, respiratory syncytial virus, HSV-1, polio virus type 1, and severe acute respiratory syndrome coronavirus (SARS-CoV).[79–81] The protein 3CLpro is known to be blocked by several flavonoid compounds such as pectolinarin, quercetin 3-β-d glucose, herbacetin, rhoifolin, and isobavachalcone.[82]

6. Conclusion

The bioactive compounds are vastly found throughout the plant kingdom, with some specific compounds present in higher quantities in particular foods. Different bioactive compounds have been classified and characterized depending on their chemical nature and structural arrangement. They are the secondary metabolites, which in lower quantities will exert multiple beneficial properties during the health disorders. During the COVID-19 pandemic, the food-based approach to combat the infection has been one of the main components, which have helped in maintaining healthy well-being. The major bioactive components, their properties, and the foods abundant in these compounds have been studied in the section; this will give an insight about the main foods containing specific bioactive compounds, which will help the community or society to understand and know the role of these specific compounds in combatting the COVID-19 infection. The food groups, which are abundant in bioactive compounds include cereals, millets, legumes, fruits, vegetables, spices, beverages, herbs, etc.; however, each type of food is having its own importance with specific bioactive principal components. Hence, these food groups can be extensively studied with incorporation of specific proportions in our daily routine diet, which will provide certain quantity of the particular bioactive compounds.

References

1. Vishwakarma S, Panigrahi C, Barua S, Sahoo M, Mandliya S. Food nutrients as inherent sources of immunomodulation during COVID-19 pandemic. *LWT*. 2022:113154.
2. Galanakis CM. The food systems in the era of the coronavirus (COVID-19) pandemic crisis. *Foods*. 2020;9(4). https://doi.org/10.3390/foods9040523.
3. Pereira JA, Berenguer CV, Andrade CF, Câmara JS. Unveiling the bioactive potential of fresh fruit and vegetable waste in human health from a consumer perspective. *Appl Sci*. 2022;12(5):2747.
4. Crous-Bou M, Minguillon C, Gramunt N, Molinuevo JL. Alzheimer's disease prevention: from risk factors to early intervention. *Alzheimer's Res Ther*. 2017;9(1):71. https://doi.org/10.1186/s13195-017-0297-z.
5. Lapuente M, Estruch R, Shahbaz M, Casas R. Relation of fruits and vegetables with major cardiometabolic risk factors, markers of oxidation, and inflammation. *Nutrients*. 2019; 11(10):2381.
6. Sharifi-Rad J, Rodrigues CF, Sharopov F, et al. Diet, lifestyle and cardiovascular diseases: linking pathophysiology to cardioprotective effects of natural bioactive compounds. *Int J Environ Res Publ Health*. 2020;17(7):2326.
7. Rizou M, Galanakis IM, Aldawoud TM, Galanakis CM. Safety of foods, food supply chain and environment within the COVID-19 pandemic. *Trends Food Sci Technol*. 2020;102: 293–299.
8. CDC. *How Coronavirus Spreads*; March 2022. https://www.cdc.gov/coronavirus/2019-ncov/prevent-getting-sick/how-covid-spreads.html? CDC_AA_refVal=https%3A%2F%2Fwww.cdc.gov%2Fcoronavirus%2F2019-ncov%2Fprepare%2Ftransmission.html.
9. FAO. *COVID-19 and the Risk to Food Supply Chains: How to Respond?*; March 2020. https://www.fao.org/documents/card/en/c/ca8388en/.
10. FDA. *Best Practices for Retail Food Stores, Restaurants, and Food Pick-Up/Delivery Services during the COVID-19 Pandemic*. FDA; March 2022. https://www.fda.gov/food/food-safety-during-emergencies/best-practices-retail-food-stores-restaurants-and-food-pick-updelivery-services-during-covid-19.
11. FDA. *Food Safety and the Coronavirus Disease 2019 (COVID-19)*. FDA; 2022. https://www.fda.gov/food/food-safety-during-emergencies/food-safety-and-coronavirus-disease-2019-covid-19.
12. WHO. *COVID-19 and Food Safety: Guidance for Food Businesses: Interim Guidance, 07 April 2020*. World Health Organization; 2022. https://www.who.int/publications/i/item/covid-19-and-food-safety-guidance-for-food-businesses.
13. EFSA. *Coronavirus: No Evidence that Food is a Source or Transmission Route*. EFSA Unión Europea, Parlamento Europeo; 2022. https://www.efsa.europa.eu/en/news/coronavirus-no-evidence-food-source-or-transmission-route#:~:text=EFSA%20is%20closely%20monitoring%20the,of%20transmission%20of%20the%20virus.
14. Chin AWH, Chu JTS, Perera MRA, et al. Stability of SARS-CoV-2 in different environmental conditions. *Lancet Microbe*. 2020;1(1):e10. https://doi.org/10.1016/S2666-5247(20)30003-3.
15. Mérillon J-M, Ramawat KG. *Bioactive Molecules in Food*. Springer; 2019.
16. Xiong Y, Zhang P, Warner RD, Shen S, Fang Z. Cereal grain-based functional beverages: from cereal grain bioactive phytochemicals to beverage processing technologies, health benefits and product features. *Crit Rev Food Sci Nutr*. 2020:1–25.

17. Fardet A, Rock E, Rémésy C. Is the in vitro antioxidant potential of whole-grain cereals and cereal products well reflected in vivo? *J Cereal Sci.* 2008;48(2):258–276.
18. Scalbert A, Manach C, Morand C, Rémésy C, Jiménez L. Dietary polyphenols and the prevention of diseases. *Crit Rev Food Sci Nutr.* 2005;45(4):287–306.
19. Masisi K, Beta T, Moghadasian MH. Antioxidant properties of diverse cereal grains: a review on in vitro and in vivo studies. *Food Chem.* 2016;196:90–97.
20. López-Alarcón C, Denicola A. Evaluating the antioxidant capacity of natural products: a review on chemical and cellular-based assays. *Anal Chim Acta.* 2013;763:1–10.
21. Rothwell JA, Urpi-Sarda M, Boto-Ordonez M, et al. Phenol-Explorer 2.0: a major update of the Phenol-Explorer database integrating data on polyphenol metabolism and pharmacokinetics in humans and experimental animals. *Database.* 2012.
22. Zhang L, Han Z, Granato D. Polyphenols in foods: classification, methods of identification, and nutritional aspects in human health. In: *Advances in Food and Nutrition Research.* Elsevier; 2021:1–33.
23. Santana-Gálvez J, Jacobo-Velázquez DA. *Classification of Phenolic Compounds. Phenolic Compounds in Food.* CRC press; 2018:3–20.
24. Khoo HE, Azlan A, Tang ST, Lim SM. Anthocyanidins and anthocyanins: colored pigments as food, pharmaceutical ingredients, and the potential health benefits. *Food Nutr Res.* 2017;61(1):1361779.
25. Langi P, Kiokias S, Varzakas T, Proestos C. Carotenoids: from plants to food and feed industries. In: *Microbial Carotenoids.* 2018:57–71.
26. Bartłomiej S, Justyna R-K, Ewa N. Bioactive compounds in cereal grains–occurrence, structure, technological significance and nutritional benefits—a review. *Food Sci Technol Int.* 2012;18(6):559–568.
27. Goudar G, Sharma P, Janghu S, Longvah T. Effect of processing on barley β-glucan content, its molecular weight and extractability. *Int J Biol Macromol.* 2020;162: 1204–1216.
28. Acosta-Estrada BA, Gutiérrez-Uribe JA, Serna-Saldívar SO. Bound phenolics in foods, a review. *Food Chem.* 2014;152:46–55.
29. Shahidi F, Yeo J. Insoluble-bound phenolics in food. *Molecules.* 2016;21(9):1216.
30. Stalikas CD. Extraction, separation, and detection methods for phenolic acids and flavonoids. *J Separ Sci.* 2007;30(18):3268–3295.
31. Marais JP, Deavours B, Dixon RA, Ferreira D. *The Stereochemistry of Flavonoids. The Science of Flavonoids.* Springer; 2006:1–46.
32. Castañeda-Ovando A, de Lourdes Pacheco-Hernández M, Páez-Hernández ME, Rodríguez JA, Galán-Vidal CA. Chemical studies of anthocyanins: a review. *Food Chem.* 2009;113(4):859–871.
33. Giada M. Food phenolic compounds: main classes, sources and their antioxidant power. *Oxidative stress and chronic degenerative diseases—a role for antioxidants.* 2013:87–112. https://doi.org/10.5772/51687.
34. Lazaridou A, Biliaderis CG, Izydorczyk MS. *Functional Food Carbohydrates.* Boca Raton, FL: CRC Press; 2007.
35. Goudar G, Manne M, Sathisha GJ, et al. Phenolic, nutritional and molecular interaction study among different millet varieties. *Food Chem Adv.* 2022, 100150. https://doi.org/10.1016/j.focha.2022.100150.
36. Sharma P, Goudar G, Longvah T, Gour VS, Kothari S, Wani IA. Fate of polyphenols and antioxidant activity of barley during processing. *Food Rev Int.* 2022;38(2):163–198.

37. Sharma H, Rao PS, Singh AK. Fifty years of research on Tinospora cordifolia: from botanical plant to functional ingredient in foods. *Trends Food Sci Technol.* 2021;118: 189–206.
38. Singh D, Chaudhuri PK. A review on phytochemical and pharmacological properties of Holy basil (Ocimum sanctum L.). *Ind Crop Prod.* 2018;118:367–382.
39. Vastrad JV, Goudar G, Byadgi SA, Devi RD, Kotur R. Identification of bio-active components in leaf extracts of Aloe vera, Ocimum tenuiflorum (Tulasi) and Tinospora cordifolia (Amrutballi). *J Med Plants Res.* 2015;9(28):764–770.
40. Ikewaki N, Iwasaki M, Kurosawa G, et al. β-glucans: wide-spectrum immune-balancing food-supplement-based enteric (β-WIFE) vaccine adjuvant approach to COVID-19. *Hum Vaccines Immunother.* 2021;17(8):2808–2813.
41. Geller A, Yan J. Could the induction of trained immunity by β-glucan serve as a defense against COVID-19? *Front Immunol.* 2020;11:1782.
42. Lakshmi S, Goudar G, Singh M, Dhaliwal H, Sharma P, Longvah T. Variability in resistant starch, vitamins, carotenoids, phytochemicals and in-vitro antioxidant properties among diverse pigmented grains. *J Food Meas Char.* 2021;15(3):2774–2789.
43. Harakotr B, Suriharn B, Tangwongchai R, Scott MP, Lertrat K. Anthocyanin, phenolics and antioxidant activity changes in purple waxy corn as affected by traditional cooking. *Food Chem.* 2014;164:510–517.
44. Mora-Rochin S, Gutiérrez-Uribe JA, Serna-Saldivar SO, Sánchez-Peña P, Reyes-Moreno C, Milán-Carrillo J. Phenolic content and antioxidant activity of tortillas produced from pigmented maize processed by conventional nixtamalization or extrusion cooking. *J Cereal Sci.* 2010;52(3):502–508.
45. Hithamani G, Srinivasan K. Effect of domestic processing on the polyphenol content and bioaccessibility in finger millet (Eleusine coracana) and pearl millet (Pennisetum glaucum). *Food Chem.* 2014;164:55–62. https://doi.org/10.1016/j.foodchem.2014.04.107.
46. Kamatar M, Brunda S, Rajaput S, Hundekar R, Goudar G. Nutritional composition of seventy five elite germplasm of foxtail millet (Setaria italica). *Int J Adv Res Technol.* 2015; 4(4):1–6.
47. Goudar G, Hemalatha S, Naik R, Kamatar M. *Recapturing nutritious millets for health and management of diseases* Naik R, ed. Evaluation of Nutritional Composition of Foxtail Millet (Setaria italica) Grains Cultivated in Agro Climatic Zones of Karnataka by NIR; 37. Dharwad: National symposium; 2011:37.
48. Goudar G, Sathisha G. Effect of processing on ferulic acid content in foxtail millet (Setaria italica) grain cultivars evaluated by HPTLC. *Orient J Chem.* 2016;32(4):2251–2258.
49. Hidalgo A, Brandolini A, Pompei C. Carotenoids evolution during pasta, bread and water biscuit preparation from wheat flours. *Food Chem.* 2010;121(3):746–751.
50. Marinella MA. Indomethacin and resveratrol as potential treatment adjuncts for SARS-CoV-2/COVID-19. *Int J Clin Pract.* 2020;74(9):e13535.
51. Horne JR, Vohl M-C. Biological plausibility for interactions between dietary fat, resveratrol, ACE2, and SARS-CoV illness severity. *Am J Physiol Endocrinol Metab.* 2020;318(5): E830–E833.
52. El-Missiry MA, Fekri A, Kesar LA, Othman AI. Polyphenols are potential nutritional adjuvants for targeting COVID-19. *Phytother Res.* 2021;35(6):2879–2889. https://doi.org/10.1002/ptr.6992.
53. Suručić R, Tubić B, Stojiljković MP, et al. Computational study of pomegranate peel extract polyphenols as potential inhibitors of SARS-CoV-2 virus internalization. *Mol Cell Biochem.* 2021;476(2):1179–1193.

54. Zhou W, Niu Y, Ding X, et al. Analysis of carotenoid content and diversity in apricots (Prunus armeniaca L.) grown in China. *Food Chem.* 2020;330:127223.
55. Dias MG, Olmedilla-Alonso B, Hornero-Méndez D, et al. Comprehensive database of carotenoid contents in ibero-American foods. A valuable tool in the context of functional foods and the establishment of recommended intakes of bioactives. *J Agric Food Chem.* 2018;66(20):5055–5107.
56. Meléndez-Martínez AJ, Mandić AI, Bantis F, et al. A comprehensive review on carotenoids in foods and feeds: status quo, applications, patents, and research needs. *Crit Rev Food Sci Nutr.* 2020:1–51.
57. Reif C, Arrigoni E, Schärer H, Nyström L, Hurrell RF. Carotenoid database of commonly eaten Swiss vegetables and their estimated contribution to carotenoid intake. *J Food Compos Anal.* 2013;29(1):64–72.
58. Gupta S, Singh AK, Kushwaha PP, et al. Identification of potential natural inhibitors of SARS-CoV2 main protease by molecular docking and simulation studies. *J Biomol Struct Dyn.* 2021;39(12):4334–4345.
59. Garg S, Anand A, Lamba Y, Roy A. Molecular docking analysis of selected phytochemicals against SARS-CoV-2 Mpro receptor. *Vegetos.* 2020;33(4):766–781.
60. Kaushik S, Jangra G, Kundu V, Yadav JP, Kaushik S. Anti-viral activity of Zingiber officinale (Ginger) ingredients against the Chikungunya virus. *VirusDisease.* 2020;31(3): 270–276.
61. Wang J, Prinz RA, Liu X, Xu X. In vitro and in vivo antiviral activity of gingerenone A on influenza a virus is mediated by targeting Janus kinase 2. *Viruses.* 2020;12(10):1141.
62. Rouf R, Uddin SJ, Sarker DK, et al. Antiviral potential of garlic (Allium sativum) and its organosulfur compounds: a systematic update of pre-clinical and clinical data. *Trends Food Sci Technol.* 2020;104:219–234.
63. Wang L, Jiao H, Zhao J, Wang X, Sun S, Lin H. Allicin alleviates reticuloendotheliosis virus-induced immunosuppression via ERK/mitogen-activated protein kinase pathway in specific pathogen-free chickens. *Front Immunol.* 2017;8:1856.
64. Chavan RD, Shinde P, Girkar K, Madage R, Chowdhary A. Assessment of anti-influenza activity and hemagglutination inhibition of plumbago indica and allium sativum extracts. *Pharmacogn Res.* 2016;8(2):105–111. https://doi.org/10.4103/0974-8490.172562.
65. Choi J-B, Cheon HS, Chung M-S, Cho W-I. Pretreatment sterilization of garlic and ginger using antimicrobial agents and blanching. *Korean J Food Sci Technol.* 2018;50(2): 172–178.
66. Ulasli M, Gurses SA, Bayraktar R, et al. The effects of Nigella sativa (Ns), Anthemis hyalina (Ah) and Citrus sinensis (Cs) extracts on the replication of coronavirus and the expression of TRP genes family. *Mol Biol Rep.* 2014;41(3):1703–1711.
67. Sharma P, Longvah T. *Nigella (Nigella sativa) Seed. Oilseeds: Health Attributes and Food Applications.* Singapore: Springer; 2021:331–350. https://doi.org/10.1007/978-981-15-4194-0_13.
68. Vastrad JV, Badanayak P, Goudar G. Phenolic compounds in tea: phytochemical, biological, and therapeutic applications. *Phenolic Compounds: Chemistry, Synthesis, Diversity, Non-Conventional Industrial, Pharmaceutical and Therapeutic Applications.* 2022; 4(4):399.
69. Ge M, Xiao Y, Chen H, Luo F, Du G, Zeng F. Multiple antiviral approaches of (−)-epigallocatechin-3-gallate (EGCG) against porcine reproductive and respiratory syndrome virus infection in vitro. *Antiviral Res.* 2018;158:52–62. https://doi.org/10.1016/j.antiviral.2018.07.012.

70. Ungarala R, Munikumar M, Sinha SN, Kumar D, Sunder RS, Challa S. Assessment of antioxidant, immunomodulatory activity of oxidised epigallocatechin-3-gallate (green tea polyphenol) and its action on the main protease of SARS-CoV-2—an in vitro and in silico approach. *Antioxidants*. 2022;11(2):294.
71. Murugesan S, Kottekad S, Crasta I, et al. Targeting COVID-19 (SARS-CoV-2) main protease through active phytocompounds of ayurvedic medicinal plants–Emblica officinalis (Amla), Phyllanthus niruri Linn.(Bhumi Amla) and Tinospora cordifolia (Giloy)–A molecular docking and simulation study. *Comput Biol Med*. 2021;136:104683.
72. Manne M, Goudar G, Varikasuvu SR, et al. Cordifolioside: potent inhibitor against Mpro of SARS-CoV-2 and immunomodulatory through human TGF-β and TNF-α. *3 Biotech*. 2021; 11(3):1–25.
73. Sinha SK, Prasad SK, Islam MA, et al. Identification of bioactive compounds from Glycyrrhiza glabra as possible inhibitor of SARS-CoV-2 spike glycoprotein and non-structural protein-15: a pharmacoinformatics study. *J Biomol Struct Dyn*. 2021;39(13):4686–4700.
74. Shree P, Mishra P, Selvaraj C, et al. Targeting COVID-19 (SARS-CoV-2) main protease through active phytochemicals of ayurvedic medicinal plants–Withania somnifera (Ashwagandha), Tinospora cordifolia (Giloy) and Ocimum sanctum (Tulsi)–a molecular docking study. *J Biomol Struct Dyn*. 2022;40(1):190–203.
75. Mehany T, Khalifa I, Barakat H, Althwab SA, Alharbi YM, El-Sohaimy S. Polyphenols as promising biologically active substances for preventing SARS-CoV-2: a review with research evidence and underlying mechanisms. *Food Biosci*. 2021;40:100891.
76. Yang F, Zhang Y, Tariq A, et al. Food as medicine: a possible preventive measure against coronavirus disease (COVID-19). *Phytother Res*. 2020;34(12):3124–3136.
77. Yahfoufi N, Alsadi N, Jambi M, Matar C. The immunomodulatory and anti-inflammatory role of polyphenols. *Nutrients*. 2018;10(11):1618.
78. Villa T, Feijoo-Siota L, Rama J, Ageitos J. Antivirals against animal viruses. *Biochem Pharmacol*. 2017;133:97–116.
79. Chiang LC, Chang JS, Chen CC, Ng LT, Lin CC. *Anti-herpes simplex virus activity of Bidens pilosa* and *Houttuynia cordata*. *Am J Chin Med*. 2003;31(3):355–362. https://doi.org/10.1142/S0192415X03001090.
80. Kaul TN, Middleton Jr E, Ogra PL. Antiviral effect of flavonoids on human viruses. *J Med Virol*. 1985;15(1):71–79.
81. Yi L, Li Z, Yuan K, et al. Small molecules blocking the entry of severe acute respiratory syndrome coronavirus into host cells. *J Virol*. 2004;78(20):11334–11339.
82. Jo S, Kim S, Shin DH, Kim M-S. Inhibition of SARS-CoV 3CL protease by flavonoids. *J Enzym Inhib Med Chem*. 2020;35(1):145–151.

MicroRNAs and COVID-19

6

Prashanth Gowda[1], *Vivek Kumar*[2], *Ashish Sarangi*[3], *Jangampalli Adi Pradeepkiran*[1], *P. Hemachandra Reddy*[4,5,6,7] and *Subodh Kumar*[8]

[1]Department of Internal Medicine, Texas Tech University Health Sciences Center, Lubbock, TX, United States; [2]Department of Biotechnology, IMS Engineering College, Ghaziabad, Uttar Pradesh, India; [3]Department of Psychiatry and Behavioral Sciences Baylor College of Medicine, Houston, TX, United States; [4]Department of Pharmacology and Neuroscience, School of Medicine, Texas Tech University Health Sciences Center, Lubbock, TX, United States; [5]Neurology, Department of School of Medicine, School of Medicine, Texas Tech University Health Sciences Center, Lubbock, TX, United States; [6]Public Health Department of Graduate School of Biomedical Sciences, School of Medicine, Texas Tech University Health Sciences Center, Lubbock, TX, United States; [7]Department of Speech, Language and Hearing Sciences, School Health Professions, School of Medicine, Texas Tech University Health Sciences Center, Lubbock, TX, United States; [8]Center of Emphasis in Neuroscience, Department of Molecular and Translational Medicine, Paul L. Foster School of Medicine, Texas Tech University, Health Sciences Center El Paso, El Paso, TX, United States

Abstract

Currently, there are no reliable biomarkers for identifying COVID-19 patients and no definite therapeutics to control this deadly disease. MicroRNAs (miRNA) have been explored in several human diseases for their potential role as biomarkers and their therapeutic potential. However, there is very little information available about the roles of miRNAs in COVID-19 infection. This chapter outlines the recent updates and developments of miRNAs in COVID-19 such as miRNAs as potential biomarkers for COVID-19, the molecular basis of miRNAs in COVID-19 infection, and the use of miRNAs as therapeutics targets for COVID-19. While a few potential miRNAs have been researched for the aforementioned reasons, more research is needed to determine the roles of individual miRNAs in COVID-19 infection.

Keywords: Biomarkers; COVID-19; MicroRNAs; MiRNAs targets; Plasma; Serum

1. Introduction

MicroRNAs (miRNA) are small, conserved noncoding RNAs averaging 22 nucleotides long that posttranscriptionally regulate gene expression through complementary binding to messenger RNA (mRNA). The biogenesis of miRNA is shown in Fig. 6.1. MiRNAs are transcribed into primary-miRNA (pri-miRNA) in the nucleus by RNA

COVID-19 in Alzheimer's Disease and Dementia. https://doi.org/10.1016/B978-0-443-15256-6.00011-8

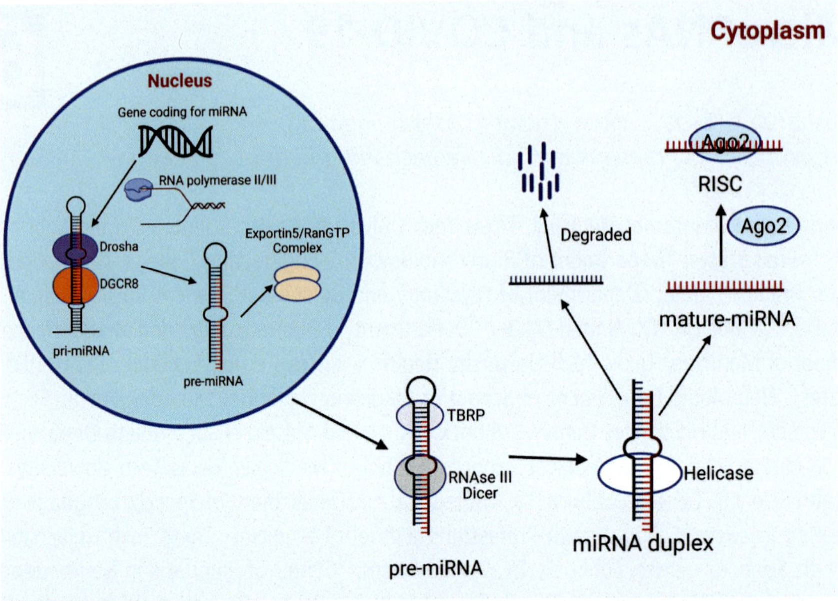

Figure 6.1 MiRNA biogenesis pathways. This figure illustrates the transcription of pri-miRNA in the nucleus, the exportation of pre-miRNA from the nucleus to the cytoplasm, and the processing of miRNA by cleavage proteins throughout this process, resulting in the formation of single-stranded mature-miRNA. Mature-miRNA can then associate with Ago2 protein to form an RNA-induced silencing complex (RISC) complex, which can silence the translation of mRNA.

polymerases II and III. DROSHA and DGCR8 proteins then cleave pri-miRNA into precursor-miRNA (pre-miRNA), which is subsequently transported to the cytoplasm by the Exportin 5/RanGTP complex. RNAse III Dicer and TBRP digest the pre-miRNA, creating a miRNA duplex that is unwound by helicase into a single-stranded mature-miRNA. The complementary strand of the mature miRNA is then degraded. The mature-miRNA interacts with Ago2 protein to form an RNA-induced silencing complex (RISC), which binds to 3′UTR of the mRNA through complementary base pairing. Consequently, this binding downregulates gene expression by degrading messenger RNA (mRNA).

MiRNAs dysregulation has been noted in several human diseases.[1–3] Similarly, several miRNAs have been found in patients with COVID-19 that may be implicated in the pathogenesis of this disease. MiRNAs may play a role as biomarkers for the detection of COVID-19 and may potentially serve as therapeutic targets in treating SARS-COV-2. Because miRNAs are present in the serum and plasma of humans and other animals such as bovine fetuses, calves, horses, mice, and rats,[4] miRNAs have a potential role as useful biomarkers of detecting COVID-19, predicting disease progression/health outcomes and as potential therapeutic targets. This chapter will analyze the current literature to explore which miRNAs are dysregulated in COVID patients, how miRNAs may contribute to the progression of COVID-19 in patients, and how miRNAs may be used as therapeutic targets or treatments.

2. MicroRNAs as biomarkers for COVID-19

Currently, there are no reliable biomarkers for identifying high-risk COVID-19 patients that require immediate medical attention. Unfortunately, high-risk patients are treated only after the appearance of severe symptoms, thereby missing the critical window for treatment and often leading to increased pressure on healthcare resources.[5] Therefore, there is an urgent need for reliable biomarkers that accurately identify high-risk COVID-19 patients by predicting the severity and prognosis of the disease.

Both negative- (Ebola and H5N1) and positive-stranded viruses (West Nile and Dengue) can encode small miRNA-like RNAs.[6–8] These cell-free RNAs are highly stable and can freely circulate in human blood.[9] Researchers[5] have tried to use these freely circulating small miRNAs-like RNAs as biomarkers for early detection of COVID-19 infection and prognosis prediction. A study of peripheral blood samples collected from COVID-19 patients by Li et al. noted that 35 miRNAs were upregulated and 38 miRNAs were downregulated compared to controls.[10]

Differential miRNA expression is mainly found in the serum and plasma; thus, this chapter focuses mainly on these two fluids. No current studies assessed miRNA dysregulation in other biofluids outside of serum and plasma. This hole in our understanding can be supplanted by further study to analyze miRNA dysregulation in other fluids such as cerebrospinal fluid.

2.1 Serum microRNAs

Numerous studies have demonstrated that serum miRNAs have the potentials to be used as biomarkers for early detection and prognosis of lung cancer,[11] colorectal cancer,[12] diabetes,[13] and COVID-19, among many other diseases. For example, miR-nsp3-3p in serum could be used to predict the severity of COVID-19 7.4 days in advance of severe symptoms at 97.1% accuracy.[5] This miRNA can also be used to accurately monitor patients' disease progression trend during the hospital stay at much higher accuracy than other indices such as D-dimer, C-reactive protein (CRP), lactate dehydrogenases (LDH), and peripheral lymphocyte count (PLC). Another study that used ferret models found that miR-423-5p, miR-23a-3p, and miR-195-5p discern COVID-19 infection with a 99.7% accuracy.[14] Moreover, hsa-let-7d, hsa-miR-17, hsa-miR-34b, hsa-miR-93, hsa-miR-200b, hsa-miR-200c, and hsa-miR-223 levels were found to be significantly decreased, while hsa-miR-190a and hsa-miR-203 were found to be significantly increased in COVID-19 patients when compared to healthy controls. Of these microRNAs, miR-190a had the greatest potential in identifying the prognosis of patients with COVID-19.[15] Some other miRNAs that can potentially be used as biomarkers include miR-146a-5p, miR-21-5p, and miR-126-3p.[16] Many more miRNAs are differentially regulated in serum, including miR-21, miR-155, miR-208a, and miR-499, which were found to be significantly increased in patients with COVID-19 compared to healthy controls.[17] Due to the substantial number of miRNAs that are differentially expressed in patients with COVID-19 compared to control, the analysis of serum miRNAs will have great importance in determining better methods of detection and treatment. Table 6.1 summarizes the miRNAs that are deregulated in the serum of COVID-19 patients.

Table 6.1 Serum miRNAs deregulated in COVID-19 patients.

Sl. No.	miRNA	Regulation	Extraction techniques	Patient sample	Age	ROC AUC	References
1	miR-msp3-3p		Small RNA deep sequencing	159 COVID-19 patients and 51 healthy controls		0.933	5
2	miR-423-5p, miR-23a-3p and miR-195-5p	miR-423-5p downregulated, miR-195-5p upregulated, and miR-23a-3p not differentially expressed	miRNeasy micro kit (Qiagen)	20 ferrets (10 male, 10 female) infected with COVID-19 and 12 healthy controls	Median age of infected ferrets was 4 months	1.0	14
3	let-7d, miR-17, miR-34b, miR-93, miR-200b, miR-200c, miR-223, miR-190a, and miR-203	All upregulated except hsa-miR-223, hsa-miR-190a, and hsa-miR-203	VNAT (viral nucleic acid buffer)	40 patients and 10 healthy controls	Mean age of patients was 55 and the mean age of controls was 36 years		15
4	miR-146a-5p, miR-21-5p, and miR-126-3p	All are downregulated		29 healthy control subjects and 29 serum samples	Mean age of healthy controls was 64.1 years		16
5	miR-21, miR-155, miR-208a and miR-499	Upregulated	miRNeasy Serum/ Plasma Advanced Kit (Qiagen)	18 covid patients (17 male, 1 female) and 15 healthy controls (14 male, 1 female)	median age of controls was 31 years, and the median age of the COVID-19 patients was 59 years	miR-155 = 1.000; miR-208a = 0.796; miR-499 = 0.865	17

2.2 Plasma microRNAs

MicroRNA expression has been studied to identify whether the degree of severity is correlated with the expression of various miRNAs in COVID-19. David de Gonzalo-Calvo et al. discovered that miR-148a-3p, miR-451a, and miR-486-5p could be used to differentiate between intensive care unit (ICU) patients and ward patients. Additionally, differences in the expression of miR-192-5p and miR-323a-3p were found when comparing ICU survivors with nonsurvivors.[18] MiR-17-5p and miR-142-5p were found to be downregulated while miR-15a-5p, miR-19a-3p, miR-19b-3p, miR-23a-3p, miR-92a-3p, and miR-320a were upregulated in patients with COVID-19.[19] Out of these miRNAs, researchers found that miR-19a-3p, miR-19b-3p, and miR-92a-3p have potential use as biomarkers and therapeutic targets.

MicroRNAs in the plasma have also been studied for their possible role in the progression of the disease. For example, miR-451a, which is discussed later in this chapter, contributes to the development of this disease.[20] Since miRNAs are dysregulated in various biofluids, understanding how miRNAs contribute to disease progression is of utmost importance. Table 6.2 summarizes the miRNAs that are deregulated in the plasma of COVID-19 patients.

3. Molecular basis of microRNAs in COVID-19 infection

Deregulated miRNAs in COVID-19 patients are associated with genes and proteins in many different organs, mainly targeting those found in the lungs, heart, and kidneys. As a result, deregulated miRNAs in COVID-19 are also involved in the development of respiratory disease, cardiac disease, diabetes, and kidney disease. Table 6.3 summarizes the target proteins of deregulated miRNAs that are involved in the pathogenesis of COVID-19. By understanding, which miRNAs contribute to organ dysfunction during this disease, researchers can develop therapeutics that target these miRNAs or associated proteins.

3.1 Lungs

The pulmonary system is the first and main target of the SARS-CoV-2 virus in COVID-19 patients. It causes acute respiratory distress syndrome (ARDS) by damaging the endothelial cell. It can also cause pulmonary embolism, a life-threatening condition in COVID-19 patients.[26] Numerous studies have shown to significantly increase proinflammatory cytokines (IL-4, IL-6, IL-8, IL-13, and TNF-α), intercellular adhesion molecule 1 (ICAM-1), and caspase-1 (CASP-1) levels in COVID-19 patients.[23,29] Therefore, these proteins have the potential to be used as biomarkers for early diagnosis of SARS-CoV-2 infection. Cytokine storms during the infection produce NF-κB,[31] which can cause dysfunction and apoptosis of endothelial cells possibly mediated by the activation of CASP-1.[23]

Table 6.2 Plasma miRNAs deregulated in COVID-19 patients.

S. No.	miRNA	Regulation	Extraction techniques	Patient samples	Age	ROC AUC	References
1	miR-148a-3p, miR-451a and miR-486-5p	miR-148a-3p is upregulated while miR-451a and miR-486-5p are downregulated	miRCURY LNA Universal RT microRNA PCR System (Qiagen)	9 patients (44 male, 35 female)	Mean age of patients was 68 years	0.89	18
2	miR-31-5p, miR-3125, miR-4742-3p, miR-1275, miR-3617-5p, and miR-500b-3p	miR-31-5p, miR-3125, and miR-4742-3p are upregulated while miR-1275, miR-3617-5p, and miR-500b-3p are downregulated	Next-generation sequencing using QIAseq miRNA Library Kit and QIAseq miRNA NGS 48 Index IL (Qiagen)	10 COVID-19 patients (4 male, 6 female) and 10 age- and gender-matched healthy controls (4 male, 6 female)	Mean age of healthy controls was 53 years and the mean age of patients was 53.5 years		14
3	miR-17-5p and miR-142-5p, miR-15a-5p, miR-19a-3p, miR-19b-3p, miR-23a-3p, miR-92a-3p and miR-320a	All upregulated except miR-17-5p and miR-142-5p	miRNeasy Serum/Plasma RNA isolation kit (Qiagen)	33 COVID-19 patients (20 male, 13 female) and 10 healthy controls	Mean age of patients was 45 years	0.815 for miR-19a-3p, 0.875 for miR-19b-3p, and 0.850 for miR-92a-3p with the combined ROC of all three miRNAs being 0.917	19
4	miR-451a	Downregulated		5 COVID-19 patients and 3 healthy donors			20

Table 6.3 MiRNAs and their COVID-19 targets proteins in lungs, heart and kidney diseases.

S. No.	Organ	miRNA	Targeted protein	Regulation	References
1	Lungs	miR-124	ACE2	Downregulated	21
2		miR-17-5p	ACE2	Downregulated	22
3		miR-181a	IL-17, ACE2	Downregulated	23
4		miR-214	ACE2, FGFR1-WNT/MAPK/AKT	Upregulated	23
5		miR-98	TMPRSS2, IL-10	Downregulated	21
6		let-7	ACE2	Downregulated	24
7		miR-15b-5p	ERK1	Downregulated	25
8		miR-195-5p	Spike protein (S)	Upregulated	25
9		miR-422a	MLH1, TP53	Downregulated	25
10		miR-26a-5p	IL-6, ICAM-1	Downregulated	26
11		miR-29b-3p	IL-4, IL-8	Downregulated	26
12		miR-34a-5p	CASP-1	Downregulated	26
13	Heart	miR-7-5p	TMPRSS2	Downregulated	27
14		miR-125b	ACE2	Downregulated	24
15		miR-145-5p	ACE2	Downregulated	27
16		miR155-5p	IL-6, IL-8	Upregulated	24
17		miR-208a-3p	βMHC	Upregulated	28
18		miR-212	KCNJ2	Downregulated	29
19		miR-214	TMPRSS2	Upregulated	24
20		miR-223-3p	NLRP3	Downregulated	24
21		miR-328	CACNA1C, CACNB1	Downregulated	29
22		miR-375	PDK1	Upregulated	30
23	Kidney	let-7	TLR4	Downregulated	24
24		miR-15b-5p	mTOR	Downregulated	25
25		miR-18	ACE2	Upregulated	21
26		miR-125b	ACE2	Downregulated	21
27		miR-141	ACE2	Downregulated	24
28		miR-143	ACE2	Downregulated	21
29		mIR-145	ACE2	Upregulated	21
30		miR-216a	TGFβ	Upregulated	24
31		miR-421	ACE2	Downregulated	21

Table 6.3 presents the list of miRNAs for the potential treatment of COVID-19 patients with respiratory diseases. Among them, literature shows the potential use of miR-17-5p for treatment of COVID-19 patients with respiratory tract infection,[26] miR-98, and miR-124 for treating COVID-19 patients with lung cancer,[22] miR-181a, and miR-214 for COVID-19 patients with non-small-cell lung cancer,[32] and let-7 for COVID-19 patients with idiopathic pulmonary fibrosis. Numerous studies have noted that miRNAs can be used to regulate cytokine levels and CASP-1 expression in COVID-19 patients. In a Brazilian study of postmortem lung biopsies from COVID-19 patients, a downregulation of miR-26a-5p, miR-29b-3p, and miR-34a-5p that target cytokine levels were appreciated when compared to controls.[23] Another example of miRNA regulation of cytokine levels is that of hsa-miR-15b-5p, a known regulator of cell proliferation and apoptosis by targeting the extracellular signal-regulated kinase 1 (ERK1). In hamster lung tissue infected with SARS-CoV-2, hsa-miR-15b-5p was downregulated, thus inhibiting apoptosis of infected cells, which in turn leads to the proliferation of SARS-CoV-2.[21]

3.2 Heart

Upregulation of miR-208a-3p and miR-375 is noted in COVID-19 patients with pre-existing cardiovascular diseases.[24,28] These miRNAs can lead to myocardial damage by inducing apoptosis in ischemic cardiomyocytes. Upregulated miRNAs such as miR-155-5p and miR-214 and downregulated miRNAs such as miR-125b and miR-223-3p are known to reduce endothelial cell injury due to high glucose levels.[33] MiR-155-5p and miR-214 are found overexpressed in heart failure patients to control the release of proinflammatory cytokines. MiR-212 and miR-328 are downregulated in the elderly and people with arrhythmia when they were infected with COVID-19.[30]

Diabetes mellitus is a metabolic disease known to increase the severity and mortality of COVID-19 patients by increasing cardiac injury and diabetic ketoacidosis.[27,34] Both diabetes and COVID-19 have an intertwined relationship as they alter the metabolism of their patients, which instigates pathological remodeling leading to cardiac arrhythmias and heart failure. For example, infection of cardiomyocytes with SARS-CoV-2 can induce cytotoxic effects as it changes the expression of genes involved in metabolism and immune response functions. Some circulating miRNAs such as miR-7-5p and miR-145-5p (Table 6.3) are predicted to have antiviral functions in diabetes patients and inhibit SARS-CoV-2 replication by directly targeting the S-protein.[25] However, levels of these miRNAs among the elderly and people with diabetes are significantly lower compared to those of healthy young adults. Therefore, adjustment of the expression of certain miRNA levels can also serve as a therapeutic strategy to treat COVID-19 patients with diabetes.

3.3 Kidney

Numerous studies have reported that kidney diseases elevate the risk of SARS-CoV-2 infection and high mortality.[35,36] Additionally, COVID-19 can exacerbate kidney damage. Dehydration is one of the major consequences of COVID-19 that can lead to acute kidney injury. The infectious damage to the kidney can also lead to increased protein loss in the urine.[37] Other possible mechanisms of COVID-19-related kidney injury include sepsis[38] and rhabdomyolysis.[33]

The downregulation of let-7 and miRNA-141 and overexpression of TGFβ via upregulation of miR-216a increase kidney fibrosis.[39] Other miRNAs that take part in the altered expression of ACE2 protein, a receptor of the spike protein of the SARS-CoV-2 virus, are miR-18, miR-125b, miR-143, miR-145, and miR-421 (Table 6.3).[22] Due to ACE2 dysfunction, the dysregulation of these miRNAs can further contribute to kidney damage.

4. MicroRNAs as therapeutic for COVID-19

Numerous therapeutic approaches such as antibiotic and antiviral therapies,[40] glucocorticoid therapy,[41] invasive and noninvasive ventilation,[42] and renal replacement therapy[43] have been evaluated to treat COVID-19 patients. However, none of these therapeutics in large-scale clinical studies have shown significant efficacy in treating COVID-19 patients.[44] There is an urgent need for finding an alternative therapeutic strategy to prevent the progression of the disease. It is now known that miRNAs play an active role in host-pathogen interactions during viral infection, and viral miRNAs have been observed to make extensive changes in the host transcriptome.[33] Therefore, in recent years, miRNAs have been used as antiviral regulators of viral genes in treating various diseases such as hepatitis B[45] and COVID-19.[33] The differential expression of many miRNAs indicates that miRNAs might play a significant role in the development and treatment of COVID-19. Therefore, analyzing which COVID-19 genes are targeted by miRNAs that are differentially expressed in COVID-19 patients provides great insight into both how miRNAs contribute to the progression of this disease and how to use these genes as potential therapeutic targets. Many studies have already identified many target genes of miRNAs in COVID-19 patients. The next step is to work to develop treatments that either directly targets these genes or maintain adequate levels of miRNAs or other regulators that play a role in controlling the expression of these genes. Table 6.3 displays these deregulated miRNAs found in COVID-19 patients with respiratory, diabetes/heart failure, kidney, and liver diseases.

ACE2 and TMPRS22 are two common protein targets of miRNAs associated with COVID-19. SAR2-CoV-2, using its spike protein, binds to the receptor ACE2 as an entry point to infect a wide range of human cells.[46] The binding of spike protein with ACE2 inhibits regulation of ANG II concentration leading to inflammation and cell death in the alveoli adversely affecting oxygen supply into the body. This causes injury to the lungs in COVID-19 patients. Although the lungs are the primary organ of injury by SARS-CoV-2 as it enters the body through the nose or mouth, it also injures other organs where tissues express ACE2 including the heart, kidney, liver, and digestive tracts.[47] Overall, ACE2-associated miRNAs play vital roles in the normal functioning of multiple organs, and their expressions are affected by SARS-CoV-2

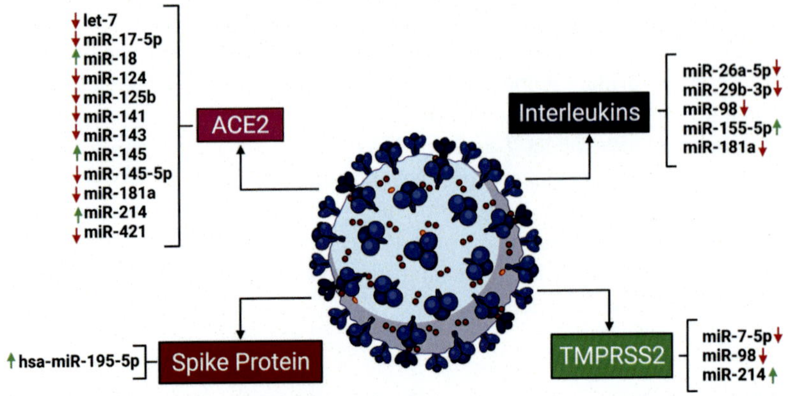

Figure 6.2 Deregulated miRNAs and their important protein targets in COVID-19 patients. Many important proteins are dysregulated in COVID-19 patients due to the differential expression of many miRNAs. This figure illustrates which deregulated miRNAs are associated with each of the important protein targets (ACE2, TMPRSS2, interleukins, and spike protein). The green arrow pointing upwards indicates that the miRNA is upregulated in COVID-19 patients compared to healthy controls while the red arrow pointing downwards indicates that the miRNA is downregulated.

infection. TMPRSS2 is a serine protease involved in the priming of the spike protein and is essential for the entry of the virus into cells.[48] Therefore, researchers can potentially look to develop treatments that inhibit TMPRSS2 or are involved with ACE2 by regulating associated miRNAs. Fig. 6.2 illustrates ways by which the miRNAs related to these proteins are deregulated.

5. Conclusion

Researchers have identified many different miRNAs that are deregulated in COVID-19 patients and involved in the pathogenesis of the disease. Many of these miRNAs were found in plasma and serum samples and can possibly be used as effective biomarkers in detecting COVID-19. Since many of these miRNAs are differentially expressed in the lungs, heart, and kidney, these miRNAs can also serve as therapeutic targets when treating COVID-19. However, the mechanism for miRNA dysregulation during COVID-19 is still largely unknown.

In this chapter, we identified 20 deregulated miRNAs in serum samples (Table 6.1), 18 differentially expressed miRNAs in plasma samples (Table 6.2), and 31 miRNAs that have altered their targeted protein expression in the lungs, heart, and kidneys (Table 6.3). These tables do not include all the miRNAs that are dysregulated in the plasma, serum, or body organs as there are significantly more miRNAs that were not included. However, the miRNAs presented in the tables provide great insight into how blood serum or plasma samples can be used to analyze miRNA to both detect the disease but also predict the severity of the disease. In addition, these tables show

how many miRNAs contribute to dysfunction in multiple organs in patients with COVID-19.

The next step for researchers is to develop treatment methods that target miRNAs found to contribute to the progression of COVID-19. By targeting these miRNAs, treatments can serve to limit the harm done to organ systems and improve health outcomes.

References

1. Kumar S, Reddy PH. The role of synaptic microRNAs in Alzheimer's disease. *Biochim Biophys Acta, Mol Basis Dis*. December 1, 2020;1866(12):165937. https://doi.org/10.1016/j.bbadis.2020.165937.
2. Kumar S, Reddy PH. A new discovery of microRNA-455-3p in Alzheimer's disease. *J Alzheimers Dis*. 2019;72(s1):S117–S130. https://doi.org/10.3233/JAD-190583.
3. Kumar S, Reddy PH. Are circulating microRNAs peripheral biomarkers for Alzheimer's disease? *Biochim Biophys Acta*. September 2016;1862(9):1617–1627. https://doi.org/10.1016/j.bbadis.2016.06.001.
4. Chen X, Ba Y, Ma L, et al. Characterization of microRNAs in serum: a novel class of biomarkers for diagnosis of cancer and other diseases. *Cell Res*. October 2008;18(10):997–1006. https://doi.org/10.1038/cr.2008.282.
5. Fu Z, Wang J, Wang Z, et al. A virus-derived microRNA-like small RNA serves as a serum biomarker to prioritize the COVID-19 patients at high risk of developing severe disease. *Cell Discov*. July 6, 2021;7(1):48. https://doi.org/10.1038/s41421-021-00289-8.
6. Chen Z, Liang H, Chen X, et al. An Ebola virus-encoded microRNA-like fragment serves as a biomarker for early diagnosis of Ebola virus disease. *Cell Res*. March 2016;26(3):380–383. https://doi.org/10.1038/cr.2016.21.
7. Li X, Fu Z, Liang H, et al. H5N1 influenza virus-specific miRNA-like small RNA increases cytokine production and mouse mortality via targeting poly(rC)-binding protein 2. *Cell Res*. February 2018;28(2):157–171. https://doi.org/10.1038/cr.2018.3.
8. Li X, Zou X. *An Overview of RNA Virus-Encoded microRNAs*. ExRNA; 2019.
9. Mitchell PS, Parkin RK, Kroh EM, et al. Circulating microRNAs as stable blood-based markers for cancer detection. *Proc Natl Acad Sci U S A*. July 29, 2008;105(30):10513–10518. https://doi.org/10.1073/pnas.0804549105.
10. Li C, Hu X, Li L, Li JH. Differential microRNA expression in the peripheral blood from human patients with COVID-19. *J Clin Lab Anal*. October 2020;34(10):e23590. https://doi.org/10.1002/jcla.23590.
11. Chen L, Liu Y, Wu J, et al. Lung adenocarcinoma patients have higher risk of SARS-CoV-2 infection. *Aging (Albany NY)*. January 10, 2021;13(2):1620–1632. https://doi.org/10.18632/aging.202375.
12. Cui X, Lv Z, Ding H, Xing C, Yuan Y. MiR-1539 and its potential role as a novel biomarker for colorectal cancer. *Front Oncol*. 2020;10:531244. https://doi.org/10.3389/fonc.2020.531244.
13. Zhao C, Dong J, Jiang T, et al. Early second-trimester serum miRNA profiling predicts gestational diabetes mellitus. *PLoS One*. 2011;6(8):e23925. https://doi.org/10.1371/journal.pone.0023925.

14. Farr RJ, Rootes CL, Rowntree LC, et al. Altered microRNA expression in COVID-19 patients enables identification of SARS-CoV-2 infection. *PLoS Pathog*. July 2021;17(7): e1009759. https://doi.org/10.1371/journal.ppat.1009759.
15. Demiray A, Sarı T, Çalışkan A, Nar R, Aksoy L, Akbubak İH. Serum microRNA signature is capable of predictive and prognostic factor for SARS-COV-2 virulence. *Turk J Biochem*. 2021;46(3):245−253. https://doi.org/10.1515/tjb-2020-0520.
16. Sabbatinelli J, Giuliani A, Matacchione G, et al. Decreased serum levels of the inflammaging marker miR-146a are associated with clinical non-response to tocilizumab in COVID-19 patients. *Mech Ageing Dev*. January 2021;193:111413. https://doi.org/10.1016/j.mad.2020.111413.
17. Garg A, Seeliger B, Derda AA, et al. Circulating cardiovascular microRNAs in critically ill COVID-19 patients. *Eur J Heart Fail*. March 2021;23(3):468−475. https://doi.org/10.1002/ejhf.2096.
18. de Gonzalo-Calvo D, Benitez ID, Pinilla L, et al. Circulating microRNA profiles predict the severity of COVID-19 in hospitalized patients. *Transl Res*. October 2021;236:147−159. https://doi.org/10.1016/j.trsl.2021.05.004.
19. Fayyad-Kazan M, Makki R, Skafi N, et al. Circulating miRNAs: potential diagnostic role for coronavirus disease 2019 (COVID-19). *Infect Genet Evol*. October 2021;94:105020. https://doi.org/10.1016/j.meegid.2021.105020.
20. Yang P, Zhao Y, Li J, et al. Downregulated miR-451a as a feature of the plasma cfRNA landscape reveals regulatory networks of IL-6/IL-6R-associated cytokine storms in COVID-19 patients. *Cell Mol Immunol*. April 2021;18(4):1064−1066. https://doi.org/10.1038/s41423-021-00652-5.
21. Kim WR, Park EG, Kang KW, Lee SM, Kim B, Kim HS. Expression analyses of microRNAs in hamster lung tissues infected by SARS-CoV-2. *Mol Cells*. November 30, 2020; 43(11):953−963. https://doi.org/10.14348/molcells.2020.0177.
22. Widiasta A, Sribudiani Y, Nugrahapraja H, Hilmanto D, Sekarwana N, Rachmadi D. Potential role of ACE2-related microRNAs in COVID-19-associated nephropathy. *Noncoding RNA Res*. December 2020;5(4):153−166. https://doi.org/10.1016/j.ncrna.2020.09.001.
23. Centa A, Fonseca AS, Ferreira S, et al. Deregulated miRNA expression is associated with endothelial dysfunction in post-mortem lung biopsies of COVID-19 patients. *Am J Physiol Lung Cell Mol Physiol*. December 2, 2020. https://doi.org/10.1152/ajplung.00457.2020.
24. Garikipati VNS, Verma SK, Jolardarashi D, et al. Therapeutic inhibition of miR-375 attenuates post-myocardial infarction inflammatory response and left ventricular dysfunction via PDK-1-AKT signalling axis. *Cardiovasc Res*. July 1, 2017;113(8):938−949. https://doi.org/10.1093/cvr/cvx052.
25. Wang Y, Zhu X, Jiang XM, et al. Decreased inhibition of exosomal miRNAs on SARS-CoV-2 replication underlies poor outcomes in elderly people and diabetic patients. *Signal Transduct Targeted Ther*. August 11, 2021;6(1):300. https://doi.org/10.1038/s41392-021-00716-y.
26. Sakr Y, Giovini M, Leone M, et al. Pulmonary embolism in patients with coronavirus disease-2019 (COVID-19) pneumonia: a narrative review. *Ann Intensive Care*. 2020;10: 124. https://doi.org/10.1186/s13613-020-00741-0.
27. Yan Y, Yang Y, Wang F, et al. Clinical characteristics and outcomes of patients with severe covid-19 with diabetes. *BMJ Open Diabetes Res Care*. April 2020;8(1). https://doi.org/10.1136/bmjdrc-2020-001343.
28. Greco S, Made A, Gaetano C, Devaux Y, Emanueli C, Martelli F. Noncoding RNAs implication in cardiovascular diseases in the COVID-19 era. *J Transl Med*. October 31, 2020;18(1):408. https://doi.org/10.1186/s12967-020-02582-8.

29. Nagashima S, Mendes MC, Camargo Martins AP, et al. Endothelial dysfunction and thrombosis in patients with COVID-19-brief report. *Arterioscler Thromb Vasc Biol.* October 2020;40(10):2404–2407. https://doi.org/10.1161/ATVBAHA.120.314860.
30. Fani M, Zandi M, Ebrahimi S, Soltani S, Abbasi S. The role of miRNAs in COVID-19 disease, 10.2217/fvl-2020-0389 *Future Virol.* March 2021. https://doi.org/10.2217/fvl-2020-0389. Epub 2021 Mar 24.; 2021.
31. Liu T, Zhang L, Joo D, Sun SC. NF-kappaB signaling in inflammation. *Signal Transduct Targeted Ther.* 2017;2doi. https://doi.org/10.1038/sigtrans.2017.23.
32. Cao Y, Zhao D, Li P, et al. MicroRNA-181a-5p impedes IL-17-induced nonsmall cell lung cancer proliferation and migration through targeting VCAM-1. *Cell Physiol Biochem.* 2017; 42(1):346–356. https://doi.org/10.1159/000477389.
33. Arghiani N, Nissan T, Matin MM. Role of microRNAs in COVID-19 with implications for therapeutics. *Biomed Pharmacother.* September 25, 2021;144:112247. https://doi.org/10.1016/j.biopha.2021.112247.
34. Kumar A, Arora A, Sharma P, et al. Is diabetes mellitus associated with mortality and severity of COVID-19? A meta-analysis. *Diabetes Metabol Syndr.* July–August 2020; 14(4):535–545. https://doi.org/10.1016/j.dsx.2020.04.044.
35. Wang Y, Lv Y, Liu Q. SARS-CoV-2 infection associated acute kidney injury in patients with pre-existing chronic renal disease: a report of two cases. *Immun Inflamm Dis.* December 2020;8(4):506–511. https://doi.org/10.1002/iid3.333.
36. Chueh TI, Zheng CM, Hou YC, Lu KC. Novel evidence of acute kidney injury in COVID-19. *J Clin Med.* November 3, 2020;(11):9. https://doi.org/10.3390/jcm9113547.
37. Su H, Yang M, Wan C, et al. Renal histopathological analysis of 26 postmortem findings of patients with COVID-19 in China. *Kidney Int.* July 2020;98(1):219–227. https://doi.org/10.1016/j.kint.2020.04.003.
38. Xu G, Mo L, Wu C, et al. The miR-15a-5p-XIST-CUL3 regulatory axis is important for sepsis-induced acute kidney injury. *Ren Fail.* November 2019;41(1):955–966. https://doi.org/10.1080/0886022X.2019.1669460.
39. Cao Q, Chen XM, Huang C, Pollock CA. MicroRNA as novel biomarkers and therapeutic targets in diabetic kidney disease: an update. *FASEB Bioadv.* June 2019;1(6):375–388. https://doi.org/10.1096/fba.2018-00064.
40. Adebisi YA, Jimoh ND, Ogunkola IO, et al. The use of antibiotics in COVID-19 management: a rapid review of national treatment guidelines in 10 African countries. *Trop Med Health.* June 23, 2021;49(1):51. https://doi.org/10.1186/s41182-021-00344-w.
41. Ferrara F, Vitiello A. Efficacy of synthetic glucocorticoids in COVID-19 endothelites. *Naunyn-Schmiedeberg's Arch Pharmacol.* May 2021;394(5):1003–1007. https://doi.org/10.1007/s00210-021-02049-7.
42. Windisch W, Weber-Carstens S, Kluge S, Rossaint R, Welte T, Karagiannidis C. Invasive and non-invasive ventilation in patients with COVID-19. *Dtsch Arztebl Int.* August 3, 2020; 117(31–32):528–533. https://doi.org/10.3238/arztebl.2020.0528.
43. Raza A, Estepa A, Chan V, Jafar MS. Acute renal failure in critically ill COVID-19 patients with a focus on the role of renal replacement therapy: a review of what we know so far. *Cureus.* June 3, 2020;12(6):e8429. https://doi.org/10.7759/cureus.8429.
44. Jean SS, Lee PI, Hsueh PR. Treatment options for COVID-19: the reality and challenges. *J Microbiol Immunol Infect.* June 2020;53(3):436–443. https://doi.org/10.1016/j.jmii.2020.03.034.
45. Huang JH, Han TT, Li LX, et al. Host microRNAs regulate expression of hepatitis B virus genes during transmission from patients' sperm to embryo. *Reprod Toxicol.* March 2021; 100:1–6. https://doi.org/10.1016/j.reprotox.2020.11.004.

46. Zou X, Chen K, Zou J, Han P, Hao J, Han Z. Single-cell RNA-seq data analysis on the receptor ACE2 expression reveals the potential risk of different human organs vulnerable to 2019-nCoV infection. *Front Med.* April 2020;14(2):185−192. https://doi.org/10.1007/s11684-020-0754-0.
47. Sriram K, Insel PA. A hypothesis for pathobiology and treatment of COVID-19: the centrality of ACE1/ACE2 imbalance. *Br J Pharmacol.* November 2020;177(21):4825−4844. https://doi.org/10.1111/bph.15082.
48. Hoffmann M, Kleine-Weber H, Schroeder S, et al. SARS-CoV-2 cell entry depends on ACE2 and TMPRSS2 and is blocked by a clinically proven protease inhibitor. *Cell.* April 16, 2020;181(2):271−280. https://doi.org/10.1016/j.cell.2020.02.052. e8.

Mechanisms and implications of COVID-19 transport into neural tissue

7

Katherine G. Holder, Bernardo Galvan and Alec Giakas
School of Medicine, Texas Tech University Health Sciences Center, Lubbock, TX, United States

Abstract

The neuropathogenicity of COVID-19 was reported shortly after detection of the virus when patients began reporting symptoms of diminished taste and smell, headaches, mental status changes, and more. As the virus spread, increasing data on viral symptoms in conjunction with novel theories on COVID-19 virulence factors indicated that the virus had neurotropic properties. Several mechanisms have been proposed detailing severe acute respiratory syndrome coronavirus disease 2019 (SARS-CoV-2) transport past the blood−brain barrier and into neural tissue. This chapter offers a comprehensive review of possible neurotropic mechanisms including transport via the angiotensin-converting enzyme 2 (ACE-2) receptor, transportation directly past or through the blood−brain barrier, transsynaptic neuronal transfer, and olfactory conduction.

Keywords: ACE-2 receptor olfactory invasion; Aseptic meningitis; Blood−brain barrier; Neural invasion; Transsynaptic transfer

1. Introduction

When reports of a novel virus first circulated from Wuhan in the last months of 2019, the illness was described primarily as a respiratory virus. As the virus spread and data on the infection became more robust, the literature reported a multitude of symptoms associated with the novel illness.[1] At the beginning of 2020, neurologic symptoms became more widely reported, and by June 2020, the first retrospective case series was published detailing neurologic manifestations of hospitalized patients who had been infected with COVID-19.[1] This early report of a 214 patient case series noted that over 36% of patients hospitalized with COVID-19 experienced neurologic symptoms.[1] Severity of neurologic symptoms seemed to correlate with severity of infection and respiratory status.[1] In the following months, an abundance of literature surrounding the neurologic manifestations of COVID-19 was published. Many theories have been proposed to analyze how the novel coronavirus penetrates neuronal tissue and moves past the human blood−brain barrier (BBB) into the central nervous system

COVID-19 in Alzheimer's Disease and Dementia. https://doi.org/10.1016/B978-0-443-15256-6.00004-0

(CNS). This chapter will discuss the prevailing theories about COVID-19's interactions with the human nervous system and briefly analyze the implications of these relationships.

2. Viruses and neurological damage

Many strains within the coronavirus family, such as HCoV-229E, HCoV-OC43, and HCoV-NL63, have been studied for their significant potential for neuro-invasion. HCoV-OC43 has been particularly well studied and has been shown to thrive in neural cell in vitro cultures. Additionally, oligodendrocytes, astrocytes, microglia, and neurons are all susceptible to infection with HCoV-43. These older coronaviruses are thought to invade the CNS intranasally and then rapidly spread throughout the CNS causing direct, virus-mediated, mild injury to neural tissue.[2]

3. SARS-CoV-2 virulence and neurologic invasion

The coronavirus family's main virulence factor is its highly specialized spike protein.[3] The spike protein is found on the virus's outer capsule and binds to human cell entry receptors to gain entry into the human cells.[4]

Once the COVID-19 virus enters human tissue, it behaves similarly to other viruses.[5] COVID-19's intracellular life cycle begins when it expresses and replicates its RNA genome to produce full-length genetic copies that are subsequently incorporated into new viral particles. These new viral particles may go on to infect other cells of the current host or may be subsequently transferred to a new, susceptible host.[3] Between the S1 and S2 proteins lies a polybasic cleavage site, which permits efficient cleavage of the spike protein and promotes viral shedding. The efficient cleavage results in enhanced infectivity and is thought to be a key event in the evolution of SARS-CoV-2.[3]

Coronavirus spike-protein-compatible cell entry receptors have been identified for several coronavirus strains and include the human aminopeptidase N (APN; HCoV-229E), angiotensin-converting enzyme 2 (ACE2; HCoV-NL63, SARS-CoV and SARS-CoV-2), and dipeptidyl peptidase 4 (DPP4; MERS-CoV).[3] In 2002, shortly after the initial SARS-CoV outbreak, ACE2 was identified as the main functional receptor that enabled SARS-CoV to infect host cells.[6] Now, almost 20 years after the initial SARS-CoV outbreak, the novel SARS-CoV-2 is thought to share an affinity for the ACE2 receptor. Both of these coronaviruses have high genomic and structural similarities between their S proteins with a 76% amino acid homogeneity between the strains. This supports the identification of ACE 2 as the preferred cell surface receptor for SARS-CoV-2, commonly called COVID-19.[7–9]

3.1 ACE2 receptors

Human cells that express high numbers of coronavirus spike-protein-compatible receptors, like ACE2, display increased susceptibility to coronavirus infection and determine viral cellular tropism.[2] In humans, ACE2 is expressed in high concentrations in cells of the airway epithelia, lung parenchyma, kidney, small intestine, vascular endothelial, and widely throughout the CNS.[7] Recent literature supports that specific neuronal cells and spatial areas of the nervous system may express higher concentrations of ACE2 and therefore have increased susceptibility to COVID-19 infection. These cells include neurons, astrocytes, and oligodendrocytes. Neuronal cell cultures found that the ACE2 receptor is expressed not only on the cells' outer membrane but also in the cytoplasm.[10–12] Additionally, ACE2 was found to be highly concentrated in local areas of the CNS: the substantia nigra, ventricles, middle temporal gyrus, posterior cingulate cortex, brainstem, and olfactory bulb.[13]

The high concentration of ACE-2 in neuronal tissue combined with the breadth of neurologic symptoms reported by patients with COVID-19 raises suspicion that the novel virus may be neurotropic.[14] While literature on CNS virtual infection suggests that all viruses can reach the CNS depending on viral and host factor conditions, neurotropic viruses actually have a preference for neuronal tissue.[13] The particular mechanism of COIVD-19's neuroinvasion is still incompletely understood, but several theories have been proposed on how the novel virus may invade neuronal tissue.[15] Mechanisms implicated in SARS-CoV-2's spread into the CNS include infection of the vascular endothelium, infection of leukocytes that migrate across the BBB, trans-synaptic transfer of viral particles across neurons, and entry into the CNS via cranial nerve fibers.[2]

3.2 Blood–brain barrier

The blood–brain barrier is the highly selective microvasculature that prevents molecules in circulating blood from crossing into the CNS. The blood vessels that supply critical oxygen and nutrients to the CNS are uniquely organized to regulate the movement of ions, molecules, and cells between circulating blood and the brain. These blood vessels are made of extremely thin, specialized endothelial cells (ECs) that are held together by tight junctions and exhibit remarkably low rates of transcytosis. These ECs are supported and regulated by an extensive network of neuronal cells including mural cells, immune cells, and glial cells, which interact in the neurovascular unit to maintain the BBB. When functioning properly, these cells prevent ion dysregulation, promote signaling homeostasis, and protect against entry of immune cells and molecules into the CNS, thereby averting neuronal dysfunction and dysregulation. Perhaps, most importantly, the BBB is critical in protecting the CNS from toxins, pathogens, injury, and disease.[16]

Spread of COVID-19 across the BBB can be explained by two distinct possible mechanisms, which we will describe as the direct and indirect mechanisms. First, in the direct mechanism, viral infection of CNS microvascular endothelial cells may disrupt the BBB.[17] Endothelial cells express high levels of the ACE-2 receptor and are therefore increasingly susceptible to COVID-19 infection. ECs of the CNS microvasculature are no exception to this. When COVID infects these epithelial cells, it may disrupt their permeability and transmembrane transport, allowing the virus to pass through the epithelial cells and into CNS tissue.[18] The virus may enter these epithelial cells, replicate inside of them, and undergo subsequent release on the CNS side. The virus may also undergo ACE-2-dependent transport across the thin epithelial cells without undergoing intracellular replication. Some literature has proposed that the virus may invade via receptor-mediated transport by a receptor other than ACE-2, transport mediated by adsorptive transcytosis, or transport mediated through interactions with the endothelial glycocalyx.[17] Animal studies have demonstrated SARS-CoV-2 RNA in the vascular and perivascular space as well as the brain microvascular endothelial cells of infected mice. The same study revealed that COVID-19 used these ECs as a conduit for transcellular spread past the BBB and disrupted the neuro microvasculature basement membrane without obviously altering endothelial tight junctions.[19] The exact cellular mechanism of this transport is still incompletely understood but likely consists of one or multiple of these possible mechanisms.

The indirect mechanism that has been proposed for COVID-19 crossing of the BBB involves movement of viral particles into the BBB without direct utilization of cerebral epithelial cells. This can be further subdivided into two categories: leaky intercellular channels and leukocyte trafficking of viral particles. The leukocyte-dependent method, commonly referred to as the trojan horse method, proposes that the virus invades human immune cells that are trafficked through the BBB and subsequently expel viral particles into the CNS.[17] Just as the wooden horse carried enemy soldiers into the city of Troy, this method of "trojan horse" spread proposes that immune cells traffic pathogens into immune privileged tissues. This mechanism has been studied extensively with human immunodeficiency virus (HIV), which infects immune cells by passing through the BBB to subsequently infect the CNS.[20] The CNS is an immune privileged environment and should typically be free of leukocytes. However, in severe infection, WBCs can invade the CNS. This may happen through several mechanisms. Increased inflammation from infection or other pathologic conditions may increase endothelial cell leakage at the level of the brain microvasculature, promoting leukocyte extravasation. WBC may also be trafficked into the brain in inflammatory states that upregulate cytokine signaling.[17] When functioning properly, these phagocytic lymphocytes are able to internalize and destroy pathogens. However, in cases of severe immune dysfunction, viruses like COVID-19 may be able to survive and even multiply within lymphocytes. This ultimately allows the virus to escape proper immune surveillance and traffics viral particles into immune protected sites, namely, the brain, using the exact mechanism that is made to protect human tissue.[21] These mechanisms are diagrammed in Fig. 7.1.

3.3 Transsynaptic neuronal transfer

SARS-CoV-2 has also been described entering the CNS via transsynaptic transfer across neurons. This theory proposes that SARS-CoV-2 first invades peripheral nerves and then travels retrograde through neural synapses of the PNS to reach the CNS. This specialized mechanism of entry has been described in several other coronaviruses, including HCoV-OC43, hemagglutinating encephalomyelitis virus 67 (HEV67), and avian bronchitis virus. The spread of HEV67 in particular has been shown to invade nasal mucosa and lung epithelium then spread retrograde through the dorsal root ganglia and eventually terminate in the medullary neurons. HEV67 has been known to transfer between motor cortex neurons using a membrane mediated

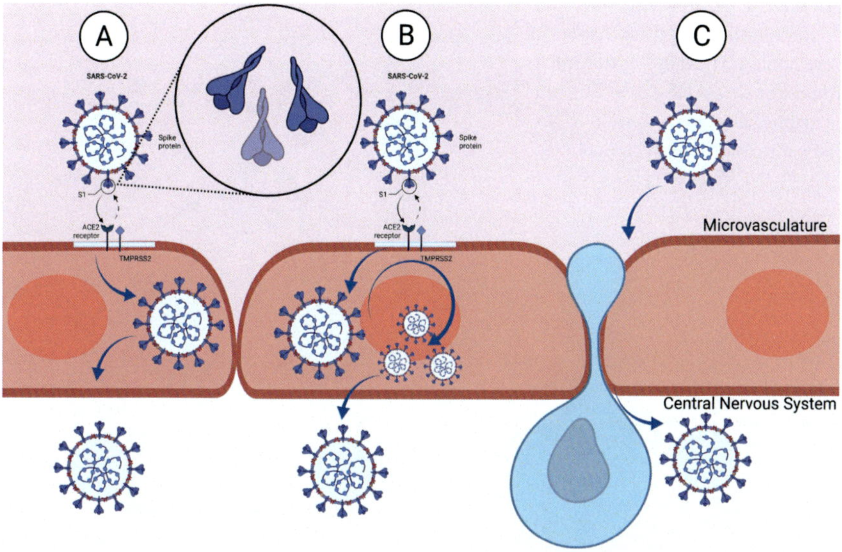

Figure 7.1 Three potential mechanisms of severe acute respiratory syndrome coronavirus 2 (SARS-CoV-2) crossing the blood–brain barrier (BBB). (A) demonstrates viral invasion of neuronal microvasculature via the angiotensin-converting enzyme-2 (ACE-2) receptor and subsequent transfer into the central nervous system (CNS). (B) demonstrates the same mechanism as (A) with additional viral replication inside endothelial cells before transfer into the CNS. (C) demonstrates inflammation and subsequent extravasation of leukocytes and viral particles directly from the neural microvasculature into the CNS. Of note, the leukocytes themselves could be infected with the novel virus and transport it into the CNS using the Trojan horse method, or leaky endothelium could simply allow viral movement into the CNS.

endo- and exocytosis pathway; it has used a similar secretory mechanism in order to transfer from neurons to satellite cells. HCoV-OC43, like HIV and herpes simplex virus, has been described utilizing fast intracellular transport, which recruits microtubules to move molecules anterograde or retrograde within a cell to infect neurons. A similar mechanism may be used by SARS-Cov-2.[2]

SARS-CoV-2 may infect the brain through anterograde axonal transport after binding to TMPRS2 protein receptors, a spike-protein-compatible receptor similar to ACE-2. SARS-CoV-2 might also reach the CNS through first invading nonneuronal cells in the olfactory epithelium with high ACE2 expression, then passing to mature neuronal cells with lower ACE2 expression. This mechanism is displayed in Fig. 7.2. Interestingly, animal models have demonstrated an increase in ACE2 and TMPRSS2 expression with increasing age; as such, elderly patients might be more susceptible to SARS-CoV-2 infection via the transsynaptic olfactory route. It has been suggested that patients with higher olfactory dysfunction from SARS-CoV-2 might represent those with a stronger innate immune response. Contrarily, older patients with higher ACE2 and TMPRSS expression might be more likely to develop severe symptoms due to their more attenuated innate immune response.[13]

It has been reported that the brainstem is the most heavily affected area of the brain by several coronaviruses. Additionally, some studies have detected MERS-CoV virus

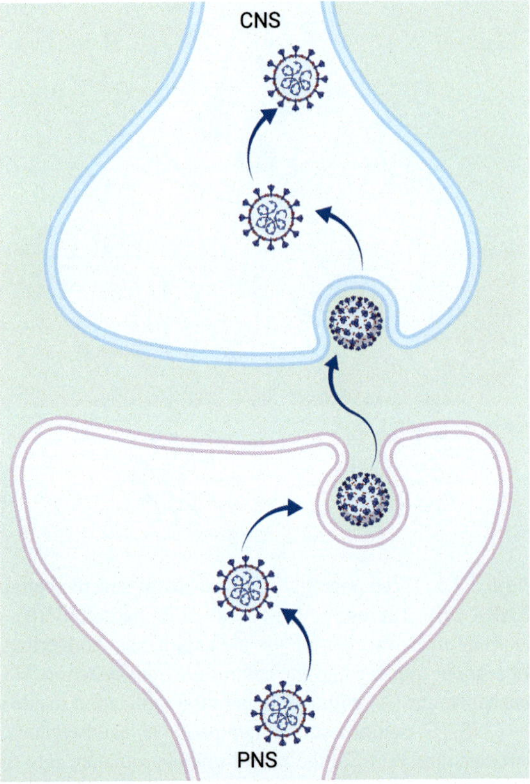

Figure 7.2 Diagram depicting retrograde transsynaptic transport of severe acute respiratory syndrome coronavirus 2 (SARS-CoV-2) from the peripheral nervous system (PNS) to the central nervous system (CNS).

in the brain without detection in the lungs. As such, it is believed that SARS-CoV-2 might spread transsynaptically from the vagus nerve to the vagal nucleus of the medulla oblongata. The vagal nucleus controls several highly integrated functions such as respiration. Therefore, it is reasonable to suggest that SARS-CoV-2 could travel from the vagus nerve through transsynaptic invasion to the vagal nuclei, affecting the respiratory centers in the medulla oblongata and, therefore, exacerbating the respiratory dysfunction induced by primary SARS-Cov-2 lung infection.[13]

3.4 Olfactory system entry

In the final proposed mechanism, SARS-CoV-2 has been widely reported to cause anosmia and ageusia; these manifestations lend support to the virus' ability to infect the CNS via the olfactory system. ACE-2, the receptor by which SARS-CoV-2 gains access to many cell types in the human body, is expressed in several nonneuronal cell types in the olfactory system, such as sustentacular and stem cells in the respiratory epithelium.[2] However, little is known about ACE-2 expression on olfactory neurons in an inflammatory setting, including during infection with SARS-CoV-2. Autopsy studies have found significant SARS-CoV-2 presence within the nasal mucosa; given the anatomic proximity of this mucosa with nerve fibers in the nasopharynx, it has been suggested that SARS-CoV-2 enters the CNS by exploiting this neural–mucosal interface.[22]

SARS-CoV-2's impact on olfactory dysfunction is not a new phenomenon within infectious disease; several other viruses exhibit this phenomenon via an inflammatory reaction of the nasal mucosa followed by development of rhinorrhea and loss of smell. Some of these viruses include rhinovirus, parainfluenza, Epstein–Barr virus, and some other coronaviruses. Previous studies have demonstrated the ability of SARS-CoV to cause neuronal death through invasion of the olfactory epithelium.[23] However, SARS-CoV-2's effect on olfactory dysfunction is unique in that it is often not associated with rhinorrhea.[24] Additionally, the olfactory manifestations of SARS-CoV-2 can present before, during, or after the onset of other otolaryngologic symptoms.[23]

In most studies to date, olfactory dysfunction has been reported in 30%–85% of patients infected with SARS-CoV-2.[25] One study, which quantitatively assessed olfactory dysfunction using the University of Pennsylvania Smell Identification Test, found that 98% of patients exhibited some smell dysfunction, 33% of which had severe hyposmia, and 58% of which had anosmia. Several mechanisms have been proposed to elucidate SARS-CoV-2's effect on the olfactory system.[26] There is evidence that the virus is able to affect the sense of smell through invasion of the CNS; increased signal on MRI has been demonstrated in the olfactory cortex following SARS-CoV-2 infection. As discussed previously, the prevailing mechanism for this CNS invasion of the olfactory cortex is the viral internalization into nerve terminals through endocytosis, retrograde transport through the nerve, followed by transsynaptic spread to adjacent brain regions.[27]

Olfactory input plays an important role in the perception of flavor, which incorporates both olfactory and gustatory sensation.[25] Additionally, "taste" disorders commonly result from olfactory deficiencies.[28] It is important to note that the olfactory mucosa contains over 350 specialized olfactory receptors that recognize specific smells; however, these receptors are localized on the olfactory epithelium and stem cells, instead of the neurons themselves.[29] As such, infections in nonneuronal cells can cause anosmia and other disturbances in odor perception, such as parosmia and cacosmia.[30,31] SARS-CoV-2 has been demonstrated to contribute to chronic anosmia and hyposmia.[32]

The sense of smell contributes to many physical, mental, and social components of a healthy lifestyle, such as food and nutrient selection, enjoyment of food, socialization, overall quality of life, and detection of safety hazards.[33] Additionally, several studies have demonstrated a link between olfactory dysfunction and depression.[34] As such, besides the long-term physical, neurologic, and systemic implications of SARS-CoV-2 infection, the virus can negatively affect the quality of life through the chronic alteration of taste and smell. Indeed, one study reported that 85.6% of patients infected with SARS-CoV-2 reported an olfactory dysfunction with an impact on quality of life.[23]

Although olfactory dysfunction has detrimental implications, it is hardly the most severe neurologic implication of CNS invasion by SARS-CoV-2. Viral spread to the CNS has been implicated in a plethora of symptoms that contribute to acute illness and may progress to long-term disease in patients infected with COVID-19. These symptoms range from headache and anosmia to impaired consciousness, seizures, cerebral vascular events, paralytic peripheral nerve disorders, meningitis, encephalitis, or even death.[13] The mechanisms of these presentations vary dramatically, some being caused by immune dysregulation or vascular disruption and others implicating the SARS-CoV-2 directly.

4. Conclusion

Understanding the mechanism of neuroinvasion by SARS-CoV-2 is critical to preventing disease progression and identifying new treatments for the novel disease. While the exact mechanism of neuroinvasion is still incompletely understood, it probably involves a combination of cerebral microvascular changes, leukocyte invasion, transsynaptic spread, and olfactory migration. Clinicians should closely monitor neurologic function of patients who have been infected with COVID-19 as early detection of deficits may be critical in improving treatment and developing therapeutic algorithms. Long-term assessment of these patients will be essential for understanding the implications of SARS-CoV-2 on neurologic function.

References

1. Mao L, Jin H, Wang M, et al. Neurologic manifestations of hospitalized patients with coronavirus disease 2019 in Wuhan, China. *JAMA Neurol.* 2020;77(6):683–690. https://doi.org/10.1001/jamaneurol.2020.1127 ([doi]).
2. V'kovski P, Kratzel A, Steiner S, Stalder H, Thiel V. Coronavirus biology and replication: implications for SARS-CoV-2. *Nat Rev Microbiol.* 2021;19(3):155–170. https://doi.org/10.1038/s41579-020-00468-6 ([doi]).
3. Zubair AS, McAlpine LS, Gardin T, Farhadian S, Kuruvilla DE, Spudich S. Neuropathogenesis and neurologic manifestations of the coronaviruses in the age of coronavirus disease 2019: a review. *JAMA Neurol.* 2020;77(8):1018–1027. https://doi.org/10.1001/jamaneurol.2020.2065 ([doi]).
4. Buzhdygan TP, DeOre BJ, Baldwin-Leclair A, et al. The SARS-CoV-2 spike protein alters barrier function in 2D static and 3D microfluidic in-vitro models of the human blood-brain barrier. S0969-9961(20)30406-X ([pii]) *Neurobiol Dis.* 2020;146:105131.
5. Sanyal S. How SARS-CoV-2 (COVID-19) spreads within infected hosts - what we know so far. *Emerg Top Life Sci.* 2020;4(4):371–378. https://doi.org/10.1042/ETLS20200165 ([doi]).
6. Li W, Moore MJ, Vasilieva N, et al. Angiotensin-converting enzyme 2 is a functional receptor for the SARS coronavirus. *Nature.* 2003;426(6965):450–454. https://doi.org/10.1038/nature02145 ([doi]).
7. Hamming I, Timens W, Bulthuis ML, Lely AT, Navis G, van Goor H. Tissue distribution of ACE2 protein, the functional receptor for SARS coronavirus. A first step in understanding SARS pathogenesis. *J Pathol.* 2004;203(2):631–637. https://doi.org/10.1002/path.1570 ([doi]).
8. Shieh WJ, Hsiao CH, Paddock CD, et al. Immunohistochemical, in situ hybridization, and ultrastructural localization of SARS-associated coronavirus in lung of a fatal case of severe acute respiratory syndrome in Taiwan. *Hum Pathol.* 2005;36(3):303–309. S0046817704006240 [pii].

9. Leung GM, Hedley AJ, Ho LM, et al. The epidemiology of severe acute respiratory syndrome in the 2003 Hong Kong epidemic: an analysis of all 1755 patients, 141/9/662 ([pii]) *Ann Intern Med.* 2004;141(9):662−673.
10. Chen R, Wang K, Yu J, et al. The spatial and cell-type distribution of SARS-CoV-2 receptor ACE2 in the human and mouse brains. *Front Neurol.* 2021;11:573095. https://doi.org/10.3389/fneur.2020.573095 ([doi]).
11. Xia H, Lazartigues E. Angiotensin-converting enzyme 2 in the brain: properties and future directions. *J Neurochem.* 2008;107(6):1482−1494. https://doi.org/10.1111/j.1471-4159.2008.05723.x ([doi]).
12. Doobay MF, Talman LS, Obr TD, Tian X, Davisson RL, Lazartigues E. Differential expression of neuronal ACE2 in transgenic mice with overexpression of the brain renin-angiotensin system, 00292.2006 ([pii]) *Am J Physiol Regul Integr Comp Physiol.* 2007; 292(1):373.
13. Yachou Y, El Idrissi A, Belapasov V, Ait Benali S. Neuroinvasion, neurotropic, and neuroinflammatory events of SARS-CoV-2: understanding the neurological manifestations in COVID-19 patients. *Neurol Sci.* 2020;41(10):2657−2669. https://doi.org/10.1007/s10072-020-04575-3 ([doi]).
14. Zhou Z, Kang H, Li S, Zhao X. Understanding the neurotropic characteristics of SARS-CoV-2: from neurological manifestations of COVID-19 to potential neurotropic mechanisms. *J Neurol.* 2020;267(8):2179−2184. https://doi.org/10.1007/s00415-020-09929-7 ([doi]).
15. Najjar S, Najjar A, Chong DJ, et al. Central nervous system complications associated with SARS-CoV-2 infection: integrative concepts of pathophysiology and case reports ([doi]) *J Neuroinflammation.* 2020;17(1). https://doi.org/10.1186/s12974-020-01896-0, 231-0.
16. Daneman R, Prat A. The blood-brain barrier. *Cold Spring Harbor Perspect Biol.* 2015;7(1): a020412. https://doi.org/10.1101/cshperspect.a020412 ([doi]).
17. Erickson MA, Rhea EM, Knopp RC, Banks WA. Interactions of SARS-CoV-2 with the blood-brain barrier. *Int J Mol Sci.* 2021;22(5):2681. https://doi.org/10.3390/ijms22052681 ([pii]).
18. MacLean MA, Kamintsky L, Leck ED, Friedman A. The potential role of microvascular pathology in the neurological manifestations of coronavirus infection, 55-1 *Fluids Barriers CNS.* 2020;17(1). https://doi.org/10.1186/s12987-020-00216-1 ([doi]).
19. Zhang L, Zhou L, Bao L, et al. SARS-CoV-2 crosses the blood-brain barrier accompanied with basement membrane disruption without tight junctions alteration. *Signal Transduct Targeted Ther.* 2021;6(1):337. https://doi.org/10.1038/s41392-021-00719-9 ([doi]).
20. Kim WK, Corey S, Alvarez X, Williams K. Monocyte/macrophage traffic in HIV and SIV encephalitis. *J Leukoc Biol.* 2003;74(5):650−656. https://doi.org/10.1189/jlb.0503207 ([doi]).
21. Santiago-Tirado FH, Doering TL. False friends: phagocytes as trojan horses in microbial brain infections. *PLoS Pathog.* 2017;13(12):e1006680. https://doi.org/10.1371/journal.ppat.1006680.
22. Meinhardt J, Radke J, Dittmayer C, et al. Olfactory transmucosal SARS-CoV-2 invasion as a port of central nervous system entry in individuals with COVID-19. *Nat Neurosci.* 2021; 24(2):168−175. https://doi.org/10.1038/s41593-020-00758-5 ([doi]).
23. Lechien JR, Chiesa-Estomba CM, De Siati DR, et al. Olfactory and gustatory dysfunctions as a clinical presentation of mild-to-moderate forms of the coronavirus disease (COVID-19): a multicenter European study. *Eur Arch Oto-Rhino-Laryngol.* 2020;277(8): 2251−2261. https://doi.org/10.1007/s00405-020-05965-1 ([doi]).

24. Netland J, Meyerholz DK, Moore S, Cassell M, Perlman S. Severe acute respiratory syndrome coronavirus infection causes neuronal death in the absence of encephalitis in mice transgenic for human ACE2. *J Virol.* 2008;82(15):7264–7275. https://doi.org/10.1128/JVI.00737-08 ([doi]).
25. Izquierdo-Dominguez A, Rojas-Lechuga MJ, Mullol J, Alobid I. Olfactory dysfunction in the COVID-19 outbreak. *J Investig Allergol Clin Immunol.* 2020;30(5):317–326. https://doi.org/10.18176/jiaci.0567 ([doi]).
26. Moein ST, Hashemian SM, Mansourafshar B, Khorram-Tousi A, Tabarsi P, Doty RL. Smell dysfunction: a biomarker for COVID-19. *Int Forum Allergy Rhinol.* 2020;10(8):944–950. https://doi.org/10.1002/alr.22587 ([doi]).
27. Iadecola C, Anrather J, Kamel H. Effects of COVID-19 on the nervous system. S0092-8674(20)31070-9 ([pii]) *Cell.* 2020;183(1):16–27. e1.
28. Doty RL. Olfactory dysfunction and its measurement in the clinic. *World J Otorhinolaryngol Head Neck Surg.* 2015;1(1):28–33. https://doi.org/10.1016/j.wjorl.2015.09.007 ([doi]).
29. Garcia-Esparcia P, Schluter A, Carmona M, et al. Functional genomics reveals dysregulation of cortical olfactory receptors in Parkinson disease: novel putative chemoreceptors in the human brain. *J Neuropathol Exp Neurol.* 2013;72(6):524–539. https://doi.org/10.1097/NEN.0b013e318294fd76 ([doi]).
30. Yan CH, Faraji F, Prajapati DP, Ostrander BT, DeConde AS. Self-reported olfactory loss associates with outpatient clinical course in COVID-19. *Int Forum Allergy Rhinol.* 2020;10(7):821–831. https://doi.org/10.1002/alr.22592 ([doi]).
31. Brann DH, Tsukahara T, Weinreb C, et al. Non-neuronal expression of SARS-CoV-2 entry genes in the olfactory system suggests mechanisms underlying COVID-19-associated anosmia. *Sci Adv.* 2020;6(31). https://doi.org/10.1126/sciadv.abc5801. Epub 2020 Jul 24. doi: eabc5801 [pii].
32. Fiani B, Covarrubias C, Desai A, Sekhon M, Jarrah R. A contemporary review of neurological sequelae of COVID-19. *Front Neurol.* 2020;11:640. https://doi.org/10.3389/fneur.2020.00640 ([doi]).
33. Boesveldt S, Yee JR, McClintock MK, Lundstrom JN. Olfactory function and the social lives of older adults: a matter of sex. *Sci Rep.* 2017;7:45118. https://doi.org/10.1038/srep45118 ([doi]).
34. Hummel T, Whitcroft KL, Andrews P, et al. Position paper on olfactory dysfunction. *Rhinol Suppl.* 2017;54(26):1–30. https://doi.org/10.4193/Rhino16.248 ([doi]).

Further reading

1. Uversky VN, Elrashdy F, Aljadawi A, Ali SM, Khan RH, Redwan EM. Severe acute respiratory syndrome coronavirus 2 infection reaches the human nervous system: how? *J Neurosci Res.* 2021;99(3):750–777. https://doi.org/10.1002/jnr.24752 ([doi]).
2. Ke Z, Oton J, Qu K, et al. Structures and distributions of SARS-CoV-2 spike proteins on intact virions. *Nature.* 2020;588(7838):498–502. https://doi.org/10.1038/s41586-020-2665-2 ([doi]).

Immunogenetic landscape of COVID-19 infections related neurological complications

8

Balakrishnan Karuppiah[1], *Rathika Chinniah*[1], *Sasiharan Pandi*[1], *Vandit Sevak*[1], *Padma Malini Ravi*[2] *and Dhinakaran Thadakanathan*[1]
[1]Madurai HLA Centre, Madurai Kidney Centre & Transplantation Research Institute, Madurai, Tamil Nadu, India; [2]Department of Immunology, School of Biological Sciences, Madurai Kamaraj University, Madurai, Tamil Nadu, India

Abstract

The human leukocyte antigen (HLA) is a critical component of antigen presentation and plays crucial role in conferring differential susceptibility and severity of diseases caused by viruses such as COVID-19. The immunogenetic profile of populations, BCG vaccination status, and a host of lifestyle factors might contribute to the observed variations in mortality rates due to COVID-19. These genetic, epigenetic, and environmental factors could widely influence infection dynamics and immune responses against COVID-19. The aim of this review is to provide an update on HLA association with SARS-CoV-2 infection in global populations and to highlight the possible neurological involvements. We also set out to explore the HLA immunogenetic markers related to COVID-19 infections that can be used in screening high-risk individuals for personalized therapies and in community-based vaccine development.

Keywords: CNS; COVID-19; Cytokines; Haplotypes ACE2 receptor; HLA alleles; MHC; Neuroinvasive; Neurological diseases; Neurotropic; VIPs.

1. Introduction

Many central nervous system (CNS) infections are predominantly caused by viruses. The respiratory viruses with neuroinvasive and neurotropic potential include: the influenza virus, human meta-pneumovirus, members of the enterovirus/rhinovirus genus, coxsackieviruses, respiratory syncytial virus, Henipavirus genus, Hendra, and Nipah viruses. The neuroinvasive coronaviruses associated with neurological involvement include circulating endemic SARS coronovirus strains such as OC43 and 229E, severe acute respiratory syndrome coronavirus (SARS-CoV), and Middle East respiratory syndrome coronavirus (MERS-CoV).[1] The mortality rates of COVID-19 vary across countries (0.2%−15%) depending on a number of demographic factors such as age, smoking habit, pre-existing co-morbidities and type of

COVID-19 in Alzheimer's Disease and Dementia. https://doi.org/10.1016/B978-0-443-15256-6.00009-X

medication, and geo-physical factors such as temperature and humidity and a number of socio-biological factors (life expectancy, average income, social norms, Bacillus Calmette-Guérin (BCG) vaccination status, etc) and genetic factors among others.[2,3] In general, countries in the Northern hemisphere have faced the maximum incidence of COVID-19 disease, while those in the Southern hemisphere have only been moderately affected.

The impact of COVID-19 at a population level is highly complex. The clinical variations in COVID-19 severity and symptomatic presentation could be attributed to the differences in host immunologic/immunogenetic factors. Through the process of evolution, viruses have exerted selective pressures on the human populations to develop genetic variations for resistance or susceptibility toward pathogens. Thus, it is possible that the observed regional variations of impact of COVID-19 may reflect, alongside other factors, the host genetic variations.[4]

2. HLA immunogenetic variations

The major histocompatibility complex (MHC) is approximately four mega base regions located on the short arm of human chromosome 6 (6p21). MHC encodes the classical transplantation HLA genes and many other (non-MHC) genes with essential roles for immune function and cellular processes such as nervous system development and neuro-physiological function. The HLA region comprises six major loci that encode structurally homologous proteins classified into HLA class I (HLA-A,-B,-Cw) and class II (HLA-DR,-DQ,-DP) antigens as a function of their biology, structure, tissue distribution, and peptide presentation to T-cells.[5,6] The genes of HLA antigens are the most polymorphic in human genome with a total of 31,675 alleles reported and still evolving in synchronization with pathogen evolution. Among these, 23,002 alleles were in HLA class I loci and 8673 alleles were in HLA class II loci.[7]

Common genetic polymorphisms of MHC genes, structural variations such as copy number variation (CNVs), and many other variations resulting from the presence or absence of retroviral sequences (e.g., Alu, LINE, HERV, LTR, MER, and SVA) have also been observed among MHC haplotypes and implicated in health and diseases. Although many human gene sequences appear to be "trans-species" and highly conserved, the MHC region too have changed considerably during primate evolution, and it is now realized that genomic structure has been affected more by periodic duplications, "indels" and transposable elements such as "Alus" and human endogenous retrovirus (HERVs).

Human leukocyte antigen (HLA) genes are primarily implicated in host resistance or susceptibility to a range of pathogens. The presentation of viral antigens by different HLA proteins has shown to impart differential viral resistance, disease prevalence, and transmission dynamics. Each HLA genotype can stimulate unique type of T cell mediated antiviral responses and could possibly alter the symptoms and transmission of the disease.[8] The repertoire of the HLA alleles that forms novel, highly conserved and ancestral extended haplotypes (AEH) is thought to

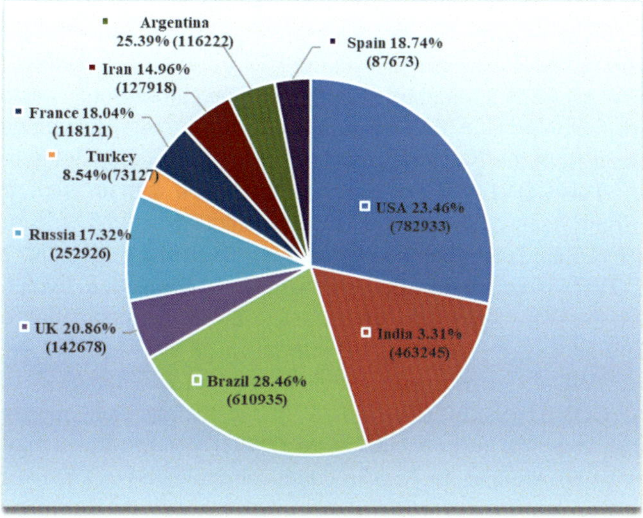

Figure 8.1 Number of deaths in top 10 countries with coronavirus disease 2019 (COVID-19)-related mortality.

contribute to better adaptation and facilitated better survival of species (natural selection) during evolutionary process (Fig. 8.1).

HLA noncoding region variations such as SNPs or small insertion/deletions (indels), as well as largerscale copy number variants (CNV) present in regulatory regions, could decide the course of viral diseases. For example, variation in the 3′UTR of HLA DPB1 is linked with spontaneous clearance of hepatitis B virus in both Japanese and US populations.[9,10] The proposed mechanism for enabling viral clearance might be linked to the rs9277534 A/G SNP that determines HLA DP cell-surface expression.[10] Investigating the variations in both coding and noncoding regions in the MHC using high-throughput and/or high-resolution technologies in population groups with diverse ancestries will be required to fully understand their role in infection induced neurological diseases.

3. HLA immunogenetics and COVID-19

Epidemiological studies from various groups have reported different HLA alleles as either protective against or susceptible toward COVID-19 infection.[11–13] The discrepancies between the studies can be attributed to the study design and the sample size, as well as ethnic variations of populations studied. In one case, the populations in South Asia, such as the Andaman Islanders, carried longer and larger "Runs of Homozygosity" (RoH) (a condition of carrying huge lethal genes in the population), primarily because of their long isolation and intense inbreeding practices.[14] Such population isolates with their low level of heterozygosity and longer/larger ROH may have a higher risk of COVID-19. It has been shown that a polymorphism rs2285666 ($G > A$) of

ACE2 gene of X chromosome may increase the expression level of ACE 2 receptor up to 50%.[15,16] This polymorphism was widespread in South Asia, and the haplotype associated with this SNP was shared with the East Eurasian populations.[14,17]

A recent GWAS study has identified several loci on chromosome 3 associated with the severe risk factor for COVID-19 among Europeans.[3,8] A risk haplotype of 50 kb size introgressed from the Neanderthals, the "Neanderthal core haplotype" has been documented.[16] This COVID-19 risk haplotype presented with an allele frequency of 30% among South Asians, 8% among Europeans, and 4% among African Americans. However, 63% of Bangladesh populations carried at least one copy of this Neanderthal haplotype. COVID-19 mortality among Bangladesh population living in the UK revealed twice the risk of mortality compared to the native population of Britannic pedigree nicely attested this finding.[18] While much of the world has been severely impacted by COVID-19, Africa, with a population of >1.2 billion people, has a lower percentage of COVID-19 deaths especially in the malaria-endemic region.[19] One possible explanation for lower incidence of COVID-19 in Africa may be due to high diversity of HLA alleles in African continent compared to other regions of the world.

HLA antigens and the "Original antigenic sin" (OAS) phenomenon could be an important proposition to understand SARS-CoV-2 infection or vaccination modalities.[20,21] The OAS refers to the activation of a more vigorous immune reaction to the priming antigen (wild/original antigenic type of the virus) than an immunogenic booster (mutated antigenic type) that binds poorly, if at all, to antibodies caused by the priming antigen type/dose.[20,22,23] Extreme diversity of HLA alleles within the MHC, frequency variations of alleles of different HLA loci and differences in HLA-peptide binding efficiencies and interactions could all contribute for such an immunologically distorted event.[13]

3.1 Population size and increased immunogenetic variations

In general, the greater the population size, the greater the genetic diversity. Furthermore, high levels of genetic diversity in a population imply higher heterozygosity in the context of infectious epidemics.[24,25] Thus, when the genes/alleles of individuals/groups in a population vary greatly, it favors the populations for better fitness, including survival against infectious diseases and epidemic outbreaks.[26,27] Of the three groups of HLA loci (HLA-A/-B/-C), HLA-B locus revealed an unexplained level of diversity: it appears that the HLA-B locus alleles are diversifying more rapidly than the alleles of any other loci. The molecular-level significance of which is poorly understood.

The differences in immunogenetic landscape due to HLA allele frequency variations in populations might influence the variations of disease outcome. Top 10 countries with the highest number of COVID-19 cases are compared for the number of cases and deaths (Table 8.1). India, considering its huge population size, revealed comparatively smaller number of COVID-19 deaths. With a total of 34,426,036 COVID-19 cases (as of November 2021) and with a total of 463,245 COVID-19 deaths reported so-far, India with a population of 1,398,494,658 (1.39 billion; second highly populous country in the world) stands second based on number of cases and

Table 8.1 Top 10 countries with COVID-19 cases and deaths.[a]

S.No	Population	Population size	Total cases[b]	Total deaths	Rank based no. of deaths
1.	USA	33,36,48,943	4,78,34,810	782,933	III
2.	India	139,84,94,658	3,44,26,036	463,245	X
3.	Brazil	21,46,19,177	2,19,40,950	610,935	I
4.	UK	6,83,72,226	94,87,302	142,678	IV
5.	Russia	14,60,19,815	89,92,595	252,926	VII
6.	Turkey	8,55,75,739	83,63,959	73,127	IX
7.	France	6,54,70,493	72,60,503	118,121	VI
8.	Iran	8,54,57,576	60,27,269	127,918	VIII
9.	Argentina	4,57,61,013	53,04,059	116,222	II
10.	Spain	4,67,79,479	50,47,156	87,673	V

[a]Cf: https://www.worldometers.info/coronavirus/(as of November 2021).
[b]Top 10 ranked countries based on total COVID-19 cases.

10th based on deaths (a simple calculation of total cases and/or deaths versus total population of the country) among top 10 countries with COVID-19 infections.[28] Other countries at the top of the ladder are the United States (cases vs. deaths: I/III), Brazil (III/I), the United Kingdom (IV/IV), Russia (V/VII), Turkey (VI/IX), France (VII/VI), Iran (VIII/VIII), Argentina (IX/II), and Spain (X/V).

The observed lower mortality rate in India could be attributed to factors such as food habits, lifestyle, community-level disease burden, BCG vaccination practice, environmental pathogen load, and HLA immunogenetic diversity among others. The existence of broad-based natural immunity in the general population as a consequence of the extensive microbial load in the natural environment and higher level of immunogenetic diversity could be the possible explanations for this reduced mortality. None the less, higher level of HLA diversity could in turn, favor viral mutations—a co-evolutionary mechanism promoted by the process of natural selection.[29] This molecular evolutionary phenomenon explains clearly the emergence of new infectious strains of viruses (such as Delta variant) in countries like India.

3.2 Specific HLA variants and population associations

A number of studies have reported the association of HLA variants (alleles) in host susceptibility/resistance to SARS (Lin et al., 2003), MERS,[30] and influenza.[31,32] A number of alleles of MHC class I loci (HLA-A/-B/-C) such as A*25, A*11, B*27:07, B*15:27, B*08, B*44, B*15:01, B*51, B*14, B*18, B*49, C*01, C*03, C*04:01, and C*07:29, and class II loci (HLA-DRB1/DQB1) such as DRB1*03, DRB1*04:01, DRB1*04:02, DRB1*04:05, DRB1*08:01, DRB1*15:01, DQB1*04, and DQB1*06:02 were reported to be associated positively (susceptible) with COVID-19 infections in global populations (Table 8.2). A three-locus haplotype, "A*30:02-B*14:02-C*08:02" revealed strong association with moderate to severe COVID-19 disease in Sardinian (Italy) cohort.[33]

Table 8.2 HLA class I/II alleles and haplotypes association with COVID-19 among global populations.

Population	Susceptible alleles					Haplotypes	References*
	Class I			Class II			
	A	B	C	DRB1	DQB1		
Canary Islands (Spain)	A*11	B*39	C*16, C*04		DQB1*04		33
Chinese	A*11:01	B*15:27 B*51:01	C*07:29 C*14:02				34,35
Fujian, China		B*15:27		DRB1*04:06			36
72 different countries (AFND Data base)	A*01	B*07, B*08, B*44	C*05:01 (C*05 + KIR2DS4)				37
Italy	A*25	B*08, B*44, B*15:01 B*51	C*01 and C*03				38
Italy		B*27:07		DRB1*15:01	DQB1*06:02		39
Sardania (Italy)	A*23:01			DRB1*08:01		A*30:02-B*14:02-C*08:02	40
Italy						A*02:01-B*18:01-C*07:01-DRB1*11:04	12

South Han Chinese	A*24:02			DRB4*01:01	DQB1*03:01	DPA1*02:02-DPB1*05:01	41
Hong Kong Chinese patients		B22 serotype (B*54:01, B*55:01, B*55:07, B*55:12 and B*56:01) B*27					42
GWAS study (Italy)		B*27:07		DRB1*15:01	DQB1*06:02		3
Japanese	A*11:01:01:01	B*52:01:02:02	C*12:02:02:01				43
Iran	A*01, A*03	B*07, B*38					44
Saudi patients	A*01		C*03				45

HLA alleles DRB1*03:01, HLA-Cw*15:02, and HLA-A*02:01 decrease susceptibility to SARS-CoV-1 infection in the Chinese population.[34] HLA-DRB1*04:01 allele was found to offer protective effect from the development of respiratory failure following exposure to COVID-19 in a population group of European descent.[46] In the Japanese population, allele HLA-A*11:01, together with the HLA-A*02:06 and HLA-B*54:01 alleles, could protect against COVID-19 infection.[47] More interestingly, HLA-DRB1*13:02 has shown to confer protection against various illnesses ranging from hepatitis B and C,[48] influenza,[49] HIV,[50] malaria,[51] and a number of autoimmune disorders.[52] Further, DRB1*13:02, has been shown to not only protect against age-related deterioration in neural network functioning, but also to negate the deleterious effects induced by apoE4 gene on neural network functioning.[53] However, the molecular aspects of these HLA alleles affording protection toward COVID-19 remains elusive.

3.3 Delta variant

South East Asia is being devastated with three highly transmissible and immune-evasive variants of concern such as B.1.1.7 (alpha), B.1.351 (beta), and B.1.617.2 (delta). The SARS-CoV-2 delta variant (B.1.617.2) that evolved from the outbreak in India has created widespread pandemic and re-emergence of a highly infectious strain.[54] South East Asia, with more than 655 million people (two thirds of the world's population), could be the next hotspot due to geographical and socioeconomic factors, such as sporadic malnourishment and the emergence of new variants. The main reasons attributed for this highly devastating pathogen emergence, across a large swath of land mass of South East Asia, includes illegal movement of workers/general population between countries, limited vaccine rollout, poor socio-economic conditions, and highly turbulent political climate in many countries of this region. On the basis of available genome sequences, the Delta variant seems to have successfully overtaken the Alpha variant in many of the South East Asian nations. Most importantly, the viral genomic data have confirmed criss-cross transportation of these variants as a consequence of uncontrolled human migrations between countries. Thus, the whole of South East Asia is being considered as a new regional hotspot of COVID-19. Thus, the immunogenetics of the populations and other socio-economic conditions prevailing in this region together orchestrated the emergence of newer strains and presented highly complicated infection dynamics. Hence, a multicentric approach to study these critical aspects in relation to COVID-19 emergence/reemergence is the need of the hour.

4. HLA associations in COVID-19 induced neurological disorders

The HLA genes play a role in those autoimmune neurological disorders triggered by infectious agents.[55] The allele HLA-A*33 was associated with acute inflammatory demyelinating polyradiculoneuropathy (AIDP) in China.[56] HLA-DRB1*03:01 was associated with Gullien Bar Syndrome (GBS) in Iraq[57] and DQB1*05:01 in German[58] populations. In COVID-19 brain autopsies, microglia nodules and increased expression of microglial hyper-activity markers such as IBA1, HLA-DR, and CD68 were reported in recent and older hypoxic/ischemic injuries.[59–61] Further, marked neuronal loss among the cerebral cortex, hippocampus, medulla, and cerebellar Purkinje cell layer were all demonstrated.[59,62] A large number of published literature have reported the development of a number of neurological complications such as headache and neuroinflammatory or cerebrovascular complications among COVID-19 patients.

The analysis of cerebrospinal fluid collected from neuro-inflammatory COVID-19 patients have shown the expansion of dedifferentiated monocytes and exhausted CD4+ T cells.[63] It has been demonstrated that the antigen presentation by CD4+ T cells to cells expressing HLA-DQB1*06:02 leads to increased IL-17 proinflammatory cytokine levels, that in turn confers increased risk to the development of anti-myelin directed autoimmune responses.[64,65] Interestingly, in a previous study, the haplotypes HLA-DR2-DQ6, DR4-DQ8, and DR3-DQ2 (most predominant two-locus haplotypes reported in different populations) have shown to accommodate peptides from infectious pathogens to CD4+ T-cells in Europeans who survived the bottleneck of different, life-threatening infections prevalent in Europe.[66] Thus, Human leukocyte antigen (HLA) genes play a critical role in immune protection from foreign antigens including viruses, bacteria, and parasites.[67] Recent studies have implicated the HLA region genes/alleles in neurodegenerative diseases including AD.[68,69]

4.1 Opportunities for future study

Recent lines of evidences have demonstrated the fixation/selection of human gene variants encoding "virus interacting proteins" (VIPs), which continued for at least 900 generations (or approx. 25,000 years) by analyzing "selective sweep signals" (a skewed selection event that favors enrichment of beneficial variant in populations) that clustered around those host proteins that are preferentially hijacked by RNA viruses as receptors for entry into the human host.[70–72] Similarly, from among 5291 VIPs, 42 coronavirus interacting—VIPs (CoV-VIPs) were functionally shortlisted by using high-throughput molecular methods.[73] The relationships between HLA and these COV-VIPs must be elucidated.

5. Conclusions

HLA class I molecules present pathogen-derived peptides on the surface of infected cells for recognition by T cells. The extreme polymorphism of HLA alleles and highly complex HLA-mediated immune responses orchestrated by cytokines and other mediators have complicated our understanding of diseases caused by infectious viruses such as COVID-19. More particularly, the polymorphism observed in HLA-A, -B, and -C loci alleles is suggested to provide an advantage in pathogen defense mediated by CD4+ and CD8+ T-cells. Thus, investigating HLA markers and COVID-19 disease associations at molecular level can help us better understand the molecular etiologies of infections caused by novel viruses such as SARS-CoV-2.

References

1. Desforges M, Le Coupanec A, Dubeau P, et al. Human coronaviruses and other respiratory viruses: underestimated opportunistic pathogens of the central nervous system? *Viruses*. 2019;12(1):1–28. https://doi.org/10.3390/v12010014.
2. Manolis AS, Manolis AA, Manolis TA, Apostolopoulos EJ, Papatheou D, Melita H. *Since January 2020 Elsevier Has Created a COVID-19 Resource Centre with Free Information in English and Mandarin on the Novel Coronavirus COVID- 19. The COVID-19 Resource Centre Is Hosted on Elsevier Connect, The Company ' S Public News and Information*. 2020 January.
3. Severe Covid-19 GWAS Group, Ellinghaus D, Degenhardt F, et al. Genomewide association study of severe covid-19 with respiratory failure. *N Engl J Med*. 2020;383(16): 1522–1534. https://doi.org/10.1056/nejmoa2020283.
4. Goeury T, Creary LE, Brunet L, et al. Deciphering the fine nucleotide diversity of full HLA class I and class II genes in a well-documented population from sub-Saharan Africa. *Hla*. 2018;91(1):36–51. https://doi.org/10.1111/tan.13180.
5. *The Human Leucocyte Antigens and Clinical Medicine: An Overview*. 1999:11256570 (Cdc).
6. Klein JAN, Sato A. The HLA system: first of two parts. *N Engl J Med*. 2000;343(10): 702–709. https://doi.org/10.1056/NEJM200009073431006.
7. http://www.ebi.ac.uk/imgt/hla/stats.html.
8. Ganna A, Unit TG, General M. The COVID-19 Host Genetics Initiative, a global initiative to elucidate the role of host genetic factors in susceptibility and severity of the SARS-CoV-2 virus pandemic. *Eur J Hum Genet*. 2020;28(6):715–718. https://doi.org/10.1038/s41431-020-0636-6.
9. Kamatani Y, Wattanapokayakit S, Ochi H, et al. A genome-wide association study identifies variants in the HLA-DP locus associated with chronic hepatitis B in Asians. *Nat Genet*. 2009;41(5):591–595. https://doi.org/10.1038/ng.348.
10. Thomas R, Thio CL, Apps R, et al. A novel variant marking HLA-DP expression levels predicts recovery from hepatitis B virus infection. *J Virol*. 2012;86(12):6979–6985. https://doi.org/10.1128/jvi.00406-12.
11. Iturrieta-zuazo I, Geraldine C, García-soidán A, Pintos-fonseca ADM, Alonso-alarcón N, Pariente-rodríguez R. *Since January 2020 Elsevier Has Created a COVID-19 Resource Centre with Free Information in English and Mandarin on the Novel Coronavirus COVID-19. The COVID-19 Resource Centre Is Hosted on Elsevier Connect, the Company ' S Public News and Information*. 2020 January.
12. Pisanti S, Deelen J, Gallina AM, et al. Correlation of the two most frequent HLA haplotypes in the Italian population to the differential regional incidence of Covid-19. *J Transl Med*. 2020;18(1):1–16. https://doi.org/10.1186/s12967-020-02515-5.
13. Nguyen A, David JK, Maden SK, et al. Human leukocyte antigen susceptibility map for SARS-CoV-2. *J Virol*. 2020;94(13):1–12. https://doi.org/10.1128/JVI.00510-20.

14. Singh PP, Suravajhala P, Basu Mallick C, et al. COVID-19: impact on linguistic and genetic isolates of India. *Gene Immun*. 2021 February:1–4. https://doi.org/10.1038/s41435-021-00150-8.
15. Chen YY, Zhang P, Zhou XM, et al. Relationship between genetic variants of ACE2 gene and circulating levels of ACE2 and its metabolites. *J Clin Pharm Therapeut*. 2018;43(2): 189–195. https://doi.org/10.1111/jcpt.12625.
16. Zeberg H, Pääbo S. The major genetic risk factor for severe COVID-19 is inherited from Neanderthals. *Nature*. 2020;587(7835):610–612. https://doi.org/10.1038/s41586-020-2818-3.
17. Srivastava A, Pandey RK, Singh PP, et al. Most frequent South Asian haplotypes of ACE2 share identity by descent with East Eurasian populations. *PLoS One*. 2020;15(9 September): 1–7. https://doi.org/10.1371/journal.pone.0238255.
18. Public Health England. Disparities in the Risk and Outcomes of COVID-19. PHE Publ. Published online 2020:89. https://www.gov.uk/government/publications/covid-19-review-of-disparities-in-risks-and-outcomes.
19. Stower H. Spread of SARS-CoV-2. *Nat Med*. 2020;26(4):465. https://doi.org/10.1038/s41591-020-0850-3.
20. Roncati L, Palmieri B. What about the original antigenic sin of the humans versus SARS-CoV-2? *Med Hypotheses*. 2020;142(April):109824. https://doi.org/10.1016/j.mehy.2020.109824.
21. Shi Y, Wang Y, Shao C, et al. COVID-19 infection: the perspectives on immune responses. *Cell Death Differ*. 2020;27(5):1451–1454. https://doi.org/10.1038/s41418-020-0530-3.
22. Mongkolsapaya J, Dejnirattisai W, Xu X, et al. Original antigenic sin and apoptosis in the pathogenesis of dengue hemorrhagic fever. *Nat Med*. 2003;9(7):921–927.
23. Yewdell JW, Santos JJS. Original antigenic sin: how original? how sinful? *Cold Spring Harb Perspect Med*. 2021;11(5):1–16. https://doi.org/10.1101/cshperspect.a038786.
24. Reed DH, Frankham R. Correlation between fitness and genetic diversity. *Conserv Biol*. 2003;17(1):230–237. https://doi.org/10.1046/j.1523-1739.2003.01236.x.
25. Tishkoff SA, Verrelli BC. Patterns of human genetic diversity: implications for human evolutionary history and disease. *Annu Rev Genom Hum Genet*. 2003;4:293–340. https://doi.org/10.1146/annurev.genom.4.070802.110226.
26. G.S. Cooke and A.V. S. Hill. No Title.; 2001. https://www.nature.com/articles/35103577.
27. Lyons EJ, Frodsham AJ, Zhang L, Hill AVS, Amos W. Consanguinity and susceptibility to infectious diseases in humans. *Biol Lett*. 2009;5(4):574–576. https://doi.org/10.1098/rsbl.2009.0133.
28. https://www.worldometers.info/coronavirus/.
29. Klepiela P, Leslie AJ, Honeyborne I, et al. Dominant influence of HLA-B in mediating the potential co-evolution of HIV and HLA. *Nature*. 2004;432(7018):769–774. https://doi.org/10.1038/nature03113.
30. Hajeer AH, Balkhy H, Johani S, Yousef MZ, Arabi Y. Association of human leukocyte antigen class II alleles with severe Middle East respiratory syndrome-coronavirus infection. *Ann Thorac Med*. 2016;11(3):211–213. https://doi.org/10.4103/1817-1737.185756.
31. Falfán-Valencia R, Narayanankutty A, Reséndiz-Hernández JM, et al. An increased frequency in HLA class I alleles and haplotypes suggests genetic susceptibility to influenza A (H1N1) 2009 pandemic: a case-control study. *J Immunol Res*. 2018;2018. https://doi.org/10.1155/2018/3174868.
32. Luckey D, Weaver EA, Osborne DG, Billadeau DD, Taneja V. Immunity to Influenza is dependent on MHC II polymorphism: study with 2 HLA transgenic strains. *Sci Rep*. 2019; 9(1):1–10. https://doi.org/10.1038/s41598-019-55503-1.

33. Lorente L, Martín MM, Franco A, et al. HLA genetic polymorphisms and prognosis of patients with COVID-19. *Med Intensiva*. 2021;45(2):96–103. https://doi.org/10.1016/j.medin.2020.08.004.
34. Wang F, Huang S, Gao R, et al. Initial whole-genome sequencing and analysis of the host genetic contribution to COVID-19 severity and susceptibility. *Cell Discov*. 2020;6(1). https://doi.org/10.1038/s41421-020-00231-4.
35. Wang D, Hu B, Hu C, et al. Clinical characteristics of 138 hospitalized patients with 2019 novel coronavirus-infected pneumonia in Wuhan, China. *J Am Med Assoc*. 2020;323(11): 1061–1069. https://doi.org/10.1001/jama.2020.1585.
36. Yu X, Ho K, Shen Z, et al. The association of human leukocyte antigen and COVID-19 in southern China. *Open Forum Infect Dis*. 2021;8(9):1–5. https://doi.org/10.1093/ofid/ofab410.
37. Sakuraba A, Haider H, Sato T. Population difference in allele frequency of hla-c*05 and its correlation with COVID-19 mortality. *Viruses*. 2020;12(11). https://doi.org/10.3390/v12111333.
38. Correale P, Mutti L, Pentimalli F, et al. Hla-b*44 and c*01 prevalence correlates with COVID-19 spreading across Italy. *Int J Mol Sci*. 2020;21(15):1–12. https://doi.org/10.3390/IJMS21155205.
39. Novelli A, Andreani M, Biancolella M, et al. HLA allele frequencies and susceptibility to COVID-19 in a group of 99 Italian patients. *Hla*. 2020;96(5):610–614. https://doi.org/10.1111/tan.14047.
40. Littera R, Campagna M, Deidda S, et al. Human leukocyte antigen complex and other immunogenetic and clinical factors influence susceptibility or protection to SARS-CoV-2 infection and severity of the disease course. The Sardinian experience. *Front Immunol*. 2020;11(December):1–14. https://doi.org/10.3389/fimmu.2020.605688.
41. Warren RL, Birol I. HLA predictions from the bronchoalveolar lavage fluid samples of five patients at the early stage of the Wuhan seafood market COVID-19 outbreak. *ArXiv*. 2020: 4–7. https://doi.org/10.1093/bioinformatics/btaa756 (class I).
42. Yung YL, Cheng CK, Chan HY, et al. Association of HLA-B22 serotype with SARS-CoV-2 susceptibility in Hong Kong Chinese patients. *Hla*. 2021;97(2):127–132. https://doi.org/10.1111/tan.14135.
43. Khor SS, Omae Y, Nishida N, et al. HLA-A*11:01:01:01, HLA-C*12:02:02:01-HLA-B*52:01:02:02, age and sex are associated with severity of Japanese COVID-19 with respiratory failure. *Front Immunol*. 2021;12(April). https://doi.org/10.3389/fimmu.2021.658570.
44. Saadati M, Chegni H, Ghaffari AD, Mohammad Hassan Z. The potential association of human leukocyte antigen (Hla)-a and-b with COVID-19 mortality: a neglected risk factor. *Iran J Public Health*. 2020;49(12):2433–2434. https://doi.org/10.18502/ijph.v49i12.4837.
45. Naemi FMA, Al-adwani S, Al-khatabi H, Al-nazawi A. Association between the HLA genotype and the severity of COVID-19 infection among South Asians. *J Med Virol*. 2021; 93(7):4430–4437. https://doi.org/10.1002/jmv.27003.
46. Langton DJ, Bourke SC, Lie BA, et al. The influence of HLA genotype on the severity of COVID-19 infection. *Hla*. 2021;98(1):14–22. https://doi.org/10.1111/tan.14284.
47. Toyoshima Y, Nemoto K, Matsumoto S, Nakamura Y, Kiyotani K. SARS-CoV-2 genomic variations associated with mortality rate of COVID-19. *J Hum Genet*. 2020;65(12): 1075–1082. https://doi.org/10.1038/s10038-020-0808-9.
48. Singh R, Kaul R, Kaul A, Khan K. A comparative review of HLA associates with hepatitis B and C viral infections across global populations. *World J Gastroenterol*. 2007;13(12):

1770−1787. https://www.ncbi.nlm.nih.gov/pmc/articles/PMC4149952/%0Ahttp://ovidsp.ovid.com/ovidweb.cgi?T=JS&PAGE=reference&D=emed8&NEWS=N&AN=2007225108.
49. Posteraro B, Pastorino R, Di Giannantonio P, et al. The link between genetic variation and variability in vaccine responses: systematic review and meta-analyses. *Vaccine*. 2014; 32(15):1661−1669. https://doi.org/10.1016/j.vaccine.2014.01.057.
50. International HIV Controllers Study, Pereyra F, Jia X, et al. The major genetic determinants of HIV-1 control affect HLA class I peptide presentation. *Science*. 2010;330(6010): 1551−1557. https://doi.org/10.1126/science.1195271.
51. Hill AVS, Allsopp CEM, Kwiatkowski D, et al. Common West African HLA antigens are associated with protection from severe malaria. *Nature*. 1991;352(6336):595−600. https://doi.org/10.1038/352595a0.
52. Bettencourt A, Carvalho C, Leal B, et al. The protective role of HLA-DRB1 13 in autoimmune diseases. *J Immunol Res*. 2015;2015:3−8. https://doi.org/10.1155/2015/948723.
53. James LM, Dolan S, Leuthold AC, Engdahl BE, Georgopoulos A, Georgopoulos AP. The effects of human leukocyte antigen DRB1*13 and apolipoprotein E on age-related variability of synchronous neural interactions in healthy women. *EBioMedicine*. 2018;35: 288−294. https://doi.org/10.1016/j.ebiom.2018.08.026.
54. Campbell F, Archer B, Laurenson-Schafer H, et al. Increased transmissibility and global spread of SARSCoV- 2 variants of concern as at June 2021. *Euro Surveill*. 2021;26(24): 1−6. https://doi.org/10.2807/1560-7917.ES.2021.26.24.2100509.
55. Muñiz-Castrillo S, Vogrig A, Honnorat J. Associations between HLA and autoimmune neurological diseases with autoantibodies. *Autoimmun Highlights*. 2020;11(1):1−13. https://doi.org/10.1186/s13317-019-0124-6.
56. Guo L, Wang W, Li C, Liu R, Wang G. The association between HLA typing and different subtypes of Guillain Barré syndrome. *Zhonghua Nei Ke Za Zhi*. 2002;41(6):381−383.
57. Hasan ZN, Zalzala HH, Mohammedsalih HR, et al. Association between human leukocyte antigen-DR and demylinating Guillain-Barré syndrome. *Neurosciences*. 2014;19(4): 301−305.
58. Schirmer L, Worthington V, Solloch U, et al. Higher frequencies of HLA DQB1*05:01 and anti-glycosphingolipid antibodies in a cluster of severe Guillain−Barré syndrome. *J Neurol*. 2016;263(10):2105−2113. https://doi.org/10.1007/s00415-016-8237-6.
59. Al-Dalahmah O, Thakur KT, Nordvig AS, et al. Neuronophagia and microglial nodules in a SARS-CoV-2 patient with cerebellar hemorrhage. *Acta Neuropathol Commun*. 2020;8(1): 1−7. https://doi.org/10.1186/s40478-020-01024-2.
60. Kantonen J, Mahzabin S, Mäyränpää MI, et al. Neuropathologic features of four autopsied COVID-19 patients. *Brain Pathol*. 2020;30(6):1012−1016. https://doi.org/10.1111/bpa.12889.
61. Matschke J, Lütgehetmann M, Hagel C, et al. Neuropathology of patients with COVID-19 in Germany: a post-mortem case series. *Lancet Neurol*. 2020;19(11):919−929. https://doi.org/10.1016/S1474-4422(20)30308-2.
62. Zhao Q, Meng M, Kumar R, et al. Correspondence crystallopathies. *J Med Virol*. 2020; 69(1):2016−2017. http://www.ncbi.nlm.nih.gov/pubmed/32293753.
63. Heming M, Li X, Räuber S, et al. Neurological manifestations of COVID-19 feature T cell exhaustion and dedifferentiated monocytes in cerebrospinal fluid. *Immunity*. 2021;54(1): 164−175. https://doi.org/10.1016/j.immuni.2020.12.011. e6.
64. Mangalam A, Luckey D, Basal E, Behrens M, Rodriguez M, David C. HLA-DQ6 (DQB1*0601)-restricted T cells protect against experimental autoimmune encephalomyelitis in HLA-DR3.DQ6 double-transgenic mice by generating anti-inflammatory IFN-γ. *J Immunol*. 2008;180(11):7747−7756. https://doi.org/10.4049/jimmunol.180.11.7747.

65. Kaushansky N, Ben-Nun A. DQB1*06: 02-associated pathogenic anti-myelin autoimmunity in multiple sclerosis-like disease: potential function of DQB1*06:02 as a disease-predisposing allele. *Front Oncol.* 2014;4(Oct):2−7. https://doi.org/10.3389/fonc.2014.00280.
66. Matzaraki V, Kumar V, Wijmenga C, Zhernakova A. The MHC Locus and Genetic Susceptibility to Autoimmune and Infectious Diseases. Published online 2017. doi:10.1186/s13059-017-1207-1.
67. Mhc II . *Surface Structures Involved in Target Recognition by Human Cytotoxic T Lymphocytes.* 1982 (Cml).
68. Lambert JC, Ibrahim-Verbaas CA, Harold D, et al. Meta-analysis of 74,046 individuals identifies 11 new susceptibility loci for Alzheimer's disease. *Nat Genet.* 2013;45(12):1452−1458. https://doi.org/10.1038/ng.2802.
69. Carr JS, Steele NZ, Bonham LW, et al. [P2−118]: fine-mapping of the human leukocyte antigen (Hla) locus as a risk factor for Alzheimer'S disease. *Alzheimer's Dementia.* 2017;13(7S_Part_13):1−25. https://doi.org/10.1016/j.jalz.2017.06.768.
70. Sawyer SL, Wu LI, Emerman M, Malik HS. Positive selection of primate TRIM5α identifies a critical species-specific retroviral restriction domain. *Proc Natl Acad Sci U S A.* 2005;102(8):2832−2837. https://doi.org/10.1073/pnas.0409853102.
71. Uricchio LH, Petrov DA, Enard D. Exploiting selection at linked sites to infer the rate and strength of adaptation. *Nat Ecol Evol.* 2019;3(6):977−984. https://doi.org/10.1038/s41559-019-0890-6.
72. Enard D, Petrov DA. Ancient RNA virus epidemics through the lens of recent adaptation in human genomes: Ancient RNA virus epidemics. *Philos Trans R Soc B Biol Sci.* 2020;375(1812). https://doi.org/10.1098/rstb.2019.0575rstb20190575.
73. Speidel L, Forest M, Shi S, Myers SR. A method for genome-wide genealogy estimation for thousands of samples. *Nat Genet.* 2019;51(9):1321−1329. https://doi.org/10.1038/s41588-019-0484-x.

Impact of COVID-19 on ischemic stroke condition

9

Tochi Eboh[1], *Hallie Morton*[2], *P. Hemachandra Reddy*[2,3,4,5,6] *and Murali Vijayan*[2]

[1]School of Medicine, Texas Tech University Health Sciences Center, Lubbock, TX, United States; [2]Department of Internal Medicine, Texas Tech University Health Sciences Center, Lubbock, TX, United States; [3]Department of Pharmacology and Neuroscience, School of Medicine, Texas Tech University Health Sciences Center, Lubbock, TX, United States; [4]Department of Neurology, Texas Tech University Health Sciences Center, Lubbock, TX, United States; [5]Public Health Department of Graduate School of Biomedical Sciences, School of Medicine, Texas Tech University Health Sciences Center, Lubbock, TX, United States; [6]Department of Speech, Language and Hearing Sciences, School Health Professions, School of Medicine, Texas Tech University Health Sciences Center, Lubbock, TX, United States

Abstract

Coronavirus disease 2019 (COVID-19) results from the infection by severe acute respiratory syndrome coronavirus 2 (SARS-CoV-2). The disease was first reported in Wuhan, China, when patients were found to be suffering from severe pneumonia and acute respiratory distress syndrome. It has now grown to be the first global pandemic since 1920. Patients infected with SARS-CoV-2 develop a multitude of ailments, including arterial thrombosis, which leads to acute conditions like stroke. Stroke in COVID-19 cannot be explained by a single mechanism but instead is defined by the interplay of many mechanisms, including the development of cytokine storms resulting in activation of the innate immune system, thrombotic microangiopathy, endothelial disruption, and the multifactorial activation of the coagulation cascade. Thromboprophylaxis in low—molecular-weight heparin has been shown to affect severely ill patients infected with COVID-19 beneficially. However, patients who develop stroke because of COVID-19 have poorer outcomes despite maximal medical, endovascular, and microsurgical treatment compared with non-COVID-19-infected patients. A significant challenge in managing stroke during the pandemic is maintaining high-quality care for stroke patients while protecting healthcare team members and staff.

Keywords: Cerebrovascular accident; COVID-19; Cytokine storm; Endothelial disruption; Hypercoagulability; Revascularization; Thrombosis

1. Introduction

1.1 Stroke

Cerebrovascular accident (CVA), commonly known as stroke, can be caused by two mechanisms: ischemia and hemorrhagia. Ischemic stroke is responsible for 85% of

COVID-19 in Alzheimer's Disease and Dementia. https://doi.org/10.1016/B978-0-443-15256-6.00018-0

CVAs. It is characterized by the sudden loss of circulation to a specific brain region, which results in a corresponding loss of neurological function that lasts greater than 24 h.[1–4] Subtypes of ischemic stroke include thrombotic, embolic, lacunar strokes, and cryptogenic. Thrombotic stroke generally refers to local obstruction of an artery due to the rupture of an atherosclerotic plaque, thrombus. Cerebral thrombus usually develops at branch points, most commonly the bifurcation of the middle cerebral artery in the circle of Willis.[4–6]

Embolic stroke is due to a thrombo emboli, a clot that travels from the site where it formed to another body location.[4] The most common source of cerebral emboli is the left side of the heart. Embolus suddenly occludes recipient sites' circulation, leading to abrupt and usually maximal symptoms at the start. This differs from ischemic strokes caused by thrombosis, in which multiple sites within different vascular regions may be affected.[4]

A lacunar stroke occurs secondary to hyaline arteriolosclerosis, which is a complication of hypertension.[7–10] Hyaline arteriolosclerosis is small vessel atherosclerosis caused by protein leakage of the vessel walls, which produces vascular hardening and thickening.[4] Lacunar strokes most commonly involve the lenticulostriate vessels, small branches of the MCA, resulting in small cystic areas of infarction.

Cryptogenic stroke is defined as a brain infarction not attributable to a definite cardioembolism, artery atherosclerosis, or small artery disease. According to the American Stroke Association, one in four ischemic strokes are classified as cryptogenic.

2. Coronavirus and SARS CoV-2

Coronavirus is a family of viruses named because of their crown-shaped spikes on their surface. The coronavirus family is further sub-classified into four genera: alpha, beta, gamma, and delta coronavirus. Only the alpha and beta genera can infect humans.[11] The alpha genera include HCoV-229E and HCoV-NL63, while the beta genera include HCoV-HKU1, HCoV-OC43, Middle East respiratory syndrome coronavirus (MERS-CoV), the severe acute respiratory syndrome coronavirus (SARS-CoV), and SARS-CoV-2. Both SARS-CoV and MERS-CoV have caused outbreaks of potentially fatal respiratory tract infections in 2003 and 2012, respectively.[12]

Coronavirus is a medium-sized, enveloped, single-stranded plus sense RNA virus. They have four important structural components': the spike surface protein (S), membrane protein (M), nuclear capsid protein (N), and the envelope protein (E).[13] The spike surface protein is heavily glycosylated and mediates receptor binding and fusion with the host cell membrane. In the case of SARS-Cov 2, the S protein infects the human cell through the ACE-2 receptor. Additionally, the S protein is the main antigen that stimulates neutralizing antibodies and targets cytotoxic T-cells.[13] The M protein plays an important part in viral assembly. It has a short N-terminal domain that projects to the external surface of the envelope and a long c terminus inside the envelope. The N protein is associated with the RNA genome to form the nucleocapsid.[13]

SARS-CoV-2 is a novel coronavirus that emerged in Wuhan city, China, in December of 2019. The virus rapidly spread worldwide, causing a global pandemic.

COVID-19, the disease SARS-Cov-2 causes, is transmitted through respiratory droplets via coughing and sneezing. There is approximately a 5.2-day incubation period during which infected individuals can shed large amounts of viral particles.[14] Although nearly 33% of people infected with SARS-Cov-2 never develop symptoms, the most common reported symptoms include fever, dry cough, myalgia, fatigue, headache, and anorexia. Anosmia, dysgeusia, and gastrointestinal symptoms are other notable symptoms. In addition, the virus has been seen to cause neurological diseases such as epileptic seizure, Guillian barre, acute transverse myelitis, and cerebrovascular events.[14]

Since September 15th, 2021, the worldwide confirmed coronavirus disease 2019 (COVID-19) cases have been about 226 million, resulting in over 4.6 million deaths. The United States alone has reported more than 41 million (41,229,421) with about 660,000 deaths. Although the primary reported cause of death in patients suffering from COVID-19 is an acute respiratory syndrome (ARDS), a wide range of ailments have been reported, from asymptomatic disease to kidney damage, cardiac illness, and neurological manifestations. Thrombotic events occur in up to one-third of patients who suffer from COVID-19. These events are predominantly pulmonary emboli, but studies have shown an increasing prevalence of acute ischemic stroke (AIS) during and following SARS-COV2 infections. This section aims to dive into the development of stroke in COVID-19-infected patients.

3. Epidemiology of stroke and COVID-19

Initially, reports of COVID-19 cases from China showed a high incidence of neurological manifestations with rates of patients experiencing neurological symptoms reaching upward of 36%, CVA occurring in about 5.7% of patients with severe disease compared with 0.8% of patients with the nonsevere disease.[1,15] A retrospective cohort study from New York City academic hospitals found that approximately 1.6% of adults with COVID-19 who visited the ED or were hospitalized experienced ischemic stroke. This rate was higher when compared to the rate of patients that were hospitalized for influenza.[16] Similarly, another study in New York in 2020 found that 0.9% of patients hospitalized with COVID-19 suffered from a stroke, with 65.6% of these patients dying during their hospital stay. It has also been reported that cryptogenic stroke occurred twice as frequently in COVID-19-positive patients compared to COVID-19-negative patients. Another study proposed that cryptogenic stroke represented a unique stroke mechanism associated with a higher probability of early mortality. Initial reports also found an increased risk for patients who suffered from COVID-19 to experience stroke months after the serological diagnosis.[17]

4. Mechanism of stroke in COVID-19

Stroke in COVID-19 cannot be explained by a single mechanism but instead is defined by the interplay of many mechanisms, including the development of cytokine storms

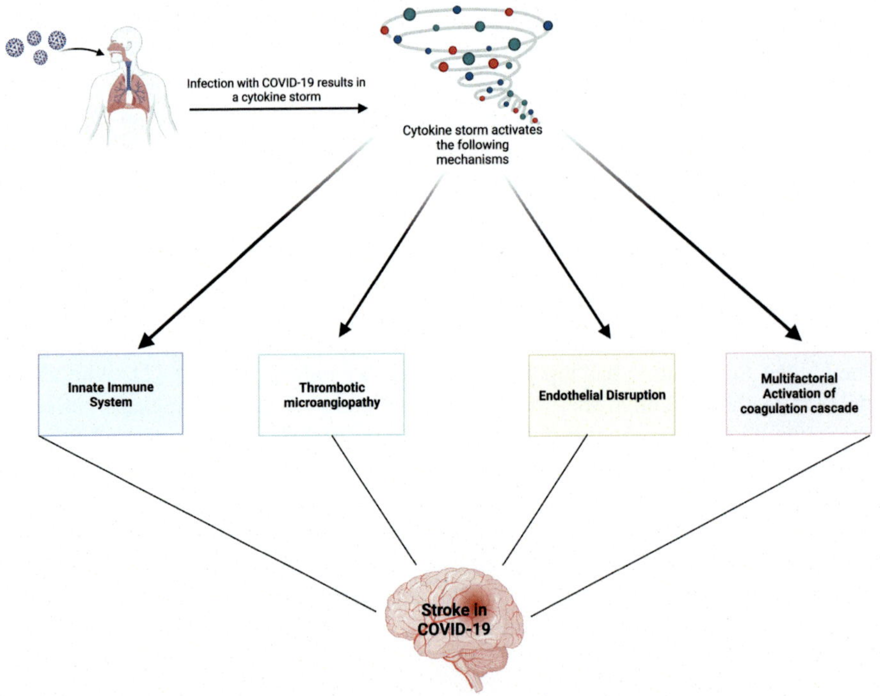

Figure 9.1 Mechanism of stroke in coronavirus disease 2019 (COVID-19): Stroke in COVID-19 is defined by the interplay of many mechanisms, including the development of cytokine storms resulting in activation of the innate immune system, thrombotic micro-angiopathy, endothelial disruption, and the multifactorial activation of the coagulation cascade.

resulting in activation of the innate immune system, thrombotic microangiopathy, endothelial disruption, and the multifactorial activation of the coagulation cascade[18] (Fig. 9.1).

One unifying variable found in COVID-19-infected patients suffering from stroke is the increase in D-dimer. D-dimer is a fibrin degradation product present in the blood after a clot has been degraded. The presence of d-dimer suggests the activation of the coagulation cascade and the innate immune system. Elevated D-dimer is directly correlated to coagulability. Among COVID-19 positive patients who suffered from large-vessel occlusion in one New York study, 90% had elevated D-dimer, while 42% presented acute stroke symptoms rather than respiratory symptoms. Recent studies have found that the increase in D-dimer levels is proportional to the severity of the disease, with D-dimer levels appearing to be more significantly elevated in patients with more severe diseases.[19]

The role of cytokine storm has proven to be another important mechanism in developing ischemic stroke in COVID-19-infected patients. Increased levels of IL-6 and C-reactive protein associated with cytokine storms are associated with an increased risk of both stroke and myocardial infarction when elevated.[20]

5. Thrombosis

COVID-19 has been associated with increased arterial and venous thrombosis rates both systemically and in the pulmonary vasculature. While numerous mechanisms of thrombosis have been proposed, activation of the innate immune system in response to the presence of a viral pathogen is thought to be the primary driver.[18] The term "immunothrombosis" is a process in which activated neutrophils and monocytes interact with platelets and the coagulation cascade leading to intravascular clot formation in blood vessels, atherosclerosis. 18

6. Cytokine storm

The initiating sequence of thrombosis in patients with COVID-19 is a hyperinflammatory response resulting in a cytokine storm.[18] Cytokines are proteins secreted by host immune cells that serve an essential role in innate defense, including the recruitment of adaptive immune cells (Fig. 9.2). In COVID-19, there is a hypersecretion of

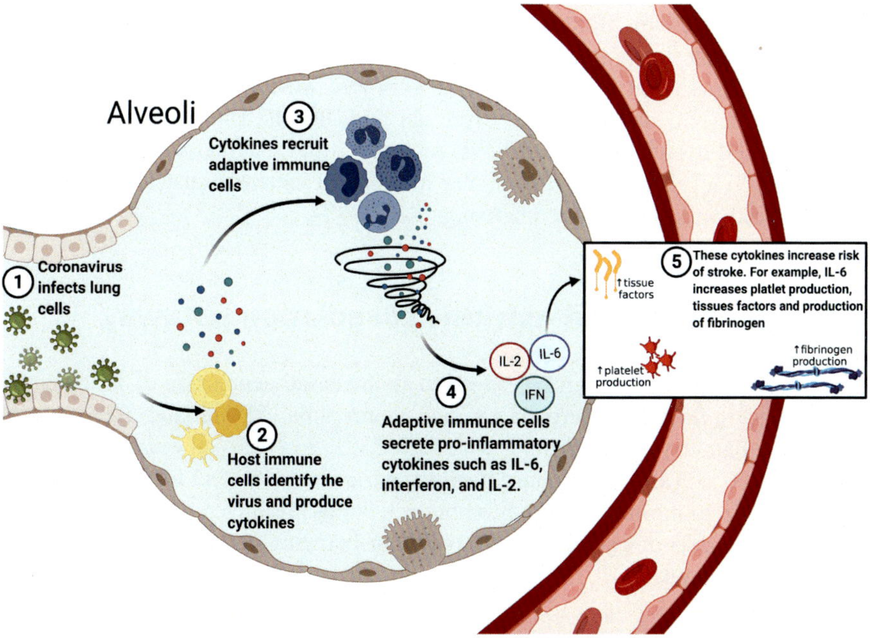

Figure 9.2 Cytokine storm: In severe coronavirus disease 2019 (COVID-19) patients, accumulation of cytokines is exaggerated resulting in a "cytokine storm," in which aberrant cytokine expression and disproportionate inflammation result in persistent acute lung injury extending beyond the time of peak viral load. In the figure, it illustrates that, (1) coronavirus infects lung cells, (2) immune cells, including macrophages, identify the virus and produce cytokines, (3) cytokines attract more immune cells, such as white blood cells, which in turn produce more cytokines, creating a cycle of inflammation that damages the lung cells, and (4) damage can occur through the formation of fibrin, (5) weakened blood vessels allow fluid to seep in and fill the lung cavities, leading to respiratory failure.

proinflammatory cytokines such as IL-6, interferon (IFN), and IL-2. IL-6 perpetuates the hypercoagulable state associated with COVID-19 through (1) increasing platelet production, (2) increasing the expression of tissue factors on endothelial cells and monocytes, which can give rise to endothelial dysfunction, and (3) activation of acute-phase response resulting in increased production of fibrinogen by hepatocytes. IFN-y can similarly increase platelet production and impair vascular endothelium (Fig. 9.2).

7. Endothelium disruption

The endothelium is a single cellular layer lining blood vessels that provide a mechanical barrier between the circulating blood and the underlying basement membrane. The endothelium also controls the vascular tone and is a strong immunomodulator shown to have antithrombotic activity.[21] Endothelial dysfunction associated with COVID-19 has been thought to occur by direct cytotoxic damage and indirect inflammatory effects. Direct cytotoxic damage occurs via infection of vascular endothelial cells, leading to cellular damage and apoptosis, thereby decreasing the antithrombotic activity of the normal endothelium. Moreover, elevated levels of Von Willebrand Factor (VWF), an endothelial adhesion protein, with an associated reduction in levels of ADAMTS13, a metalloproteinase that regulates the size of VWF, were reported in both patients admitted to the ICU and those who were not critically ill. Thus, the summation of endothelial cell dysfunction, large VWF multimers, and insufficient VWF cleavage due to reduced levels of ADAMTS13 may account for increased platelet and wall interactions leading to thrombotic microangiopathy.[22]

8. Tissue factor and extrinsic coagulation pathway

Elevated proinflammatory cytokines lead to an increased expression of TF on endothelial cells, infiltrating macrophages, and neutrophils. TF initiates the extrinsic coagulation pathway by combining with factor VIIa to create an enzyme that activates factor X to Xa.[23] Subsequent downstream activations lead to the conversion of prothrombin to thrombin, with thrombin activating fibrinogen to fibrin, fibrin being the final product of a stable clot. The previously mentioned process can also lead to platelet activation.[23]

9. Treatment of acute ischemic stroke in COVID-19

9.1 *Thromboprophylaxis*

Every patient admitted to the hospital should receive deep vein thrombosis prophylaxis regardless of the hospitalization reason. This normally comes in low-molecular-weight heparin, but other anticoagulants like apixaban can also be used.[24] These drugs act to inhibit the coagulation cascade at different points. For example, heparin inhibits the activation of factor X to Xa and the activation of prothrombin to thrombin. Low-molecular-weight heparin has been shown to have a better prognosis in severely ill COVID-19-infected patients as well as those with elevated d-dimer.[25] However, some studies have found patients admitted with COVID-19 still experienced ischemic stroke despite anticoagulation therapy. Another found that anticoagulation therapy was identified as an independent risk factor for intracranial bleeding in some COVID-19 patients. Additionally, the role of increased platelet activation and aggregation via inflammatory responses in COVID-19 could support the use of antiplatelet therapy like clopidogrel. Lastly, the targeting of the inflammatory response through the use of immunomodulatory therapy may also provide some efficacy in the prevention of acute stroke.[26]

9.2 *Intravenous thrombolysis*

The current national guidelines recommend intravenous thrombolysis for select stroke patients who fall in a specific category. Patients must be able to be treated within 3–4.5 h after the onset of symptoms and be assessed via CT scan to rule out hemorrhagic stroke.[27] This recommendation has proven to be paradoxical when used to assess COVID-19-infected patients. These patients elevated inflammatory markers have been associated with an increased risk of death or disability related to post-tPA intracranial hemorrhage. Therefore, special evaluation should be made when deciding to push tPA in COVID-19-positive patients.[28]

10. Anesthesia for mechanical thrombectomy

Mechanical thrombectomy is an interventional procedure that removes blood clots from blood vessels.[29] This procedure may be used during an acute ischemic stroke due to large artery intracranial occlusions. First, using a continuous X-ray or fluoroscopy, the physician can guide a catheter through a small incision in either the wrist or the abdomen, giving access to the artery.[27] Next, a net-like device called a stent retriever is inserted into the catheter. The stent retriever is then pushed through the clot to capture it, allowing the physician to pull the catheter out, removing the clot.

Interventionalists have preferred routine intubation because it reduces pain and discomfort for the patient while improving the ease of performing the procedure, minimizing the potential for iatrogenic vessel dissection.[30] Many centers have adopted a policy of intubating or sedating every patient prior to mechanical thrombectomy; however, some studies have suggested that the use of general anesthesia and intubation results in the increased door to perfusion time (138 vs. 100 min), which was associated with a higher probability of in-hospital mortality.[31] Interestingly, COVID-19 infection was not associated with statistically significant in-hospital mortality.[31]

11. Clinical characteristics of patients with COVID-19 and stroke

Symptoms of COVID-19 usually appear between 2 and 14 days after exposure to the virus. The most common presenting symptoms include cough (79.4%), fever (77.1%), dyspnea (56.5%), myalgias (23.8%), diarrhea (23.7%), and nausea and vomiting (19.1%).[32] In addition, many patients develop lymphopenia (90%), with some developing elevated liver function values and inflammatory markers. Different strains present differing clinical presentations.[32]

Acute ischemic stroke can manifest clinically in different ways depending on the etiology of the stroke. Thrombotic stroke can develop over hours to days as large artery ischemic may evolve over longer periods of time. Embolic stroke presents with sudden onset of deficits.[4] Stroke symptoms largely vary with common presenting symptoms, including speech deficits, sudden numbness or weakness in the face, arm, or leg (usually on one side of the body), confusion, lethargy, headache, and coma.[33]

12. Major challenges of managing stroke during COVID-19 situation

A major challenge in managing stroke during the pandemic is maintaining high-quality care for stroke patients while protecting healthcare team members and staff. During the initial outbreak of COVID-19, decreased hospital admission volume was noted because patients feared infection when in the hospital. Studies found the number of acute stroke admissions reduced by between 50% and 80% during the early phases of lockdown.[34] Since then, there have been substantial adaptions of current stroke protocols in the ER to help balance COVID-19 infection control and stroke management.

Timely reperfusion is an important goal in the treatment of acute stroke.[35] Early initiation of reperfusion therapies is dependent on optimal prehospital and in-hospital chains that have been altered during the COVID-19 pandemic. So-called "protective pathways" are utilized in patients with suspected COVID-19 infections.[36] Such pathways require each patient to be assessed in screening areas before entering the ER. In these areas, health personnel must wear appropriate PPE, and the patient should always wear a surgical mask. A complete infection screen, including exposure history and travel history, should be obtained, and a nasopharyngeal swab test is recommended.[36]

Studies assessing changes in acute stroke management during the pandemic have revealed conflicting results. One cohort study in Europe in 2021 showed no change in the quality of acute stroke care service between 2020 and 2019.[37] While another study looking at 17 stroke centers in many countries, including the United States, found an estimated 54-minute increase in symptom onset to needle time in COVID-19 patients suffering from stroke.[38] The continuous rise in COVID-19 cases with associated demand for medical attention has caused a significant burden on health care systems, with increased health care utilization beyond hospital capacity.[39]

13. Conclusion and future directions

In conclusion, the acute inflammatory and innate immune response to COVID-19 infections can induce a prothrombic and hypercoagulable state, leading to acute ischemic stroke in certain patient populations. The interplay between cerebrovascular events and COVID-19 is critical in guiding the management of this specific patient population. Treatments that target these mechanisms may mitigate the adverse macrovascular and microvascular effects of COVID-19. It seems likely that anticoagulation will continue to play a substantial role in the prophylactic management of acute stroke in COVID-19 patients.

Further studies should aim to explore which patients benefit most from the use of anticoagulation as well as the efficacy of antiplatelet and anti-inflammatory therapies in AIS prophylaxis and treatment.

References

1. Vijayan M, Chinniah R, Ravi PM, Sivanadham R, Joseph AKM, Vellaiappan NA, Krishnan JI, Karuppiah B. MTHFR (C677T) CT genotype and CT-apoE3/3 genotypic combination predisposes the risk of ischemic stroke. *Gene*. 2016;591(2):465–470. https://doi.org/10.1016/j.gene.2016.06.062.
2. Vijayan M, Kumar S, Bhatti JS, Reddy PH. Molecular links and biomarkers of stroke, vascular dementia, and Alzheimer's disease. *Mole Biol Aging*. 2017;146:95–126. https://doi.org/10.1016/bs.pmbts.2016.12.014.
3. Vijayan M, Kumar S, Yin XL, Zafer D, Chanana V, Cengiz P, Reddy PH. Identification of novel circulatory microRNA signatures linked to patients with ischemic stroke. *Human Mole Genet*. 2018;27(13):2318–2329. https://doi.org/10.1093/hmg/ddy136.
4. Sattar HA. *Fundamentals of Pathology: Medical Course and Step 1 Review*. Chicago, IL: Pathoma.com; 2021.
5. Murali V, Rathika C, Ramgopal S, Padma Malini R, Arun Kumar MJ, Neethi Arasu V, Jeyaram Illiayaraja K, Balakrishnan K. Susceptible and protective associations of HLA DRB1*/DQB1* alleles and haplotypes with ischaemic stroke. *Int J Immunogenet*. 2016; 43(3):159–165. https://doi.org/10.1111/iji.12266.
6. Vijayan M, Alamri FF, Al Shoyaib A, Karamyan VT, Reddy PH. Novel miRNA PC-5P-12969 in ischemic stroke. *Mol Neurobiol*. 2019;56(10):6976–6985. https://doi.org/10.1007/s12035-019-1562-x.
7. Vijayan M, Chinniah R, Ravi PM, Mosses Joseph AK, Vellaiappan NA, Krishnan JI, Karuppiah B. ACE-II genotype and I allele predicts ischemic stroke among males in south India. *Meta Gene*. 2014;2:661–669. https://doi.org/10.1016/j.mgene.2014.09.003.
8. Vijayan M, Reddy PH. Peripheral biomarkers of stroke: Focus on circulatory microRNAs. *Biochim Et Biophys Acta-Mole Bas Dis*. 2016;1862(10):1984–1993. https://doi.org/10.1016/j.bbadis.2016.08.003.
9. Vijayan M, Reddy PH. Stroke, vascular dementia, and alzheimer's disease: molecular links. *J Alzheimers Dis*. 2016;54(2):427–443. https://doi.org/10.3233/JAD-160527.
10. Vijayan M, Reddy PH. Non-coding RNAs based molecular links in type 2 diabetes, ischemic stroke, and vascular dementia. *J Alzheimers Dis*. 2020;75(2):353–383. https://doi.org/10.3233/JAD-200070.

11. McIntosh K, Peiris JSM. Coronaviruses. In: Richman DD, Whitley RJ, Hayden FG, eds. *Clinical Virology*. 3rd ed. Washington, DC: ASM Press; 2009:1155. https://www.uptodate.com/contents/coronaviruses/abstract/6. Accessed October 3, 2021.
12. Petrosillo N, Viceconte G, Ergonul O, Ippolito G, Petersen E. COVID-19, SARS and MERS: are they closely related? *Clin Microbiol Infect*. 2020;26(6):729–734. https://doi.org/10.1016/j.cmi.2020.03.026.
13. Enjuanes L, Smerdou C, Castilla J, et al. Development of protection against coronavirus induced diseases. A review. *Adv Exp Med Biol*. 1995;380:197–211. https://doi.org/10.1007/978-1-4615-1899-0_34.
14. Lauer SA, Grantz KH, Bi Q, et al. The incubation period of coronavirus disease 2019 (COVID-19) from publicly reported confirmed cases: estimation and application. *Ann Intern Med*. 2020;172(9):577–582. https://doi.org/10.7326/M20-0504.
15. Mao L, Jin H, Wang M, et al. Neurologic manifestations of hospitalized patients with coronavirus disease 2019 in Wuhan, China. *JAMA Neurol*. 2020;77(6):683–690. https://doi.org/10.1001/jamaneurol.2020.1127.
16. Merkler AE, Parikh NS, Mir S, et al. Risk of ischemic stroke in patients with coronavirus disease 2019 (COVID-19) vs patients with influenza. *JAMA Neurol*. 2020;77(11):1366–1372. https://doi.org/10.1001/jamaneurol.2020.2730.
17. Tu TM, Seet CYH, Koh JS, et al. Acute ischemic stroke during the convalescent phase of asymptomatic COVID-2019 infection in men. *JAMA Network Open*. 2021;4(4):e217498. https://doi.org/10.1001/jamanetworkopen.2021.7498.
18. Loo J, Spittle DA, Newnham M. COVID-19, immunothrombosis and venous thromboembolism: biological mechanisms. *Thorax*. 2021;76(4):412–420. https://doi.org/10.1136/thoraxjnl-2020-216243.
19. Shah S, Shah K, Patel SB, et al. Elevated d-dimer levels are associated with increased risk of mortality in coronavirus disease 2019, 10.1097/CRD.0000000000000330 *Cardiol Rev*. 2020. https://doi.org/10.1097/CRD.0000000000000330. Published online August 18.
20. Borish LC, Steinke JW. 2. Cytokines and chemokines. *J Allergy Clin Immunol*. 2003;111(2 Suppl):S460–475. https://doi.org/10.1067/mai.2003.108.
21. Hadi HAR, Carr CS, Al Suwaidi J. Endothelial dysfunction: cardiovascular risk factors, therapy, and outcome. *Vasc Health Risk Manag*. 2005;1(3):183–198.
22. Ackermann M, Verleden SE, Kuehnel M, et al. Pulmonary vascular endothelialitis, thrombosis, and angiogenesis in Covid-19. *New England J Med*. 2020;383(2):120–128. https://doi.org/10.1056/NEJMoa2015432.
23. Gale AJ. Continuing education course #2: current understanding of hemostasis. *Tox Path*. 2011;39(1):273–280. https://doi.org/10.1177/0192623310389474.
24. Nutescu EA, Burnett A, Fanikos J, Spinler S, Wittkowsky A. Pharmacology of anticoagulants used in the treatment of venous thromboembolism. *J Thromb Thrombolysis*. 2016;41(1):15–31. https://doi.org/10.1007/s11239-015-1314-3.
25. Al-Samkari H, Gupta S, Leaf RK, et al. Thrombosis, bleeding, and the observational effect of early therapeutic anticoagulation on survival in critically ill patients with COVID-19. *Ann Intern Med*. 2021;174:622.
26. Landmesser U. Effect of anticoagulation therapy on clinical outcomes in moderate to severe coronavirus disease 2019 (COVID-19). *Clinicaltrials*; 2021. https://clinicaltrials.gov/ct2/show/NCT04416048. Accessed October 3, 2021.
27. Oliveira-Filho J, Martins SCO, Pontes-Neto OM, et al. Guidelines for acute ischemic stroke treatment: part I. *Arq Neuropsiquiatr*. 2012;70(8):621–629. https://doi.org/10.1590/s0004-282x2012000800012.

28. Qureshi AI, Abd-Allah F, Al-Senani F, et al. Management of acute ischemic stroke in patients with COVID-19 infection: Report of an international panel. *Int J Stroke*. 2020; 15(5):540–554. https://doi.org/10.1177/1747493020923234.
29. Smith WS, Sung G, Saver J, et al. Mechanical thrombectomy for acute ischemic stroke: final results of the Multi MERCI trial. *Stroke*. 2008;39(4):1205–1212. https://doi.org/10.1161/STROKEAHA.107.497115.
30. Takahashi C, Liang CW, Liebeskind DS, Hinman JD. To tube or not to tube? the role of intubation during stroke thrombectomy. *Front Neurol*. 2014;5:170. https://doi.org/10.3389/fneur.2014.00170.
31. Kasab SA, Almallouhi E, Alawieh A, et al. International experience of mechanical thrombectomy during the COVID-19 pandemic: insights from STAR and ENRG. *J NeuroIntervent Surg*. 2020;12(11):1039–1044. https://doi.org/10.1136/neurintsurg-2020-016671.
32. Goyal P, Choi JJ, Pinheiro LC, et al. Clinical characteristics of Covid-19 in New York city. *New England J Med*. 2020;382(24):2372–2374. https://doi.org/10.1056/NEJMc2010419.
33. Yamakawa M, Kuno T, Mikami T, Takagi H, Gronseth G. Clinical characteristics of stroke with COVID-19: a systematic review and meta-analysis. *J Stroke Cerebrovasc Dis*. 2020; 29(12):105288. https://doi.org/10.1016/j.jstrokecerebrovasdis.2020.105288.
34. Markus HS, Brainin M. COVID-19 and stroke-a global world stroke organization perspective. *Int J Stroke*. 2020;15(4):361–364.
35. Saver JL, Goyal M, van der Lugt A, et al. Time to treatment with endovascular thrombectomy and outcomes from ischemic stroke: a meta-analysis. *JAMA*. 2016;316(12): 1279–1288. https://doi.org/10.1001/jama.2016.13647.
36. Venketasubramanian N, Anderson C, Ay H, et al. Stroke care during the COVID-19 pandemic: international expert panel review. *CED*. 2021;50(3):245–261. https://doi.org/10.1159/000514155.
37. Altersberger VL, Stolze LJ, Heldner MR, et al. Maintenance of acute stroke care service during the covid-19 pandemic lockdown. *Stroke*. 2021;52(5):1693–1701. https://doi.org/10.1161/STROKEAHA.120.032176.
38. Hajdu SD, Marto JP, Saliou G. Response by Hajdu et al to letter regarding article, "acute stroke management during the COVID-19 pandemic: does confinement impact eligibility for endovascular therapy?". *Stroke*. 2020;51(11):e340–e341. https://doi.org/10.1161/STROKEAHA.120.032096.
39. Dafer RM, Osteraas ND, Biller J. Acute stroke care in the coronavirus disease 2019 pandemic. *J Stroke Cerebrovasc Dis*. 2020;29(7):104881. https://doi.org/10.1016/j.jstrokecerebrovasdis.2020.104881.

The psychiatric effects of COVID-19 in the elderly

10

Ashish Sarangi[1] *and Subodh Kumar*[2]
[1]University of Missouri, Columbia, MO, United States; [2]Center of Emphasis in Neuroscience, Department of Molecular and Translational Medicine, Paul L. Foster School of Medicine, Texas Tech University Health Sciences Center El Paso, El Paso, TX, United States

Abstract

The coronavirus (COVID-19) pandemic has exposed and highlighted pre-existing psychiatric illness in the elderly as well as predisposed them to new and emerging psychiatric pathology. The impact of this devastating illness has been felt in each setting including nursing homes and prisons and has been a barrier toward healthy aging. Despite the many challenges faced by our elderly, resilience and wisdom have served as protective factors in our fight against the pandemic. This chapter highlights the psychiatric effects of the illness and ways to manage the burden associated with psychopathology.

Keywords: COVID-19; Geriatric mental health; Neurocognitive disorder; Pandemic; Psychiatric illness; Psychopharmacology

1. Introduction

The geriatric age group is a vulnerable population with a greater risk of cognitive decline, chronic medical illness, and falls. From a mental health standpoint, this population is exposed to the uncertainty that comes with end of life and post demise. The elderly is not immune to psychiatric conditions. In fact, management of major depressive disorder and anxiety disorders can often be complicated by treatment resistance. Given how susceptible this group is, it is of no surprise that the current coronavirus disease 2019 (COVID-19) pandemic has wreaked havoc in the mental well-being of this population.

The elderly is the most affected by COVID-19 infection in terms of mortality, reaching 14.8% for individuals over 80 versus 0.2% for individuals under 40 years.

They are prone to COVID-19 infection more than the general population. They are also less able to follow the CDC-recommended preventive guidelines like wearing masks, hand hygiene, or social distancing/isolation because of cognitive issues. Loneliness and psychiatric disorders are more prevalent in this age group further enhanced because of the COVID-19 pandemic. Moreover, they are less accustomed to the use of mobile phones or telepsychiatry services, and with restrictions to public transport due to COVID-19, it has become challenging for this population to seek professional mental health care. There is an imminent need to create more awareness among not

COVID-19 in Alzheimer's Disease and Dementia. https://doi.org/10.1016/B978-0-443-15256-6.00015-5

only psychiatrists, psychologists, but also public health officials and infectious disease physicians on post-COVID posttraumatic stress disorder (PTSD) in the elderly population. It is essential to effectively identify these vulnerable individuals at greater risk of developing PTSD due to the current pandemic and the associated quarantine. We need to build more and better psychological support systems, in-person or virtual, crisis interventions that can help reduce anxiety and posttraumatic stress in this population.

2. Elderly isolation during COVID-19

The elderly face a significant risk of developing severe illness from COVID-19 due to the physiologic changes associated with advanced age as well as underlying multimorbidity in this age group. Frailty and chronic medical conditions make the nursing home elderly one of the most vulnerable populations in society. Older adults with mental illness are also at higher risk for psychiatric sequelae such as increased anxiety, depression, and stress. Older adults with pre-existing mental illness or cognitive impairment are particularly susceptible to behavioral and mental health consequences stemming from the heightened isolation and uncertainty about the state of the outside world and the health of their family and friends.

Self-isolation will disproportionately affect elderly individuals whose only social contact is out of the home such as at daycare venues, community centers, and places of worship. In many nursing homes, caregivers are unable to visit their loved ones for fear of spreading this disease among at-risk adults. Furthermore, restrictions on group activities such as playing board games, watching TV in a common space, and engaging in art therapy, may harm residents' mental and physical well-being.

Quarantine and isolation, while recognized as highly effective tools in the control of infectious diseases, have been difficult to implement effectively in nursing homes. Dementia, present in many nursing home residents, has complicated the effective handling of COVID-19 in nursing homes. Dementia may impair individuals' ability to understand and acknowledge the necessity of isolation and voluntarily comply with isolation procedures. People living with dementia, who are inexperienced in telecommunication and depend primarily on in-person support, might feel lonely and abandoned and become withdrawn.[1]

It is the author's opinion that more resources need to be allocated to geriatric nursing homes to adequately address the mental health of not only patients but also their caregivers. Some of the strategies that may be beneficial include frequent testing for COVID-19, aggressive vaccination protocols, and utilization of exercise and music programs. Exercise programs can include a mix of aerobic and strength training as tolerated. These programs have already been implemented to some extent and have shown early success.

3. Elderly health care during COVID-19

Nursing home staff are also under immense strain. They have experienced ethical and moral distress associated with balancing the potential harms associated with the isolation of residents versus the severe consequences of inadequate infection control measures.

Zonghua Liu et al. observed that under the stress of fear of infection coupled with worries about the residents' condition, the level of anxiety among staff in nursing homes increased, causing them to develop signs of exhaustion and burnout after a month of full lockdown of the facilities.[2] Ultimately, COVID-19 has taught painful lessons associated with the difficulties of isolating the elderly. The need for controlling infection may come at a high cost to the mental health and well-being of patients and staff, which is a critical public health issue.

Regular health maintenance visits once provided an opportunity for these adults to leave the facility and enjoy some time outdoors. As physical visits to the doctor virtually ceased, residents lost out on this valuable opportunity that many found therapeutic. These visits have been replaced with Telemedicine. Telemedicine consultations and other technological tools can aid geriatric patients to dispel some of the quarantine-induced loneliness. Using technology for doctor's visits reduces the risk of infections that could occur with physical visits and is proving to be useful in facilitating the continued care of the elderly.[1] Nursing homes have also utilized tools, such as Facetime, to allow residents to communicate with their families, providing a much-needed break from isolation and a regained sense of connectedness with the outside world.

4. COVID-19-associated psychiatric disorders

4.1 An elderly focus

The pandemic has brought forth pre-existing physical and psychological issues that have increased depression and suicide. Before the COVID-19 pandemic, about 20% of individuals aged 55 years and older experienced either psychological or drug abuse concerns. While no statistics are yet available to estimate the number of suicides among the elderly due to COVID-19, adults 75 years and older have some of the highest suicide rates, and although they comprise only 12% of the U.S. population, older adults account for 18% of all suicides. Strict lockdowns, physical and social distancing, and a sense of being disconnected are factors that could increase rates of depression and anxiety in older adults. Resource rationing in some healthcare facilities has further compounded the sense of feeling like a burden on the healthcare system in the elderly population.

During the COVID-19 pandemic, it has been well publicized that the severity of COVID-19 often depends on chronic health conditions with mortality rates increasing with increasing ages (odds ratio: 1.1, 95% confidence interval: 1.03−1.17, $P = 0.0043$).[3] Thus, for older adults with pre-existing conditions, fear of contracting COVID-19 and likely requiring a higher level of care potentially increases concerns over being a burden (Table 10.1).

4.2 COVID-19-associated delirium

COVID-19 delirium can present with symptoms and signs of confusion, fluctuating mental status, and agitation. This is postulated to be secondary to neuroinflammation.

Table 10.1 summarizes the various psychiatric syndromes that the geriatric population is prone to during the pandemic and associated symptoms. It is important to recognize that patients can presents with any one of these or multiple disorders simultaneously.

Psychiatric disorder	Symptoms	References
Dementia	Altered mental status, agitation, disorientation, refusal of care, disorientation, and loss of appetite, orientation/attention disturbances	Bianchetti et al.[4]; Isaia et al.[5]; Ward et al.[6]
Delirium	Rigidity, alogia, abulia, and elevated inflammatory markers	Beach et al.[7]
Sleep disorders	Insomnia and parasomnias	Ligouri et al.[8]; Nalleballe et al.[9]; Romero-Sanchez et al.[10]
Anxiety	Increased heart rate and insomnia	Ligouri et al.[8]; Nalleballe et al.[9]; Zhang et al.[11]; Romero-Sanchez et al.[10]
Depression	Social isolation, loneliness, insomnia, decreased appetite, and anhedonia	Ligouri et al.[8]; Nalleballe et al.[9]; Zhang et al.[11]; Romero-Sanchez et al.[10]
Post-traumatic stress disorder	Having recurrent intrusive thoughts	Cai et al.[12]
Psychosis	Delusions, auditory hallucinations, and visual hallucinations	Parra et al.[13]; Varatharaj et al.[14]; Romero-Sanchez et al.[10]
Confusion		Helms et al.[15]; Ligouri et al.[8]; Nalleballe et al.[9]
Suicidal ideation	Suicidal thoughts and self-harm	Nallaballe et al.[9]
Mood disorder		Nallaballe et al.[9]
Grief disorder	Persistent and pervasive yearning, longing or preoccupying thoughts and memories of the deceased, grief-related emotional pain, and causing significant distress or impairment in functioning	Goveas and Shear[16]

Multiple studies suggest that COVID-19 may indirectly affect the central nervous system through the associated inflammatory immune response and medical interventions that are administered. Immunologic findings in patients with COVID-19 include elevated serum C-reactive protein and proinflammatory cytokines (e.g., interleukin-6) and decreased total blood lymphocyte counts.[7,17]

Delirium in the setting of COVID-19 in the ICU setting for patient who are mechanically ventilated is also associated with a high rate of mortality.[18]

Delirium in patients with a baseline history of dementia can be especially debilitating and may worsen pre-existing neurocognitive deficits as well as trigger institutionalization. Pre-COVID measure taken by healthcare staff to manage delirium in the hospital setting has been difficult. Some of these measures include frequent contact with known caregivers and exercises to get the patients moving. These have been difficult to perform as many healthcare facilities have had limited personal protective equipment and were trying to limit COVID exposure and disease transmission.

4.3 Neurocognitive disorders

Cognition and memory can be chronically affected by COVID-19. Miskowiak found an association between cognitive impairments and D-dimer levels during the acute phase of illness supporting previous studies that use of heparin and tissue plasminogen activator (tPA) may improve outcomes.[19] Cognitive recovery varies, with some studies showing significant recovery in majority of COVID-19 patients at 1 month with Rass and colleagues found that 23% of studied patients had cognitive defects (measured by Montreal Cognitive Assessment) at 3-month follow-up.[20] This was increased to 50% among those who had encephalopathy during COVID-19 illness. McLoughlin et al. found that, among previously hospitalized adult patients with COVID-19, cognition at 4-week follow-up (measured using the modified Telephone Instrument for Cognitive Status) was similar to those without delirium. However, physical function (as measured by combined Barthel Index and Nottingham Extended Activities of Living scores) was markedly worse in those with delirium (97 versus 153, $P < .01$).[21] Furthermore, higher rates of dementia were observed among COVID-19 survivors compared to those with influenza (HR 2.33; 95% CI 1.77−3.07; $P < .0001$) or other respiratory infections (HR 1.71; 95% CI 1.50−1.95). Interestingly, van der Borst and colleagues, in a study of 124 COVID-19 patients (46 with severe or critical disease), found that mental or cognitive function was not correlated with disease severity via the Clinical Frailty Scale.[22] Conversely, a study of 120 participants with mild to moderate COVID-19 observed that 98.3% (118/120) had normal cognitive functioning at 4-month follow-up. However, this group had a relatively low rate of comorbidities (15% hypertension, 8.3% obesity, 3.3% diabetes) and rarely received supplementary oxygen therapy (1.6%).[23] Finally, inpatient rehabilitation was associated with significant improvement in memory and cognitive function in a cohort ($n = 29$) of hospitalized COVID-19 patients (mean hospital and intubation length 32.2 and 18.7 days, respectively), with 90% of patients discharged home after a mean of 16.7 of inpatient rehab, suggesting the importance of such care in recovery from COVID-19.[24]

There is also concern that COVID-19 may cause or worsen neurological diseases, such as Alzheimer's. For example, apolipoprotein E (ApoE) e4e4 allele, highly associated with Alzheimer's, may increase the risk of severe COVID-19. Risk for Alzheimer's might be indirectly increased due to respiratory dysfunction, as tau hyperphosphorylation is increased by hypoxia. Alzheimer's patients could be more vulnerable to COVID-19, given increased ACE expression in brains of Alzheimer's patients has been observed; tau hyperphosphorylation increased due to hypoxia.[22]

4.4 Major depressive disorder

Significant psychosocial stressors during the pandemic including loss of a stable income, social isolation, loss of close loved ones, increased workloads, fear of infecting family members, as well as personal decline in physical health have worsened pre-existing depression and predisposed individuals to new onset depressive pathology. It is important to not attribute fatigue, insomnia, and decrease in appetite to depression as these symptoms can occur in COVID-19-related illness. It is prudent to focus on depressed mood and anhedonia as key indicators of a co-occurring depressive disorder in individuals with COVID-19 illness. A comorbid major depressive disorder may prolong recovery from illness and limit a patient's desire to seek postdischarge follow up care from the hospital. For diagnostic purposes, it is critical to tease out the chronological history of development of depressive symptoms to establish if a patient had a history of pre-existing depression versus developing new onset disease due to COVID-19.

4.5 Obsessive compulsive disorder

The COVID-19 pandemic has had a mixed impact on the population as far as obsessive-compulsive disorder (OCD) is concerned. While many patients with pre-existing disease have fared well during the pandemic as they do not have to venture out because of social isolation measures, many have developed new onset symptoms. "Germophobia" has been normalized by society during the pandemic with many conducting frequent hand-sanitizing and hand-washing measures. It is still important to remember that these measures themselves are not indicative of the disorder, but the proportion of time spent doing these activities and if they are disruptive to daily functioning may be a better measure of psychopathology.

4.6 Psychosis

Patients infected with COVID-19 may display paranoid and persecutory delusions along with auditory and visual hallucinations, which can mimic a primary psychiatric illness such as schizophrenia.[13] This can also occur in the context of a delirium. Although this can be scary both for the patient and family members, it is important to rule out pre-existing mental health pathology and determine if the symptoms are secondary to side effects from an iatrogenic medication or an illicit substance the patient may have consumed.

4.7 Posttraumatic stress disorder

Posttraumatic stress disorder deserves a special mention as it is often underrecognized. Experiencing near death secondary to COVID-19 infection can result in significant trauma, which may further manifest as debilitating nightmares, flashbacks of medical life-saving procedures, as well as avoidance of healthcare facilities.

5. Management of psychiatric disorders related to COVID-19

After stabilization of the patient from COVID-19-related life-threatening complications, a thorough psychiatric history and mental status exam is critical to identify harboring mental health illness. Monotonous speech, apathy, isolation, decreased appetite, and hopelessness are all suggestive of a depressive disorder. A thorough and complete suicide risk assessment needs to be conducted to ensure safety. Patients who have been mechanically ventilated may develop debilitating nightmares and flashbacks after being readjusted to room air. Hallucinations and paranoia can indicate a resulting psychotic illness secondary to COVID-19 encephalopathy. Fluctuating consciousness, disorientation, and agitation can indicate delirium, which is life-threatening.

A thorough basic laboratory panel and urine drug screen are critical to rule out secondary causes of psychiatric symptoms. A lumbar puncture for CSF analysis and brain imaging should be done as indicated especially in patients with suspected delirium and prominent psychotic symptoms. Immunologic findings in patients with COVID-19 include elevated serum C-reactive protein and proinflammatory cytokines (e.g., interleukin-6) and decreased total blood lymphocyte counts.

6. Pharmacological agents

Low-dose selective serotonin receptor inhibitors (SSRIs) such as *Citalopram* or *Sertraline* may be useful to address depressive symptoms, agitation, and features of PSTD in patient with current or prior COVID-19 infection. Care should be taken to address common side effects such as nausea and vomiting as such patients may already be struggling with same from the infection or may be on other medications, which can contribute to gastrointestinal side effects.

Low-dose antipsychotic agents such as *Haloperidol* may be needed in patients with severe agitation during delirium although antipsychotics should generally be avoided in the elderly due to the risk of sudden death secondary to cardiac arrhythmias. Antipsychotics also have the risk of lowering the seizure threshold, which can be dangerous in this population already at risk of seizures secondary to brain inflammation.

Benzodiazepines such as *Lorazepam* and *Diazepam* should be avoided as these agents can result in respiratory depression, which can add to COVID-19-induced hypoxia. Benzodiazepines can also result in paradoxical agitation in patients, which can result in self-harm or harm to others. Low-dose melatonin may be beneficial in regulating the sleep—wake cycle in patients who are delirious secondary to COVID-19.

There is minimal evidence currently available to support the use of acetylcholinesterase inhibitors who demonstrate cognitive deficits secondary to COVID-19; however, further research is ongoing in this area.

After discharge from the hospital, a biopsychosocial and holistic approach should be employed in caring for the elderly with COVID-19-related psychiatric disorders. Stress reduction techniques, individual psychotherapy employing techniques of cognitive behavioral therapy, and trauma-focused therapies can be useful to help mitigate long-term psychiatric sequelae. Mindfulness, yoga, and physical exercise can be helpful in regulating mood and well-being.

7. Conclusion

Despite many challenges the geriatric population faced during the ongoing pandemic, the goal for healthy aging can still be achieved through utilization of available cognitive reserve and resources from the community. This includes both biological and nonbiological treatments in the management of COVID-19-associated psychiatric illness. Various stakeholders need to join hands and develop screening tools and policies that address the long-term mental health outcomes of the pandemic.

References

1. Su Z, Meyer K, Li Y, et al. Technology-based interventions for nursing home residents: implications for nursing home practice amid and beyond the influence of COVID-19: a systematic review protocol. *Res Sq*. 2020. https://doi.org/10.21203/rs.3.rs-56102/v2. Published online December.
2. [The epidemiological characteristics of an outbreak of 2019 novel coronavirus diseases (COVID-19) in China]. *Zhonghua Liuxingbingxue Zazhi*. 2020;41(2):145–151. https://doi.org/10.3760/cma.j.issn.0254-6450.2020.02.003.
3. Moreno-Pérez O, Merino E, Leon-Ramirez J-M, et al. Post-acute COVID-19 syndrome. Incidence and risk factors: a Mediterranean cohort study. *J Infect*. 2021;82(3):378–383. https://doi.org/10.1016/j.jinf.2021.01.004.
4. Bianchetti A, Rozzini R, Guerini F, et al. Clinical presentation of COVID19 in dementia patients. *J Nutr Health Aging*. 2020;24(6):560–562. https://doi.org/10.1007/s12603-020-1389-1.
5. Isaia G, Marinello R, Tibaldi V, Tamone C, Bo M. Atypical presentation of covid-19 in an older adult with severe Alzheimer disease. *Am J Geriatr Psychiatry*. 2020;28(7):790–791. https://doi.org/10.1016/j.jagp.2020.04.018.
6. Ward CF, Figiel GS, McDonald WM. Altered mental status as a novel initial clinical presentation for COVID-19 infection in the elderly. *Am J Geriatr Psychiatry*. 2020;28(8):808–811. https://doi.org/10.1016/j.jagp.2020.05.013.
7. Beach SR, Praschan NC, Hogan C, et al. Delirium in COVID-19: a case series and exploration of potential mechanisms for central nervous system involvement. *Gen Hosp Psychiatr*. 2020;65:47–53. https://doi.org/10.1016/j.genhosppsych.2020.05.008.
8. Liguori C, Pierantozzi M, Spanetta M, et al. Subjective neurological symptoms frequently occur in patients with SARS-CoV2 infection. *Brain Behav Immun*. 2020;88:11–16. https://doi.org/10.1016/j.bbi.2020.05.037.
9. Nalleballe K, Reddy Onteddu S, Sharma R, et al. Spectrum of neuropsychiatric manifestations in COVID-19. *Brain Behav Immun*. 2020;88:71–74. https://doi.org/10.1016/j.bbi.2020.06.020.

10. Romero-Sánchez CM, Díaz-Maroto I, Fernández-Díaz E, et al. Neurologic manifestations in hospitalized patients with COVID-19: the ALBACOVID registry. *Neurology*. 2020; 95(8):e1060−e1070. https://doi.org/10.1212/WNL.0000000000009937.
11. Zhang J, Lu H, Zeng H, et al. The differential psychological distress of populations affected by the COVID-19 pandemic. *Brain Behav Immun*. 2020;87:49−50. https://doi.org/10.1016/j.bbi.2020.04.031.
12. Cai X, Hu X, Ekumi IO, et al. Psychological distress and its correlates among COVID-19 survivors during early convalescence across age groups. *Am J Geriatr Psychiatry*. 2020; 28(10):1030−1039. https://doi.org/10.1016/j.jagp.2020.07.003.
13. Parra A, Juanes A, Losada CP, et al. Psychotic symptoms in COVID-19 patients. A retrospective descriptive study. *Psychiatr Res*. 2020;291:113254. https://doi.org/10.1016/j.psychres.2020.113254.
14. Varatharaj A, Thomas N, Ellul MA, et al. Neurological and neuropsychiatric complications of COVID-19 in 153 patients: a UK-wide surveillance study. *Lancet Psychiatr*. 2020;7(10): 875−882. https://doi.org/10.1016/S2215-0366(20)30287-X.
15. Helms J, Kremer S, Merdji H, et al. Neurologic features in severe SARS-CoV-2 infection. *N Engl J Med*. 2020;382(23):2268−2270. https://doi.org/10.1056/NEJMc2008597.
16. Goveas JS, Shear MK. Grief and the COVID-19 pandemic in older adults. *Am J Geriatr Psychiatry*. 2020;28(10):1119−1125. https://doi.org/10.1016/j.jagp.2020.06.021.
17. Thakur KT, Miller EH, Glendinning MD, et al. COVID-19 neuropathology at Columbia University Irving medical center/New York Presbyterian hospital. *Brain*. 2021;144(9): 2696−2708. https://doi.org/10.1093/brain/awab148.
18. Patel U, Malik P, Usman MS, et al. Age-adjusted risk factors associated with mortality and mechanical ventilation utilization amongst COVID-19 hospitalizations—a systematic review and meta-analysis. SN Compr Clin Med. Published online August 2020:1−10. http://doi/10.1007/s42399-020-00476-w.
19. Miskowiak KW, Johnsen S, Sattler SM, et al. Cognitive impairments four months after COVID-19 hospital discharge: pattern, severity and association with illness variables. *Eur Neuropsychopharmacol*. 2021;46:39−48. https://doi.org/10.1016/j.euroneuro.2021.03.019.
20. Pendlebury ST, Cuthbertson FC, Welch SJ, Mehta Z, Rothwell PM. Underestimation of cognitive impairment by Mini-Mental State Examination versus the Montreal Cognitive Assessment in patients with transient ischemic attack and stroke: a population-based study. *Stroke*. June 2010;41(6):1290−1293. https://doi.org/10.1161/STROKEAHA.110.579888. Epub 2010 Apr 8. PMID: 20378863.
21. Mcloughlin BC, Miles A, Webb TE, et al. Functional and cognitive outcomes after COVID-19 delirium. *Eur Geriatr Med*. 2020;11(5):857−862. https://doi.org/10.1007/s41999-020-00353-8.
22. Peterson CJ, Sarangi A, Bangash F. Neurological sequelae of COVID-19: a review. *Egypt J Neurol Psychiatry Neurosurg*. 2021;57(1):122. https://doi.org/10.1186/s41983-021-00379-0.
23. Sanyaolu A, Okorie C, Marinkovic A, et al. Comorbidity and its impact on patients with COVID-19. *SN Compr Clin Med*. 2020;1−8. https://doi.org/10.1007/s42399-020-00363-4. Advance online publication.
24. Olezene CS, Hansen E, Steere HK, et al. Functional outcomes in the inpatient rehabilitation setting following severe COVID-19 infection. *PLoS One*. 2021;16(3):e0248824. https://doi.org/10.1371/journal.pone.0248824.

Section Two

Alzheimer's disease and dementia during COVID-19

Blood brain barrier disruption following COVID-19 infection and neurological manifestations

11

Sonam Deshwal, Neha Dhiman and Rajat Sandhir
Department of Biochemistry, Panjab University, Chandigarh, Punjab, India

Abstract

Neurological manifestations have been reported following infection with severe acute respiratory syndrome coronavirus 2 (SARS-CoV-2). The presence of SARS-CoV-2 in brains of affected individuals has been documented. However, the exact route of entry into the brain and subsequent post-infection consequences are not fully understood. Blood—brain barrier (BBB) is an interface between systemic circulation and central nervous system (CNS) that strictly regulates entry of specific molecules from blood to the brain. The functional component of BBB is neurovascular unit (NVU) and any alterations in the structure or function of BBB is detrimental to the CNS functions. Evidence suggests that SARS-CoV-2 infection disrupts BBB integrity and functions directly or indirectly. This chapter highlights the likely mechanisms involved in entry of SARS-CoV-2 into the brain. Further, the alterations in BBB have been implicated in neurological symptoms observed in SARS-CoV-2 patients. Moreover, systemic inflammation and other peripheral factors post infection also contribute to the disruption of BBB. The key protein of SARS-CoV-2, spike protein (S1) induces significant alterations in BBB properties. Entry of S1 protein into brain triggers a proinflammatory cascade that affects BBB integrity. Therefore, understanding the pathophysiological mechanisms in BBB dysfunction and subsequent neurological manifestations along with long-term effects on brain particularly Alzheimer's disease (AD) following coronavirus disease 2019 (COVID-19) is of utmost importance.

Keywords: Alzheimer's disease; Blood—brain barrier; COVID-19; Inflammation; Neurovascular unit; SARS-CoV-2

1. Introduction

Severe acute respiratory syndrome coronavirus 2 (SARS-CoV-2) mediated neurological manifestations range from mild headache, dizziness, anosmia, and ageusia to serious complications such as encephalopathy, seizures, Guillain-Barre syndrome, and acute cerebrovascular diseases, ischemic stroke, and cerebral hemorrhage.[1] Autopsy reports have shown the presence of SARS-CoV-2 RNA in the brains of coronavirus disease 19 (COVID-19) patients.[2,3] Encephalitis has been reported following

COVID-19 in Alzheimer's Disease and Dementia. https://doi.org/10.1016/B978-0-443-15256-6.00005-2

COVID-19 infection suggesting that viral protein is present in the brain.[4] Neurological effects occur either from direct viral entry into the brain to cause infection of brain cells or by indirect systemic inflammatory response. Primarily, SARS-CoV-2 invades the host via axonal route from olfactory epithelium and spreads through the circulatory system and across the blood-brain barrier (BBB). Both the pathways ultimately lead to the activation of immune system in the brain.[5]

BBB is an extremely selective barrier that isolates central nervous system (CNS) from the circulatory system. BBB consists of endothelial cells joined by tight junctions (TJs) that express angiotensin-converting enzyme-2 (ACE2) receptors.[6] In addition, the neuronal cells also express ACE2 receptors.[7] The virus primarily enters the CNS from the circulatory system by using two types of host cells; endothelial cells and leukocytes.[8] The virus enters by binding to the ACE2 receptors on the basolateral surface of the endothelial cells via transcellular route between these cells. The virus replicates in the brain parenchyma to infect neuronal cells.[9] The presence of other receptors such as neuropilin-1 (NRP-1) and basigin (BSG) has also been suggested on endothelial cells of neurovascular unit (NVU).[10] Therefore, the entry of virus via NRP-1 and BSG receptors expressed in cerebral vasculature cannot be ruled out.[11] Once inside the brain, virus multiplies and impairs neuronal functions and ultimately leads to neuron cell death. The infection of CNS endothelial cells by virus impacts BBB integrity. The leaky BBB due to TJ disruption allows the virus entry via paracellular route.[12] Further, the presence of SARS-CoV-2 mRNA in the cerebrospinal fluid (CSF) suggests that virus crosses the BBB. A more recent ultrastructural autopsy of SARS-CoV-2 infected individuals showed the presence of virus in the frontal lobe, particularly in the neurons, and spreading to the capillary endothelium.[13] These data support the view that once in circulation, SARS-CoV-2 is expected to spread via the bloodstream to the brain. Thus, BBB could be an interface for direct and indirect connection of SARS-CoV-2 with the brain. This chapter describes the various mechanisms involved in the invasion of SARS-CoV-2 into the brain. Further, the consequences of direct or indirect SARS-CoV-2 infection on BBB disruption are highlighted. The clinical findings have been explained in context of BBB disruption by the SARS-CoV-2. Moreover, the disruption of BBB by SARS-CoV-2 and associated neurological conditions particularly Alzheimer's disease (AD) have been discussed. Lastly, the possible strategies to prevent BBB damage have also been compiled.

2. Structure and function of BBB

BBB is a selectively permeable barrier that separates blood vessels from the brain parenchyma. The primary cells that build vasculature of BBB is the brain endothelial cells (BECs) that control the passage of ions and small hydrophilic molecules in and out of the brain. These BECs are surrounded by other cell types that are involved in the formation of BBB including pericytes and astrocytes.[9] These cells provide mechanical support and maintain the integrity of the BBB. The pericytes are the closest type of cells associated with BECs. The major role of pericytes at BBB is to induce and maintain BBB phenotype and to regulate microvascular tone.[14,15] The pericytes along with BECs are responsible for regulating the infiltration of immune cells. The astrocytes are glial cells that surround the blood vessels with their long elongated cellular processes. They are crucial for maintaining BBB phenotype and provide regulation to capillary tone via cross-talk with the pericytes. The astrocytes also control the blood flow and action of vascular smooth muscle cells. The cells of innate immune response such as macrophages, microglia, and leukocytes affect the function of BBB in response to pathogenic infection.[9,16–18] Together, BECs, their supporting cells such as pericytes, parenchymal cells including neurons, interneurons, and astrocytes, and extracellular matrix substances and associated TJ proteins are known as NVU.[19] A typical structure of BBB is presented in Fig. 11.1.

The barrier property of BBB is conferred by the presence of TJ proteins on BECs. The TJ proteins comprise of claudins, occludens, and junctional adhesion proteins like zona occludens (ZO).[20] The claudins mainly control the paracellular passage of molecules at the BBB.[21] The TJs not only control the paracellular diffusion of molecules between BECs but also regulate the lateral diffusion of proteins in the membrane. Thus, TJ contribute to the polarity of the BECs.[19] Additionally, BECs express efflux transporters like P-glycoprotein and neurotransmitter-metabolizing enzymes such as catechol O-methyl transferase (COMT), monoamine oxidases (MAO), gamma-amino butyric acid (GABA) transaminase, cholinesterases, aminopeptidases, and endopeptidases. These help in inhibiting the entry of endogenous substances and xenobiotics into the brain.[22,23] The brain requires nourishment and tropic support from the circulation, and certain substances are required to be transported to the brain. In

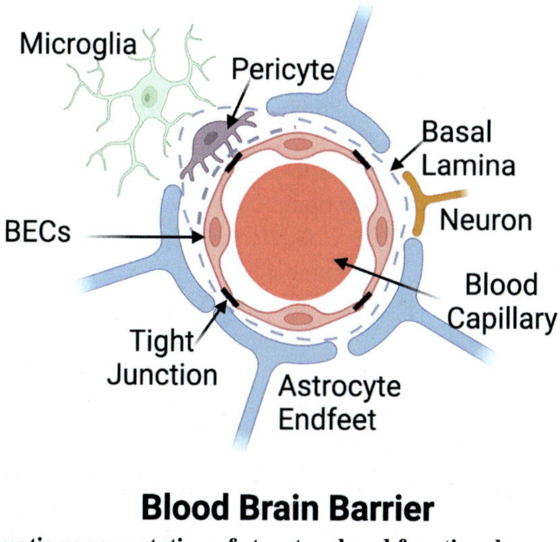

Figure 11.1 Schematic representation of structural and functional components of blood–brain barrier (BBB): The figure shows blood capillary surrounded by brain endothelial cells (BECs) and associated TJs, basal lamina along with pericytes, microglia, astrocytes (foot proc ess) that constitute the BBB.

addition, some proteins and peptides utilize special kind of transporters to cross the BBB. At BBB, different types of transport systems including energy-independent, energy-dependent, receptor-mediated transcytosis, and adsorptive endocytosis are present. In addition, BBB also acts as signaling and secretory link that allows bidirectional communication between blood and the brain.[19]

3. Mechanisms of SARS-CoV-2 entry into the brain

Till date, seven types of coronaviruses that naturally infect humans are known, out of which two are known to enter CNS.[24] In an autopsy report, coronavirus RNA was detected in the CNS of 48% of the cases.[25] The neuroinvasive potential of SARS-CoV and middle east respiratory syndrome coronavirus (MERS-CoV) have been previously investigated[26–28]. SARS-CoV-1 and MERS-CoV are known to infect immune cells and enter the brain using the hematogenous route.[24] Infected immune cells cross BBB in a tightly regulated process involving cross-talk with the BECs and subsequent diapedesis for entry.[29,30] SARS-CoV-2 being closely related to these viruses provides strong evidence in support of its neuroinvasive potential. Table 11.1 lists the seven coronaviruses and their probable mechanisms of entry into the brain.

The presence of SARS-CoV-2 in the human brain tissues or CSF is described in various studies.[38,39] Intranasal administration of SARS-CoV-2 results in entry of virus into brain.[39,40] However, these studies do not define whether the virus enters the brain via olfactory bulb or trigeminal nerve, or enters into the blood via nasal turbinates, or moved into lung and then reach brain by hematogenous route.

Moreover, other findings suggest that the olfactory epithelium was severely damaged with immune cell penetration in Syrian hamsters infected with SARS-CoV-2. However, these results showed the infection of SARS-CoV-2 in sustentacular cells but not of neurons in the olfactory bulb. Also, the presence of virus in the olfactory bulb was not detected.[41] A small amount of spike (S1) protein was detected in the brain following nasal administration at position of cribriform plate with sign of limited entry into the blood.[42] However, blood-to-brain transport of substances is more efficient than the nasal transport route to brain.

It has been established that SARS-CoV-2 entry into various tissues including CNS and brain-barrier tissues depends predominantly on the presence of ACE2 receptors.[33] Several studies also indicate that S1 may use other receptors and co-receptor glycoproteins, including cyclophilins, NRP-1, BSG, and dipeptidyl peptidase-4 (DPP4) and GRP78.[43–45] ACE2 has much greater role in viral uptake by lungs. However, limited uptake by other tissues indicates other receptors could be more critical in the uptake by the rest of the organs. The studies using S1 protein also indicate the involvement of other binding sites in S1 uptake by the brain.[42] Therefore, it is interpreted from the available data that ACE2 might be important in SARS-CoV-2 entry into the brain but may not be the sole mechanism in entry. The experimental study using S1 protein[42] suggests that vascular BBB is possibly the interface of entry into the brain. It is demonstrated that ACE2 expressed on choroid plexus epithelium can be infected by SARS-

Table 11.1 List of coronaviruses known to infect humans. SARS-CoV-1 and MERS-CoV enter the CNS. Experimental data suggest that SARS-CoV-1 and SARS-CoV-2 potentially use common mechanisms for CNS entry.

Virus	Discovered	Origin	Genus	CNS entry	Entry mechanism	Pathology	References
SARS-CoV-1	Guangdong, China, (April 2003)	Bats	Betacoronavirus	Yes	ACE2- and CD147-mediated entry, endocytosis	Mild to severe lower respiratory tract illness	31,32
SARS-CoV-2	Wuhan, China, (December 2019)	Bats	Betacoronavirus	Yes	ACE2-, TMPRSS2-, TIM1-, AXL-, CD147-mediated entry, endocytosis	Mild to severe lower respiratory tract illness	33,34,35
MERS-CoV	Saudi Arabia, (April 2012)	Bats	Betacoronavirus	Yes	DPP4-mediated entry	Mild to severe lower respiratory tract illness	34,36
HCoV-OC43	May have caused 1889–1890 pandemic; first isolated in 1967	Uncertain; may be rodents	Betacoronavirus	Uncertain	Neu5-mediated entry	Severe respiratory tract illness, pneumonia	31,37
HCoV-229E	First isolated in 1966	Bats	Alphacoronavirus	Uncertain	APN-mediated entry	Mild upper respiratory tract illness	37

Continued

Table 11.1 List of coronaviruses known to infect humans. SARS-CoV-1 and MERS-CoV enter the CNS. Experimental data suggest that SARS-CoV-1 and SARS-CoV-2 potentially use common mechanisms for CNS entry.—cont'd

Virus	Discovered	Origin	Genus	CNS entry	Entry mechanism	Pathology	References
HCoV-NL63	Netherlands, (2004)	Bats	Alphacoronavirus	Uncertain	ACE2-mediated entry	Mild to moderate upper respiratory tract illness, severe lower respiratory tract illness, croup, bronchiolitis	37
HCoV-HKU1	Hong Kong, (January 2004)	Rodents	Betacoronavirus	Uncertain	–	Mild upper respiratory tract illness, pneumonia	37

CoV-2 in vitro.[38] This indicates that virus particle could enter the brain via choroid plexus and vascular BBB.

Two major pathways have been suggested for entry of neurotropic viruses into the brain: hematogenous (blood) and retrograde neuronal routes (Fig. 11.2). The hematogenous route is considered as the main pathway for virus entry into the brain. However, receptor-mediated endocytosis is another route for entry of SARS-CoV-2 into the brain.

3.1 Hematogenous route

In the hematogenous route, virus enters via infected cells of BBB, epithelium cells of blood−CSF barrier of choroid plexus or through inflammatory cells as Trojan horse to reach the brain.[46] The virus enters the brain from blood by using two cell types; endothelial cells and leukocytes.[8] The virus enters the brain parenchyma by binding to ACE2 receptors on the basolateral membrane of the BECs via the transcellular route to infect the neuronal cells.[47] The infection of BECs may impact BBB integrity. Therefore, leaky BBB because of disrupted TJ permits the virus entry via the paracellular route.[12,47] Furthermore, SARS-CoV-2 has been suggested to infect choroid plexus

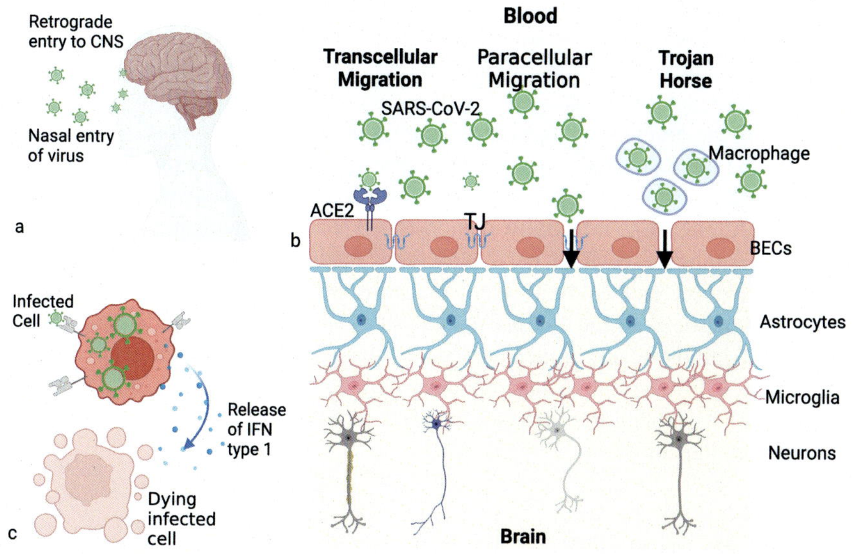

Figure 11.2 Possible routes of SARS-CoV-2 entry into the brain and immune activation. (a) SARS-CoV-2 enters central nervous system (CNS) via retrograde transport bypassing blood−brain barrier (BBB). (b) Subsequent respiratory infection of SARS-CoV-2 propagates virus in systemic circulation. SARS-CoV-2 virus penetrates BECs by interacting with ACE2 or by disrupting TJs formed by BECs or by phagocytosis that termed as transcellular migration, paracellular migration and Trojan Horse mechanism respectively. (c) In both the passages, SARS-CoV-2 infected cells release proinflammatory mediators like interferon-1 (IFN1) that alter neighboring epithelium cells and immune cells.

epithelium. The expression of ACE2 receptors in cells of choroid plexus allows virus entry into the brain and CSF.[38] These reports suggest different mechanisms by which virus enters the brain from blood.

3.2 Retrograde nerve transmission

Some of the viruses use peripheral nerves for entry into the brain and then spread throughout the CNS via process known as retrograde nerve transmission. Rabies and alpha herpes viruses follow retrograde nerve transmission to enter the brain. Another example is of gastrointestinal reovirus that uses vagal nerve to enter the brain.[48–50] In the olfactory route, virus enters brain directly from the nasal passage via the olfactory nerve and then spreads beyond the cribriform plate inside the nasal passages and into the trigeminal nerve that reaches the walls of nasal passages. The mouse hepatitis coronavirus can enter brain via olfactory and trigeminal nerves.[51] Therefore, it is suggested that the retrograde transport of viral particles in olfactory sensory neuronal axons may be an alternate entry route for SARS-CoV-2 entry.[27,52,53] This alternate hypothesis is supported by evidence of isolated anosmia symptoms in some SARS-CoV-2 infected patients.[1,54]

3.3 Receptor-mediated entry

SARS-CoV-2 uses its spike protein to interact with receptors on the host cell.[55] The most plausible entry mechanism for SARS-CoV-2 entry appears to be ACE2 receptor.[56] ACE2 receptor is expressed in almost all regions of the brain with highest expression in the brain stem that controls cardio-respiratory functions.[57] It is hypothesized that the binding affinity of ACE2 receptor for the viral spike protein S1 may be the mediating factor for its passage across the BBB.[56–58] A study using S1 protein model of SARS-CoV-2 infectivity demonstrated that on interaction with receptors on plasma membrane, SARS-CoV-2 undergoes clathrin-mediated endocytosis. This provides evidence for the involvement of endosomal system in transfer of SARS-CoV-2 RNA particles to the cytoplasm. The study also showed that deleting clathrin heavy chain inhibits clathrin-mediated endocytosis, thereby reducing infectivity.[55] Additionally, CD169-positive macrophages may facilitate viral invasion as well.[59] The ability of S1 protein to cross-murine BBB demonstrates the presence of virus in murine brain after infection.[39,40,60]

4. BBB disruption

BBB disruption is often suggested as a mechanism for the entry of virus and immune cells into the brain. BBB disruption is usually referred loosely to describe different types of BBB dysfunctions. The extent of BBB disruption differs considerably between viruses. The herpes simplex virus (HSV) being related with significantly high degree of BBB disruption results in abnormal increase in albumin levels in CSF, while human immunodeficiency virus-1 (HIV-1) results in mild disruption.[61] The BBB endothelial bed is the primary locus of attack for various neuroinvasive viruses including rabies,[62] HIV-1,[63] and Zika virus.[64] Viral pathogens have direct interactions with BECs causing infection along with inflammatory responses resulting in the loss of structural and functional integrity of BBB.[65] A disrupted BBB allows free passage of viral particles such as SARS-CoV-2 spike protein as well as infected immune cells into the brain ultimately leading to cell death.[66,67] In a study, infiltration of perivascular inflammatory cells into the brain was observed in SARS-CoV-2 infected K18-hACE2 mice.[68] Once a neurotropic virus enters brain, structural and functional deregulation occurs in BBB components (endothelial cells, pericytes, and astrocytes) stimulating innate immune response. All these cell components of BBB express various pathogen recognition receptors (PRRs) that triggers immune response in presence of virus. The proinflammatory response following infection intensifies the damage to BBB. On the other hand, type 1 IFN-mediated anti-inflammatory immune response strengthens the stability of BBB. Therefore, understanding the mechanisms involved in interaction between host and viral particles at BBB would provide novel therapeutic targets to prevent BBB disruption induced neuroinflammation. Furthermore, enhancing the protective function of BBB by inhibiting the proinflammatory response at BBB and preserving integrity of BBB will provide more opportunities to inhibit invasion by virus into the brain.[65]

4.1 Zonulin hypothesis: mechanism for SARS-CoV-2 mediated BBB disruption

Zonulin, a 47 KDa protein originally characterized by Wang in 2000.[69] Zonulin functions as endogenous regulator of intercellular TJs, is a biomarker of the intestinal barrier integrity of the small intestine. In addition to gastrointestinal (GI) tract, the upregulation of Zonulin protein has been described in lung as well as in brain tissues.[70,71] Zonulin receptors have been reported in brain tissues.[72] It has been described that Zonulin protein is also capable of reaching brain and therefore can result in increased BBB permeability.[71] Interestingly, Zonulin antagonist AT-1001 has been suggested as an anti-SARS-CoV-2 drug.[73]

Zonulin hypothesis for SARS-CoV-2 is illustrated in Fig. 11.3. According to this hypothesis, SARS-CoV-2 enters intestine via mucus from infected lungs through pulmonary mucus clearance system. The viral particle trapped in the mucus layer is carried by cilia upward to trachea via vocal chords that is removed by the GI tract. During cough, clumps of mucus are expelled directly to pharynx at high speed, these clumps of mucus get mixed with saliva and are engulfed into the esophagus.[74] Subsequently, the virus enters enterocytes via the ACE2 receptor and replicates. Although, the virus can bind and stimulate the Toll-like receptor (TLR4) that activate Zonulin protein via MyD-88 and boost proinflammatory response. Zonulin produced in lumen binds to protease-activated receptor 2 (PAR2) stimulating the destruction of TJ in the intestinal epithelial cells. At this stage, the virus uses the paracellular route to enter bloodstream and reach the brain. The virus can induce expression of the Zonulin and its receptor

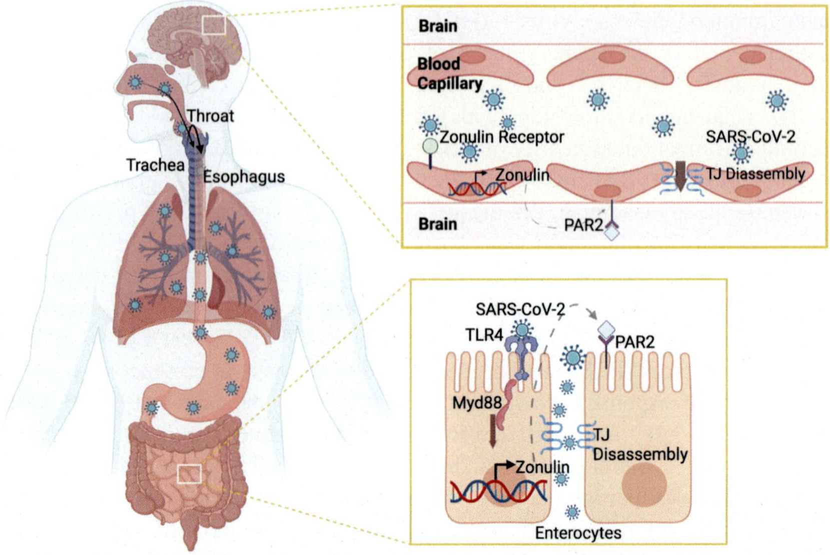

Figure 11.3 Zonulin hypothesis. SARS-CoV-2 enters intestine from mucus originating from the infected lungs via pulmonary mucus clearance system. The virus binds to TLR4 and activate Zonulin via MyD-88 that induces proinflammatory cytokine response. Activated Zonulin in lumen binds to PAR2 initiating TJ disruption (gut inset). Consequently, using paracellular route virus through systemic circulation reaches brain. Further, the virus gains access to the brain by binding to Zonulin receptor in brain and stimulate Zonulin expression. Zonulin overexpression in the brain enhances BBB permeability by TJ breakdown (brain inset).

thereby enhancing BBB permeability by disassembly of TJs. Consequently, the activated complement system by Zonulin and cytokine storm stimulated by viral infection may enhance the disruption of BBB and result in neurological manifestations observed in SARS-CoV-2 infected persons.[75]

4.2 Interaction of SARS-CoV-2 with BECs and other components of NVU

Many mechanisms have been proposed for SARS-CoV-2 entry into the BBB. One of the mechanisms is ACE2-dependent infection and replication inside BECs with subsequent release of virus in the brain. Another is ACE2-dependent delivery of SARS-CoV-2 across the BECs without replication. Further, the virus may enter via receptor-mediated transport using receptors other than ACE2. Studies have also suggested transport of virus by adsorptive transcytosis. Another suggested mode of transport is mediated by interactions with glycocalyx expressed on the surface of BECs.[56]

As evident from the studies, SARS-CoV-2 may infect brain directly, and the presence of SARS-CoV-2 has been detected in the endothelial cells of brain and in neurons.[9] It is suggested that SARS-CoV-2 virus led to BBB dysfunction by directly infecting endothelial cells or by interacting with other components of BBB. A study has shown that SARS-CoV-2 spike proteins and receptor binding domain can lead to BBB leakage in vitro of human BBB models.[66] Increased permeability after SARS-CoV-2 infection suggests the possibility of SARS-CoV-2 crossing BBB.[68]

Studies on in vitro BBB models using primary brain microvascular endothelial cells and astrocytes from hamsters or K18-hACE2 mice showed that SARS-CoV-2 replicates and infects endothelial cells. In K18-hACE2 mice, SARS-CoV-2 co-localization with ACE2 receptor in vascular endothelial cells and in vessels indicates the importance of ACE2 receptors in invasion of virus via BBB.[68] Some of the events that are triggered by SARS-CoV-2 associated endothelial dysfunctions are illustrated in Fig. 11.4. It has been shown that direct destruction caused by interaction of SARS-CoV-2 with BECs at BBB leads to leakage of non-specific serum factors into the brain. Interaction of SARS-CoV-2 proteins with endothelial cells of BBB may result in release of proteases, clotting factors, and cytokines into the brain (Fig. 11.4, **events 1–5**). Moreover, it may cause an increase in expression of cell adhesion molecules that may cause trafficking of leukocytes. However, it has been proposed that indirect effects of SARS-CoV-2 on BBB include increase in concentration of circulating proinflammatory factors and decrease in oxygen levels and clotting factors. This could

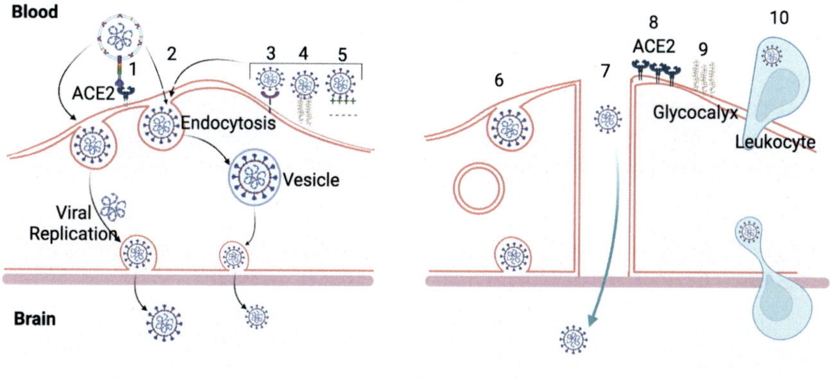

Figure 11.4 Events triggered by SARS-CoV-2 entry across BECs. Left panel (1–5) represents direct entry of virus into the brain. (1) ACE2 mediated entry with replication across BECs; (2) ACE2 mediated entry across BECs without replication; (3) Receptor -mediated entry other than ACE2; (4) Endothelial glycocalyx mediated entry; (5) Adsorptive transcytosis mediated transport. Events in the right panel (6–10) represent indirect factors such as inflammation and other pathological changes that enhance entry of SARS-CoV-2 into brain. Inflammation enhances BECs vesicular leakage. (6) induction of transendothelial channels that allows; (7) Viral leakage from blood; (8) Inflammation mediated upregulation of ACE2 and other receptors that increase viral transport; (9) Reduced or compositional change in endothelial glycocalyx results in SARS-CoV-2 interactions with brain endothelium; (10) Inflammation and other pathological conditions induced trafficking of cells through BBB induces entry of virus engulfed in immune cells by Trojan Horse mechanism.

disrupt BBB and allow virus entry through transcellular or paracellular routes. Brain endothelium results in increased release of cytokines, proteases, upregulated cell adhesion molecules, and trafficking of leukocyte to the brain (Fig. 11.4, **events 6–10**).[9] SARS-CoV-2 causes infection of vascular endothelial cells causing increased vascular permeability that leads to damage to the brain in animal models. Currently, only NVU component that has shown to be infected with SARS-CoV-2 is neurons that have been linked with neurovascular readjustment and neurodegeneration.[39]

Infection in primary human endothelial cells overexpressing ACE2 with SARS-CoV-2 induces expression of proinflammatory cytokines, clotting factors, adhesion molecules, endothelial cell lysis, and formation of multinucleate syncytia.[76] A key mechanism of action for SARS-CoV-2 pathology involves invasion through vasculature. Since ACE2 is expressed throughout the vasculature of the body,[77] it allows the virus access to multiple organs. With respect to CNS, ACE2 is also expressed on the human cerebral vasculature.[68,78] Despite the knowledge of its presence, the potential engagement of ACE2 in alterations of the BBB and its contribution to subsequent neurological complications remains to be fully understood. These studies suggest that infection of SARS-CoV-2 could cause brain endothelial dysfunctions and damage.

The diameter of virus suggests that vesicles are involved in entry across BBB. Viruses have been shown to use many kinds of vesicular systems to enter nonbarrier cells.[79] A study identified the potential of viral attachment protein (VAP), similar to S1 protein to cross-murine BBB. VAP on the virus surface and phospholipids, glycolipids, and glycoproteins on the surface of cell being attacked is involved in vesicular pathways. For instance, spike protein of SARS-CoV-2 is VAP. However, the mechanism of cell membrane binding does not involve receptor but involves adsorptive transcytosis.[80] The phenomenon of adsorptive endocytosis was initially reported for BBB as a mode for adsorbing lectin glycoproteins on luminal membrane of endothelial cells in brain, therefore, using VAPs virus can cross brain barriers.[80]

4.3 SARS-CoV-2 infection: indirect effects on BBB

Instead of directly infecting the BBB or its components, SARS-CoV-2 infection indirectly contributes to BBB dysfunctions. The conditions that result in BBB disruption include inflammation or other pathological situations such as hypoxia, thrombosis, and coagulation. These may increase endothelial vesicular system permeability, which induces the formation of transendothelial route that allows viral entry from blood into the brain. Certain disease conditions and inflammation may up-regulate ACE2 receptors that enhance the transport of virus, or alter the composition of glycocalyx on BECs, which could change the communication between endothelium and SARS-CoV-2. Ultimately, accumulation and transport of activated immune cells are enhanced through BBB in conditions involving inflammation. Further, it may allow viral entry through Trojan horse mechanism in case of infected immune cells, thereby enhancing SARS-CoV-2 uptake into the brain.[9]

4.3.1 Hyperinflammation induced increase in BBB permeability

Inflammation may occur on viral exposure or exist before the viral exposure in case of chronic conditions. As discussed in earlier section, inflammation causes various BBB dysfunctions via distinct mechanisms.[62] These alterations lead to various CNS-associated symptoms observed in viral infection such as fever, malaise, cognitive deficits, and depression-like behavior. Inflammation-mediated changes in BBB allow enhanced entry of virus into the brain. In condition of infected immune cells, diapedesis is increased while adsorptive transcytosis is increased in case of free virus entry.[81,82] As an example, HIV-1-induced inflammation induces mitogen-activated perotein kinases (MAPK) pathways thereby leading to production G-CSF and IL-6 from the luminal side of BECs that leads to transcytosis of free virus.[83] It is well established that SARS-CoV-2 infection causes inflammation at the systemic level. Systemic inflammatory response leads to increase in proinflammatory cytokines, complement proteins, acute phase proteins, chemokines, and alteration in leukocyte profiles in blood and brain.[84,85] Systemic inflammation is a characteristic feature of SARS-CoV-2 infection. An increase in plasma levels of cytokines (TNF-α, IFN-γ, IL-1β, IL-1RA, IL-7, IL-8, IL-9, IL-10, IP-10, GCSF, GM-CSF); growth factors (bFGF, PDGF, VEGF); and chemokines (CCL2, CCL3, CCL4) have been reported in SARS-CoV-2 infected individuals. The levels of IL-2, IL-7, IL-10, IP-10, TNF-α, CCL2, CCL3, and G-CSF were also higher in individuals admitted to ICU.[84] In addition, the levels of IL-6 and D-dimer were found to be higher in SARS-CoV-2 non-survivors as compared to survivors.[86] Elevated levels of cytokines in CSF have been reported in COVID-19 patients with neurological manifestations suggestive of neuroinflammation.[87] SARS-CoV-2 mediated endothelial inflammation may induce clot formation in the vasculature of peripheral tissues and CNS.[88]

BBB has the potential to react and regulate signals from immune system that emerged from the blood compartments and the brain. Therefore, conditions of BBB dysfunctions that occur in response to COVID-19 infection could revolve around interactions with immune response instead of directly interacting with virus or its components. Fig. 11.5 depicts the mechanisms of action of inflammatory response and associated processes of hypoxia and hypercoagulation (described below) involved in disrupting the BBB integrity and functioning. The neuroimmune axes events; (1) regulate BBB leakage; (2) regulation of secretion and transport across BBB by immune system; (3) secretion of immunoactive substances at BBB; (4) BBB uptake and transport of immunoactive substances; and (5) trafficking of immune cells.[89,90] These neuroimmune axes are modulated by cytokines, chemokines, and also by acute phase proteins. As an example, cytokines (IL-1β, IL-6, IFN-γ, TNF-α) and chemokines (CCL2, and CRP acute phase protein) can result in BBB disruption. The function of BBB-like transporters such as P-glycoprotein could be modulated by these factors.[89–91] Additionally, these inflammatory factors can modulate the process of adsorptive transcytosis that is one of the mechanisms used by HIV-1 to cross the BBB.[92]

Cytokines (TNF-α, IL-1α, IL-1RA, IL-1β, and IL-6) and chemokines (CCL2 and CCL11) are also carrier for transport through the intact BBB. Further, the cytokines and chemokines including TNF-α, IL-6, IL-13, G-CSF, CXCL1, CCL2, CCL3,

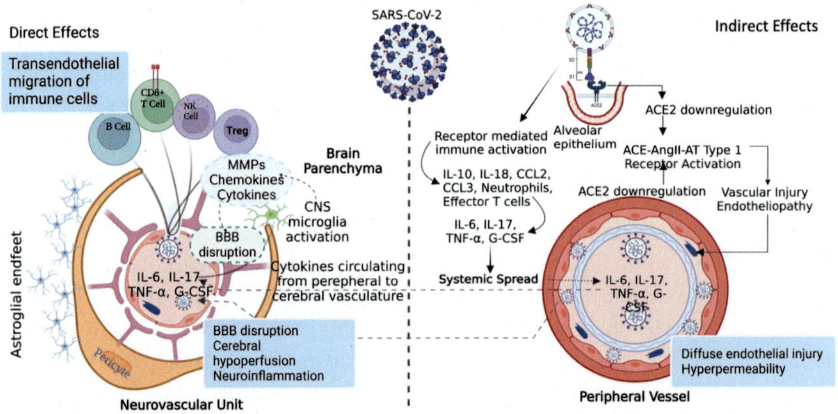

Figure 11.5 Direct and indirect mechanisms in disruption of BBB by SARS-CoV-2. The figure represents integrative clinical and experimental data linking innate immune response and systemic hyperinflammation induced by binding of S1 protein to ACE2 receptor in lungs and intestine directed toward brain vascular endothelial dysfunction, BBB disruption and subsequent brain innate immune response thereby manifesting as CNS complications. The figure also depicts the proposed action of penetrating protective immune cells from the blood into the brain via disrupted BBB and thus promoting viral clearance.

CCL4, CCL5, and CCL11 may be produced by the cells of BBB. Cytokines can freely pass-through BBB and are similar to hormones in terms of activating signaling molecules and secondary messenger. After entering the brain, they activate free calcium, thereby disrupting brain calcium homeostasis and compromising BBB integrity.[93] Leukocyte trafficking to the CSF and brain has been reported following SARS-CoV-2 infection.[94] Apparently, drugs such as IL-1 blocker anakinra and IL-6 receptor blocker tocilizumab showed significant benefits in COVID-19 patients suffering from hyperinflammation.[95] Microvascular injury in the brain of SARS-CoV-2 infected individuals have been observed. Therefore, vascular connection of BBB has been suspected based on the observations of penetrating $CD3^+$, $CD8^+$ cells, and perivascular macrophages in the perivascular spaces besides endothelial cells. Inflammatory markers like reactive gliosis have been detected in the vicinity to endothelial cells in autopsy of SARS-CoV-2 infected brain indicating that neuroinflammation may play part in brain microvascular injury.[96] The reactive gliosis and perivascular infiltrating CNS immune cells have been reported in a few SARS-CoV-2 infected patients, suggesting modulation of other non-endothelial constituents of NVU. Immune cell transport through BBB requires interaction of leukocyte and alteration of basement membrane to allow entry into the brain.[97,98] However, modulation of NVU has not been studied in condition of SARS-CoV-2 infection. Pericyte loss in lungs following SARS-CoV-2 infection is observed in patients with severe COVID-19 condition, which might appear before vasculopathy.[99] However, loss of pericytes in brain is not evaluated following COVID-19 infection.

Aberrant signaling of CNS immune pathways in response to the peripheral hyperinflammation induced by SARS-CoV-2 may disrupt BBB and increase its permeability. Effector cytokines such as IL-1β, IL-6, TNFα, and IL-17 have been shown experimentally to increase BBB permeability.[100] Proinflammatory cytokines entering the CNS activate glial cells with resultant microglial activation and proliferation (MAP), which further disrupt functional and structural integrity of BBB.[100] Studies have shown that MAP may disintegrate endothelial TJ proteins in animal models via several mechanisms like upregulation of proinflammatory cytokines and chemokines and promoting oxidative stress. The BBB breakdown creates an indefinite neuroinflammatory loop by potentially facilitating detrimental crosstalk between innate and peripheral adaptive immunity.[100,101] This crosstalk may exert adverse effects on neurotransmission causing glutamate-mediated neuronal excitotoxicity. Cerebral perfusion may reduce vasodilation and loss of endothelial dynamic autoregulatory capacity. This mechanism of BBB disruption in COVID-19 patients suffering from concurrent global encephalopathy is clinically supported by the frequent observation of cerebral hypoperfusion on perfusion imaging.[81] Upregulation of Th17 and IL-17 responses in CoV-related infections including SARS-CoV, MERS-CoV, and severe cases of SARS-Cov-2 have been recorded.[34,82,83] Th17 cell adhesion to cerebral endothelium may be supplemented by proinflammatory cytokines like TNF-α via upregulation of endothelial vascular cell adhesion molecule 1 (VCAM-1) expression. IL-17 is involved in a synergistic cross-talk with cytokines that elevates IL-6 response, exhausting the functionality of T-cells. This might be the contributing factor to lymphocytopenia observed in severe cases of COVID-19.[34,102,103]

It can be postulated that BBB is involved in neurological symptoms observed in SARS-CoV-2 infected individuals as fever is the most common symptom of COVID-19 infection that occurs in majority of cases.[104] Fever is induced by formation of prostaglandin E2 (PGE2), which is mostly synthesized by BECs in response to inflammation.[105] Upon inflammation, several symptoms induced in response to the infection may or may not have BBB involvement. Inflammation is cytokine-dependent and involves neuroimmune axis. As an example, IL-1 stimulates endothelial PGE2 to induce symptoms of malaise and discomfort through hypothalamic-pituitary adrenal (HPA) axis.[106] Blood-borne cytokines such as IL-1α cross the BBB and lead to impaired memory processing.[107] The pathways that bypass the BBB require communication with peripheral vagal nerve, which sends signals to the brain. Another route through which cytokines in blood communicate with brain is via circumventricular organs, which are the regions of brain without BBB and leaky vasculature that connect the brain by afferent and efferent pathways.[108]

5. Hypoxia

COVID-19 patients have been shown to present moderate or severe hypoxia as a clinical symptom. The asymptomatic patients show "silent hypoxemia", a condition with decreased oxygen levels, which is associated with poor survival.[109–111] Therefore, hypoxia is considered as a marker of SARS-CoV-2 severity by some research groups and clinicians. Hypoxia has subtle consequences on BBB. Various in vitro and in vivo studies suggest that decrease in oxygen levels leads to BBB disruption and may later induce the CNS disease.[112] Hypoxia brings about paracellular permeability through basement membrane breakdown and alteration in expression of TJs such as claudin-1, occludin, ZO-1, or ZO-2.[113–115] Hypoxia can enhance nonspecific vesicular transport in BECs, which is characterized by enhanced blood-borne substances in brain.[116,117] In a model of stroke, it has been seen that IL-1 causes damage to the BBB and white matter that can be suppressed by anti-IL-1 antibodies that inhibit IL-1 entry into the brain.[118–120] During hypoxic conditions, hypoxia-inducible factor-1 (HIF-1) is upregulated that increases HIF-1α protein levels thereby aggravating cytokine release following SARS-CoV-2 infection.[9] A study on human cerebral microvascular endothelial cells (hCMEC/D3) under hypoxic conditions suggests SARS-CoV-2 entry is modulated by ACE2 and transmembrane protease, serine 2 (TMPRSS2).[121]

6. Clotting and thrombosis

It has been shown that SARS-CoV-2 infection in healthy adults is linked with high chances of thrombosis and related conditions like stroke without prior comorbidities.[122] Autopsy analysis of SARS-CoV-2 infected brain showed increased occurrence of acute hypoxic/ischemic damage.[123] High prevalence of coagulation abnormalities has been described in SARS-CoV-2 cases.[124] including cerebral micro-emboli.[125] Further, the clinical data suggests that COVID-19 infection enhances the chance of immune-activated thrombotic microangiopathy (TMA) mediated by pathway.[126] TMA occurs via endothelial cell damage to small blood vessels that induces thrombocytopenia, hemolytic anemia, and lead to organ damage.[127] Neuro imaging of COVID-19 infected patients displays upregulated coagulation factors that were in accordance with TMA.[124] Immune-mediated complement system is employed in TMA, which originates from COVID-19 infection along with enhanced formation of C5b9 membrane attack complement complex discovered as possible factor.[128] However, C5b9 membrane attack complex has not yet been investigated in brain tissue of SARS-CoV-2 but linked with brain conditions like subarachnoid hemorrhage and traumatic brain injury.[129,130] An increase in other components that are engaged in clot formation and degradation like fibrinogen, thrombin, and plasmin system can also result into BBB disruption.[131–134] Studies have shown elevated fibrin degradation products and increased prothrombin time in SARS-CoV-2 in blood of infected patients, with non-survivors having higher levels in comparison to survivors.[135,136]

7. Neurological consequences of disrupted BBB post-SARS-CoV-2 infection

Human neurological disorders are a gradual process that evolves over some time; therefore, SARS-CoV-2 infected patients may produce long-term neurological effects. Very less is known in relation to the plausible long-term consequences of SARS-CoV-

2 infection on brain. Hence, SARS-CoV-2 individuals need to be followed-up to provide understanding on the effects of SARS-CoV-2 on brain.

7.1 Alzheimer's disease

Accumulating evidence reveals that coronaviruses are not all restricted to the respiratory tract but might also pervade CNS and prompt neurological disorders. Acute respiratory distress syndrome (ARDS) is a clinical characteristic of SARS-CoV-2 infected patients. ARDS is also linked with cognitive impairment and neurodegeneration.[137] It is evident that coronavirus could contribute to the pathogenesis of neurodegenerative diseases. AD is the most common form of neurodegenerative disease worldwide. Neuroimaging findings from mild cognitive impairment (MCI) and early AD individuals show BBB disruption in the hippocampus.[138] The brain of AD-affected people is distinguished by the presence of amyloid plaque aggregation and the neurofibrillary tangles that lead to neuronal cell death, synapse loss, myelin alterations, and oligodendroglia degeneration.[139] HIV-1 has been shown to cause AD in a 3D human brain organoid models.[140] Various mechanisms have been implicated in the ability of SARS-CoV-2 to induce neurological injury that could further lead to progression of AD.[141] An association between AD and COVID-19 appears to exist. It is suggested that SARS-CoV-2 infected people are more susceptible to develop AD. On the contrary, AD patients are more prone to severe SARS-CoV-2 infection.[142] COVID-19 is associated with AD neuropathology. SARS-CoV-2 may enter cognitive centers of CNS, leading AD-like phenotypes, and exaggeration of AD neuropathology in autism patients. SARS-CoV-2 infection induces cellular and molecular AD pathogenesis in iPSC-derived neurons. This study identified 24 genes that possibly regulate the pathology of AD following SARS-CoV-2 infection.[143] Autopsy analysis showed that ACE2 receptor expression is increased in brain of AD patients. Therefore, increased ACE2 receptor expression might represent a risk factor for SARS-CoV-2 infection in AD patients. SARS-CoV-2 can infect CNS through disrupted BBB, which is impaired in aging and in neurological conditions like AD.[144] BBB alterations due to direct or indirect effects of SARS-CoV-2 could lead to increased incidences of SARS-CoV-2 entry into the brain and its long-term effects. Disrupted BBB may allow the entry of immune cells into the brain, which may induce cognitive impairment and neurological effects in SARS-CoV-2 patients. Additionally, endothelial dysfunctions, which is a feature of SARS-CoV-2 infection and loss of pericytes, might lead to impaired clearance of Aβ peptides as well as of cerebral metabolites. The accumulation of excess Aβ peptides in senile plaques in hippocampus is the main pathological mechanism of AD development.[142] Aging leads to slow loss of BBB integrity,[138] and elder people are more prone to SARS-CoV-2 infection. A recent finding suggests that presence of apolipoprotein E ε4 (APOEε4) is linked with cerebrovascular dysfunctions, including pericyte degeneration and BBB disruption. Interestingly, a report from UK study showed higher prevalence of COVID-19 in APOEε4 carriers. APOEε4 expression on pericytes was demonstrated to induce leaky BBB due of impaired basement membrane.[137] Interestingly, a study highlighting the potential of SARS-CoV-2 infection-causing distortion of Tau hyperphosphorylation in brain organoids has been reported.[145] In a study from the United Kingdom of 16,749 hospitalized patients, dementia was the common symptom. However, this study suggested a link between APOEε4 gene homozygosity and SARS-CoV-2 infection susceptibility independent of dementia.[146] VCAM1 is present on surface of endothelial cells, which are stimulated by cytokines. VCAM1 is a blood and CSF biomarker of AD as its expression are increased in AD patients. VCAM1 expression is linked with the severity of dementia. BBB expression of VCAM1 is a suggestive therapy for treating age-associated neurodegenerative conditions. Serum levels of VCAM1 were remarkably increased in SARS-CoV-2 infected patients as compared to mild patients. However, the levels of VCAM1 were significantly reduced in the recovery stage as compared to severe cases. Interestingly, VCAM1 is also suggested to be involved in SARS-CoV-2 mediated vasculitis.[147]

Upon SARS-CoV-2 infection, systemic inflammation features by "cytokine storm" induce disruption of BBB and further cause damage to neuronal and glial cells that could be involved in long-term consequences. Systemic inflammation is considered as major pathophysiological mechanism behind AD development.[148] The virus-mediated systemic inflammation causes massive increase in cytokines that could cross CNS because of increased permeability of BBB. This systemic inflammation thus could enhance the neuroinflammation thereby resulting in neurodegeneration.[149] Experimental findings demonstrate that short-term mechanical ventilation stimulates the pathology of AD by causing accumulation of Aβ peptide, BBB dysfunction, systemic, and neuroinflammation.[150] The long-term existence of virus in the CNS can cause neuroinflammation and indeed initiate the development of neurological diseases.[151]

7.2 COVID-19 infection and BBB disruption in other neurodegenerative diseases

Parkinson's disease (PD) also shares risk factors with COVID-19 owing to advanced age, cardiovascular, and metabolic changes.[152,153] It is well known that viral infections may accelerate PD pathology.[154] PD patients are often immune-compromised making them vulnerable to infections including SARS-CoV-2.[155,156] Patients with advanced PD face chewing and swallowing difficulties, consequently experiencing salivary accumulation and aspiration.[157] Furthermore, chest stiffness in PD patients inhibits cough reflux that creates a favorable environment for SARS-CoV-2 infection to fester. Severe infections lead to pneumonia, which is already a leading cause of mortality in PD patients.[158] BBB disruption has been noted in PD models in rodents.[159] Positron emission tomography (PET) scans of PD patients have revealed BBB disruption.[160] Thus, BBB disruption due to neuroinflammation in PD related to COVID-19 is as a risk factor.

Ischemic stroke is the consequence of decreased blood flow to the brain results in hypoxia. Several cases have linked COVID-19 with stroke.[161–163] Acute stroke was found to be the most common neurological manifestation of COVID-19 during neuroimaging of hospitalized patients.[164] The most probable contributing factor to COVID-19 associated stroke is thought to be atherosclerosis and viral entry into the brain due to inflammation.[165] BBB damage is a well-known consequence of ischemic stroke.[166] Hence, stroke may also contribute to SARS-CoV-2 infection as a result of damaged BBB integrity.

Multiple sclerosis (MS) is also suggested as a risk factor related to COVID-19. It is an autoimmune disorder with demyelination, neuronal loss, gliosis, and inflammation in the CNS. MS patients are susceptible to viral infections and have neurological symptoms involving BBB dysfunction.[167,168] Therefore, MS patients are susceptible to SARS-CoV-2 infection as a result of disrupted BBB.

8. Treatment to prevent BBB disruption following SARS-CoV-2 infection

It has been understood from the animal studies that both systemic and neuroinflammation are involved in neurotropism following SARS-CoV-2 infection.[39,40,169] Inflammation is a critical characteristic feature of SARS-CoV-2 infection and is probably to an extent that leads to paracellular and transcellular destruction of BBB.[42] As mentioned in the earlier section, studies using S1 protein suggest that stimulation of immune cells did not increase the passage of S1 protein by adsorptive transcytosis, but increased S1 protein as a result of BBB disruption, most likely by transcellular mechanism.[42] Thus, it is believed that use of anti-inflammatory therapy could be used to suppress SARS-CoV-2 neuroinvasion. In brief, S1 protein can cross BBB and SARS-CoV-2 by vesicular mechanism by adsorptive transcytosis. Further, BBB functions are regulated by a signaling molecule 5′-adenosine monophosphate activated protein kinase (AMPK). Activation of AMPK in BECs induced BBB integrity and decreased permeability in rats.[170] AMPK activation by metformin, 5-amino imidazole-4-carboxamide ribonucleotide (AICAR) and melatonin reduced BBB impairment in rodent models of stroke.[171] The use of melatonin as adjuvant in COVID-19 is under consideration as well.[172–174] Another important signaling pathway for brain vascularization and BBB differentiation is Wnt/β-catenin pathway. In addition, dysregulated Wnt/β-catenin has been observed in several CNS diseases involving BBB dysfunction. A study showed activation with Wnt3a ligand (not expressed in BECs) or lithium chloride (GSK3β inhibitor) increased claudin-5 mRNA and occluding paracellular barrier of NUV.[175,176] Hence, Wnt activation in BECs is a likely target for BBB restoration. Therefore, molecules that can modulate inflammation and molecular pathways involved in maintaining BBB inegrity may be evaluated against COVID-19 mediated BBB disruption.

9. Conclusions

This chapter summarizes the available evidence and factors that may be involved in SARS-CoV-2 mediated BBB disruption. Further, the mechanism of entry and pathophysiology behind breaching of BBB are described. The long-term neurological implications of COVID-19 infection are understood to be a consequence of BBB disruption. Postmortem analysis of SARS-CoV-2 infected brains and animal studies provided evidence of infected endothelial cells as well as neuronal damage following COVID-19 infection. S1 protein has been suggested to cross the intact BBB via mechanism of adsorptive endocytosis indicating ACE2 receptors—independent route for entry to the brain. Thus, future studies focusing on other receptors associated with SARS-CoV-2 neuroinvasion as a therapeutic target need to be investigated as blocking these receptors may prevent viral entry and BBB disruption. It is still not well understood that neurological manifestations of SARS-CoV-2 infection are because of direct viral entry into the brain or a cause of systemic inflammation and hypoxia. Studies with more intricate in vitro models of NVU may allow us to understand the actual mechanisms of SARS-CoV-2 mediated loss of BBB integrity and consequences of altered BBB integrity. Another approach to prevent BBB disruption from SARS-CoV-2 infection may involve targeting systemic inflammation to inhibit viral infection to reach from blood to brain, which is the most common route by the virus or activated cytokines to reach BBB. Further, masking spike protein may also help in preventing spread of viral infection. As evident spike proteins are the major factors involved in BBB disruption involving various mechanisms. Inhibition of RhoA activation may also help reduce the detrimental effects of S1 on BBB. Despite limited knowledge about the implication of disrupted BBB in COVID-19 infection the disrupted BBB is probably employed by various mechanisms in expression of those neurological symptoms. Future studies need to be focus on finding the association between APOEε4, SARS-CoV-2 infection, and AD. Therefore, understanding these mechanisms may allow for better management and development of better therapeutics against long COVID associated neurological symptoms.

Abbreviations

(APOEε4) Apolipoprotein E ε4
ACE2 Angiotensin-Converting Enzyme 2
AD Alzheimer's Disease
ARDS Acute Respiratory Distress Syndrome
BBB Blood—Brain Barrier
BECs Brain Endothelial Cells
BSG Basigin
CNS Central Nervous System
COVID-19 Coronavirus Disease 2019
CSF Cerebrospinal Fluid
GI tract Gastrointestinal tract
hiPS-BCECs human-induced Pluripotent Stem cell-derived Brain Capillary Endothelial-like Cells
HIV-1 Human Immunodeficiency Virus-1
IFN Interferon
MERS-CoV Middle East Respiratory Syndrome Coronavirus
NRP-1 Neuropilin-1

NVU	Neurovascular Unit
PAR2	Protease-Activated Receptor 2
PD	Parkinson's Disease
PGE2	Prostaglandin E2
S1	Spike 1 protein
SARS-CoV-2	Severe Acute Respiratory Syndrome Coronavirus 2
TJs	Tight Junctions
TLR4	Toll-like Receptor
TMA	Thrombotic Microangiopathy
TMPRSS2	Transmembrane Protease Serine 2
VAP	Viral Attachment Protein
VCAM1	Vascular Cell Adhesion Molecule 1

References

1. Mao L, Jin H, Wang M, et al. Neurologic manifestations of hospitalized patients with coronavirus disease 2019 in Wuhan, China. *JAMA Neurol*. 2020;77(6):683–690. https://doi.org/10.1001/jamaneurol.2020.1127.
2. Serrano GE, Walker JE, Arce R, et al. Mapping of SARS-CoV-2 brain invasion and histopathology in COVID-19 disease. *medRxiv*. 2021. https://doi.org/10.1101/2021.02.15.21251511.
3. Maccio U, Zinkernagel AS, Schuepbach R, et al. Long-term persisting SARS-CoV-2 RNA and pathological findings: lessons learnt from a series of 35 COVID-19 autopsies. *Front Med*. 2022;9:778489. https://doi.org/10.3389/fmed.2022.778489.
4. Moriguchi T, Harii N, Goto J, et al. A first case of meningitis/encephalitis associated with SARS-Coronavirus-2. *Int J Infect Dis*. 2020;94:55–58. https://doi.org/10.1016/j.ijid.2020.03.062.
5. Meinhardt J, Radke J, Dittmayer C, et al. Olfactory transmucosal SARS-CoV-2 invasion as a port of central nervous system entry in individuals with COVID-19. *Nat Neurosci*. February 2021;24(2):168–175. https://doi.org/10.1038/s41593-020-00758-5.
6. Daneman R, Prat A. The blood-brain barrier. *Cold Spring Harb Perspect Biol*. 2015;7(1): a020412. https://doi.org/10.1101/cshperspect.a020412.
7. Xu J, Lazartigues E. Expression of ACE2 in human neurons supports the neuro-invasive potential of COVID-19 virus. *Cell Mol Neurobiol*. 2022;42(1):305–309. https://doi.org/10.1007/s10571-020-00915-1.
8. Li Z, Liu T, Yang N, et al. Neurological manifestations of patients with COVID-19: potential routes of SARS-CoV-2 neuroinvasion from the periphery to the brain. *Front Med*. 2020;14(5):533–541. https://doi.org/10.1007/s11684-020-0786-5.
9. Erickson MA, Rhea EM, Knopp RC, Banks WA. Interactions of SARS-CoV-2 with the blood-brain barrier. *Int J Mol Sci*. 2021;(5):22. https://doi.org/10.3390/ijms22052681.
10. Wenzel J, Lampe J, Muller-Fielitz H, et al. The SARS-CoV-2 main protease M(pro) causes microvascular brain pathology by cleaving NEMO in brain endothelial cells. *Nat Neurosci*. 2021;24(11):1522–1533. https://doi.org/10.1038/s41593-021-00926-1.
11. Cantuti-Castelvetri L, Ojha R, Pedro LD, et al. Neuropilin-1 facilitates SARS-CoV-2 cell entry and infectivity. *Science*. 2020;370(6518):856–860. https://doi.org/10.1126/science.abd2985.

12. Bohmwald K, Galvez NMS, Rios M, Kalergis AM. Neurologic alterations due to respiratory virus infections. *Front Cell Neurosci*. 2018;12:386. https://doi.org/10.3389/fncel.2018.00386.
13. Paniz-Mondolfi A, Bryce C, Grimes Z, et al. Central nervous system involvement by severe acute respiratory syndrome coronavirus-2 (SARS-CoV-2). *J Med Virol*. 2020;92(7):699–702. https://doi.org/10.1002/jmv.25915.
14. Sweeney MD, Ayyadurai S, Zlokovic BV. Pericytes of the neurovascular unit: key functions and signaling pathways. *Nat Neurosci*. 2016;19(6):771–783. https://doi.org/10.1038/nn.4288.
15. Berthiaume AA, Grant RI, McDowell KP, et al. Dynamic remodeling of pericytes in vivo maintains capillary coverage in the adult mouse brain. *Cell Rep*. 2018;22(1):8–16. https://doi.org/10.1016/j.celrep.2017.12.016.
16. Abbott NJ, Ronnback L, Hansson E. Astrocyte-endothelial interactions at the blood-brain barrier. *Nat Rev Neurosci*. 2006;7(1):41–53. https://doi.org/10.1038/nrn1824.
17. Mishra A, Reynolds JP, Chen Y, Gourine AV, Rusakov DA, Attwell D. Astrocytes mediate neurovascular signaling to capillary pericytes but not to arterioles. *Nat Neurosci*. 2016;19(12):1619–1627. https://doi.org/10.1038/nn.4428.
18. Achar A, Ghosh C. COVID-19-associated neurological disorders: the potential route of CNS invasion and blood-brain relevance. *Cells*. 2020;9(11). https://doi.org/10.3390/cells9112360.
19. Luissint AC, Artus C, Glacial F, Ganeshamoorthy K, Couraud PO. Tight junctions at the blood brain barrier: physiological architecture and disease-associated dysregulation. *Fluids Barriers CNS*. 2012;9(1):23. https://doi.org/10.1186/2045-8118-9-23.
20. Gupta S, Dhanda S, Sandhir R. 2 - Anatomy and physiology of blood-brain barrier. In: Gao H, Gao X, eds. *Brain Targeted Drug Delivery System*. Academic Press; 2019:7–31.
21. Gunzel D, Yu AS. Claudins and the modulation of tight junction permeability. *Physiol Rev*. 2013;93(2):525–569. https://doi.org/10.1152/physrev.00019.2012.
22. Loscher W, Potschka H. Blood-brain barrier active efflux transporters: ATP-binding cassette gene family. *NeuroRx*. 2005;2(1):86–98. https://doi.org/10.1602/neurorx.2.1.86.
23. Agundez JA, Jimenez-Jimenez FJ, Alonso-Navarro H, Garcia-Martin E. Drug and xenobiotic biotransformation in the blood-brain barrier: a neglected issue. *Front Cell Neurosci*. 2014;8:335. https://doi.org/10.3389/fncel.2014.00335.
24. Zubair AS, McAlpine LS, Gardin T, Farhadian S, Kuruvilla DE, Spudich S. Neuropathogenesis and neurologic manifestations of the coronaviruses in the age of coronavirus disease 2019: a review. *JAMA Neurol*. 2020;77(8):1018–1027. https://doi.org/10.1001/jamaneurol.2020.2065.
25. Arbour N, Day R, Newcombe J, Talbot PJ. Neuroinvasion by human respiratory coronaviruses. *J Virol*. 2000;74(19):8913–8921. https://doi.org/10.1128/jvi.74.19.8913-8921.2000.
26. Li K, Wohlford-Lenane C, Perlman S, et al. Middle east respiratory syndrome coronavirus causes multiple organ damage and lethal disease in mice transgenic for human dipeptidyl peptidase 4. *J Infect Dis*. 2016;213(5):712–722. https://doi.org/10.1093/infdis/jiv499.
27. Netland J, Meyerholz DK, Moore S, Cassell M, Perlman S. Severe acute respiratory syndrome coronavirus infection causes neuronal death in the absence of encephalitis in mice transgenic for human ACE2. *J Virol*. 2008;82(15):7264–7275. https://doi.org/10.1128/JVI.00737-08.
28. Glass WG, Subbarao K, Murphy B, Murphy PM. Mechanisms of host defense following severe acute respiratory syndrome-coronavirus (SARS-CoV) pulmonary infection of mice. *J Immunol*. 2004;173(6):4030–4039. https://doi.org/10.4049/jimmunol.173.6.4030.

29. Wolburg H, Wolburg-Buchholz K, Engelhardt B. Diapedesis of mononuclear cells across cerebral venules during experimental autoimmune encephalomyelitis leaves tight junctions intact. *Acta Neuropathol.* 2005;109(2):181–190. https://doi.org/10.1007/s00401-004-0928-x.
30. Greenwood J, Heasman SJ, Alvarez JI, Prat A, Lyck R, Engelhardt B. Review: leucocyte-endothelial cell crosstalk at the blood-brain barrier: a prerequisite for successful immune cell entry to the brain. *Neuropathol Appl Neurobiol.* 2011;37(1):24–39. https://doi.org/10.1111/j.1365-2990.2010.01140.x.
31. Hofmann H, Pöhlmann S. Cellular entry of the SARS coronavirus. *Trends Microbiol.* 2004;12(10). https://doi.org/10.1016/2Fj.tim.2004.08.008.
32. Wang H, Yang P, Liu K, et al. SARS coronavirus entry into host cells through a novel clathrin- and caveolae-independent endocytic pathway. *Cell Res.* 2008;18(2):290–301. https://doi.org/10.1038/cr.2008.15.
33. Jackson CB, Farzan M, Chen B, Choe H. Mechanisms of SARS-CoV-2 entry into cells. *Nat Rev Mol Cell Biol.* 2022;23(1):3–20. https://doi.org/10.1038/s41580-021-00418-x.
34. Wu D, Yang XO. TH17 responses in cytokine storm of COVID-19: an emerging target of JAK2 inhibitor Fedratinib. *J Microbiol Immunol Infect.* 2020;53(3):368–370. https://doi.org/10.1016/j.jmii.2020.03.005.
35. Murgolo N, Therien AG, Howell B, et al. SARS-CoV-2 tropism, entry, replication, and propagation: considerations for drug discovery and development. *PLoS Pathog.* 2021; 17(2). https://doi.org/10.1371/journal.ppat.1009225.
36. Millet JK, Whittaker GR. Host cell entry of Middle East respiratory syndrome coronavirus after two-step, furin-mediated activation of the spike protein. *Proc Natl Acad Sci U S A.* 2014;111(42). https://doi.org/10.1073/pnas.1407087111.
37. Matoba Y, Abiko C, Ikeda T, et al. Detection of the Human Coronavirus 229E, HKU1, NL63, and OC43 between 2010 and 2013 in Yamagata, Japan. *Jpn J Infect Dis.* 2015; 68(2). https://doi.org/10.7883/yoken.JJID.2014.266.
38. Pellegrini L, Albecka A, Mallery DL, et al. SARS-CoV-2 infects the brain choroid plexus and disrupts the blood-CSF barrier in human brain organoids. *Cell Stem Cell.* 2020;27(6): 951–961. https://doi.org/10.1016/j.stem.2020.10.001. e5.
39. Song E, Zhang C, Israelow B, et al. Neuroinvasion of SARS-CoV-2 in human and mouse brain. *J Exp Med.* 2021;(3):218. https://doi.org/10.1084/jem.20202135.
40. Winkler ES, Bailey AL, Kafai NM, et al. SARS-CoV-2 infection of human ACE2-transgenic mice causes severe lung inflammation and impaired function. *Nat Immunol.* 2020;21(11):1327–1335. https://doi.org/10.1038/s41590-020-0778-2.
41. Bryche B, St Albin A, Murri S, et al. Massive transient damage of the olfactory epithelium associated with infection of sustentacular cells by SARS-CoV-2 in golden Syrian hamsters. *Brain Behav Immun.* 2020;89:579–586. https://doi.org/10.1016/j.bbi.2020.06.032.
42. Rhea EM, Logsdon AF, Hansen KM, et al. The S1 protein of SARS-CoV-2 crosses the blood-brain barrier in mice. *Nat Neurosci.* 2021;24(3):368–378. https://doi.org/10.1038/s41593-020-00771-8.
43. Vankadari N, Wilce JA. Emerging WuHan (COVID-19) coronavirus: glycan shield and structure prediction of spike glycoprotein and its interaction with human CD26. *Emerg Microbes Infect.* 2020;9(1):601–604. https://doi.org/10.1080/22221751.2020.1739565.
44. Radzikowska U, Ding M, Tan G, et al. Distribution of ACE2, CD147, CD26, and other SARS-CoV-2 associated molecules in tissues and immune cells in health and in asthma, COPD, obesity, hypertension, and COVID-19 risk factors. *Allergy.* 2020;75(11): 2829–2845. https://doi.org/10.1111/all.14429.

45. Ibrahim IM, Abdelmalek DH, Elshahat ME, Elfiky AA. COVID-19 spike-host cell receptor GRP78 binding site prediction. *J Infect*. 2020;80(5):554−562. https://doi.org/10.1016/j.jinf.2020.02.026.
46. Desforges M, Le Coupanec A, Dubeau P, et al. Human coronaviruses and other respiratory viruses: underestimated opportunistic pathogens of the central nervous system? *Viruses*. 2019;(1):12. https://doi.org/10.3390/v12010014.
47. Alam MA, Quamri MA, Sofi G, Ayman U, Ansari S, Ahad M. Understanding COVID-19 in the light of epidemic disease described in Unani medicine. *Drug Metab Pers Ther*. 2020. https://doi.org/10.1515/dmdi-2020-0136.
48. Davis BM, Rall GF, Schnell MJ. Everything you always wanted to know about rabies virus (but were afraid to ask). *Annu Rev Virol*. 2015;2(1):451−471. https://doi.org/10.1146/annurev-virology-100114-055157.
49. Enquist LW, Husak PJ, Banfield BW, Smith GA. Infection and spread of alphaherpesviruses in the nervous system. *Adv Virus Res*. 1998;51:237−347. https://doi.org/10.1016/s0065-3527(08)60787-3.
50. Morrison LA, Sidman RL, Fields BN. Direct spread of reovirus from the intestinal lumen to the central nervous system through vagal autonomic nerve fibers. *Proc Natl Acad Sci U S A*. 1991;88(9):3852−3856. https://doi.org/10.1073/pnas.88.9.3852.
51. Perlman S, Evans G, Afifi A. Effect of olfactory bulb ablation on spread of a neurotropic coronavirus into the mouse brain. *J Exp Med*. 1990;172(4):1127−1132. https://doi.org/10.1084/jem.172.4.1127.
52. Li YC, Bai WZ, Hashikawa T. Response to commentary on "the neuroinvasive potential of SARS-CoV-2 may play a role in the respiratory failure of COVID-19 patients". *J Med Virol*. 2020;92(7):707−709. https://doi.org/10.1002/jmv.25824.
53. Hwang CS. Olfactory neuropathy in severe acute respiratory syndrome: report of a case. *Acta Neurol Taiwan*. 2006;15(1):26−28.
54. Giacomelli A, Pezzati L, Conti F, et al. Self-reported olfactory and taste disorders in patients with severe acute respiratory coronavirus 2 infection: a cross-sectional study. *Clin Infect Dis*. 2020;71(15):889−890. https://doi.org/10.1093/cid/ciaa330.
55. Bayati A, Kumar R, Francis V, McPherson PS. SARS-CoV-2 infects cells after viral entry via clathrin-mediated endocytosis. *J Biol Chem*. 2021;296:100306. https://doi.org/10.1016/j.jbc.2021.100306.
56. Baig AM. Neurological manifestations in COVID-19 caused by SARS-CoV-2. *CNS Neurosci Ther*. 2020;26(5):499−501. https://doi.org/10.1111/cns.13372.
57. Steardo L, Steardo Jr L, Zorec R, Verkhratsky A. Neuroinfection may contribute to pathophysiology and clinical manifestations of COVID-19. *Acta Physiol (Oxf)*. 2020; 229(3):e13473. https://doi.org/10.1111/apha.13473.
58. Varga Z, Flammer AJ, Steiger P, et al. Endothelial cell infection and endotheliitis in COVID-19. *Lancet*. 2020;395(10234):1417−1418. https://doi.org/10.1016/S0140-6736(20)30937-5.
59. Park MD. Macrophages: a Trojan horse in COVID-19? *Nat Rev Immunol*. 2020;20(6):351. https://doi.org/10.1038/s41577-020-0317-2.
60. Sun SH, Chen Q, Gu HJ, et al. A mouse model of SARS-CoV-2 infection and pathogenesis. *Cell Host Microbe*. 2020;28(1):124−133e4. https://doi.org/10.1016/j.chom.2020.05.020.
61. Koskiniemi M, Vaheri A, Taskinen E. Cerebrospinal fluid alterations in herpes simplex virus encephalitis. *Rev Infect Dis*. 1984;6(5):608−618. https://doi.org/10.1093/clinids/6.5.608.

62. Long T, Zhang B, Fan R, et al. Phosphoprotein gene of wild-type rabies virus plays a role in limiting viral pathogenicity and lowering the enhancement of BBB permeability. *Front Microbiol*. 2020;11:109. https://doi.org/10.3389/fmicb.2020.00109.
63. Osborne O, Peyravian N, Nair M, Daunert S, Toborek M. The paradox of HIV blood-brain barrier penetrance and antiretroviral drug delivery deficiencies. *Trends Neurosci*. 2020; 43(9):695−708. https://doi.org/10.1016/j.tins.2020.06.007.
64. Leda AR, Bertrand L, Andras IE, El-Hage N, Nair M, Toborek M. Selective disruption of the blood-brain barrier by Zika virus. *Front Microbiol*. 2019;10:2158. https://doi.org/10.3389/fmicb.2019.02158.
65. Chen Z, Li G. Immune response and blood-brain barrier dysfunction during viral neuroinvasion. *Innate Immun*. 2021;27(2):109−117. https://doi.org/10.1177/1753425920954281.
66. Buzhdygan TP, DeOre BJ, Baldwin-Leclair A, et al. The SARS-CoV-2 spike protein alters barrier function in 2D static and 3D microfluidic in-vitro models of the human blood-brain barrier. *Neurobiol Dis*. 2020;146:105131. https://doi.org/10.1016/j.nbd.2020.105131.
67. Alexopoulos H, Magira E, Bitzogli K, et al. Anti-SARS-CoV-2 antibodies in the CSF, blood-brain barrier dysfunction, and neurological outcome: studies in 8 stuporous and comatose patients. *Neurol Neuroimmunol Neuroinflamm*. 2020;7(6). https://doi.org/10.1212/NXI.0000000000000893.
68. Zhang L, Zhou L, Bao L, et al. SARS-CoV-2 crosses the blood-brain barrier accompanied with basement membrane disruption without tight junctions alteration. *Signal Transduct Target Ther*. 2021;6(1):337. https://doi.org/10.1038/s41392-021-00719-9.
69. Wang W, Uzzau S, Goldblum SE, Fasano A. Human zonulin, a potential modulator of intestinal tight junctions. *J Cell Sci*. 2000;113 Pt 24:4435−4440. https://doi.org/10.1242/jcs.113.24.4435.
70. Rittirsch D, Flierl MA, Nadeau BA, et al. Zonulin as prehaptoglobin2 regulates lung permeability and activates the complement system. *Am J Physiol Lung Cell Mol Physiol*. 2013;304(12):L863−L872. https://doi.org/10.1152/ajplung.00196.2012.
71. Skardelly M, Armbruster FP, Meixensberger J, Hilbig H. Expression of zonulin, c-kit, and glial fibrillary acidic protein in human gliomas. *Transl Oncol*. 2009;2(3):117−120. https://doi.org/10.1593/tlo.09115.
72. Lu R, Wang W, Uzzau S, Vigorito R, Zielke HR, Fasano A. Affinity purification and partial characterization of the zonulin/zonula occludens toxin (Zot) receptor from human brain. *J Neurochem*. 2000;74(1):320−326. https://doi.org/10.1046/j.1471-4159.2000.0740320.x.
73. Di Micco S, Musella S, Scala MC, et al. In silico analysis revealed potential anti-SARS-CoV-2 main protease activity by the zonulin inhibitor larazotide acetate. *Front Chem*. 2020;8:628609. https://doi.org/10.3389/fchem.2020.628609.
74. Bourouiba L, Dehandschoewercker E, Bush John WM. Violent expiratory events: on coughing and sneezing. *J Fluid Mech*. 2014;745:537−563. https://doi.org/10.1017/jfm.2014.88.
75. Llorens S, Nava E, Munoz-Lopez M, Sanchez-Larsen A, Segura T. Neurological symptoms of COVID-19: the zonulin hypothesis. *Front Immunol*. 2021;12:665300. https://doi.org/10.3389/fimmu.2021.665300.
76. Nascimento Conde J, Schutt WR, Gorbunova EE, Mackow ER. Recombinant ACE2 expression is required for SARS-CoV-2 to infect primary human endothelial cells and induce inflammatory and procoagulative responses. *mBio*. 2020;(6):11. https://doi.org/10.1128/mBio.03185-20.
77. Coronaviridae Study Group of the International Committee on Taxonomy of V. The species severe acute respiratory syndrome-related coronavirus: classifying 2019-nCoV and

naming it SARS-CoV-2. *Nat Microbiol.* 2020;5(4):536–544. https://doi.org/10.1038/s41564-020-0695-z.
78. Reynolds JL, Mahajan SD. SARS-COV2 alters blood brain barrier integrity contributing to neuro-inflammation. *J Neuroimmune Pharmacol.* 2021;16(1):4–6. https://doi.org/10.1007/s11481-020-09975-y.
79. Marsh M. The entry of enveloped viruses into cells by endocytosis. *Biochem J.* 1984;218(1):1–10. https://doi.org/10.1042/bj2180001.
80. Klasse PJ, Bron R, Marsh M. Mechanisms of enveloped virus entry into animal cells. *Adv Drug Deliv Rev.* 1998;34(1):65–91. https://doi.org/10.1016/S0169-409X(98)00002-7.
81. Helms J, Kremer S, Merdji H, et al. Neurologic features in severe SARS-CoV-2 infection. *N Engl J Med.* 2020;382(23):2268–2270. https://doi.org/10.1056/NEJMc2008597.
82. Schett G, Sticherling M, Neurath MF. COVID-19: risk for cytokine targeting in chronic inflammatory diseases? *Nat Rev Immunol.* 2020;20(5):271–272. https://doi.org/10.1038/s41577-020-0312-7.
83. Mahmudpour M, Roozbeh J, Keshavarz M, Farrokhi S, Nabipour I. COVID-19 cytokine storm: the anger of inflammation. *Cytokine.* 2020;133:155151. https://doi.org/10.1016/j.cyto.2020.155151.
84. Huang C, Wang Y, Li X, et al. Clinical features of patients infected with 2019 novel coronavirus in Wuhan, China. *Lancet.* 2020;395(10223):497–506. https://doi.org/10.1016/S0140-6736(20)30183-5.
85. Tay MZ, Poh CM, Renia L, MacAry PA, Ng LFP. The trinity of COVID-19: immunity, inflammation and intervention. *Nat Rev Immunol.* 2020;20(6):363–374. https://doi.org/10.1038/s41577-020-0311-8.
86. Zhou F, Yu T, Du R, et al. Clinical course and risk factors for mortality of adult inpatients with COVID-19 in Wuhan, China: a retrospective cohort study. *Lancet.* 2020;395(10229):1054–1062. https://doi.org/10.1016/S0140-6736(20)30566-3.
87. Garcia MA, Barreras PV, Lewis A, et al. Cerebrospinal fluid in COVID-19 neurological complications: no cytokine storm or neuroinflammation. *medRxiv.* 2021. https://doi.org/10.1101/2021.01.10.20249014.
88. Biswas S, Thakur V, Kaur P, Khan A, Kulshrestha S, Kumar P. Blood clots in COVID-19 patients: simplifying the curious mystery. *Med Hypotheses.* 2021;146:110371. https://doi.org/10.1016/j.mehy.2020.110371.
89. Erickson MA, Banks WA. Neuroimmune axes of the blood-brain barriers and blood-brain interfaces: bases for physiological regulation, disease states, and pharmacological interventions. *Pharmacol Rev.* 2018;70(2):278–314. https://doi.org/10.1124/pr.117.014647.
90. Erickson MA, Wilson ML, Banks WA. In vitro modeling of blood-brain barrier and interface functions in neuroimmune communication. *Fluids Barriers CNS.* 2020;17(1):26. https://doi.org/10.1186/s12987-020-00187-3.
91. Hsuchou H, Kastin AJ, Mishra PK, Pan W. C-reactive protein increases BBB permeability: implications for obesity and neuroinflammation. *Cell Physiol Biochem.* 2012;30(5):1109–1119. https://doi.org/10.1159/000343302.
92. Banks WA, Freed EO, Wolf KM, Robinson SM, Franko M, Kumar VB. Transport of human immunodeficiency virus type 1 pseudoviruses across the blood-brain barrier: role of envelope proteins and adsorptive endocytosis. *J Virol.* 2001;75(10):4681–4691. https://doi.org/10.1128/JVI.75.10.4681-4691.2001.
93. Yarlagadda A, Alfson E, Clayton AH. The blood brain barrier and the role of cytokines in neuropsychiatry. *Psychiatry.* 2009;6(11):18–22.

94. Tandon M, Kataria S, Patel J, et al. A comprehensive systematic review of CSF analysis that defines neurological manifestations of COVID-19. *Int J Infect Dis*. 2021;104: 390−397. https://doi.org/10.1016/j.ijid.2021.01.002.
95. Mehta P, McAuley DF, Brown M, et al. COVID-19: consider cytokine storm syndromes and immunosuppression. *Lancet*. 2020;395(10229):1033−1034. https://doi.org/10.1016/S0140-6736(20)30628-0.
96. Lee MH, Perl DP, Nair G, et al. Microvascular injury in the brains of patients with Covid-19. *N Engl J Med*. 2021;384(5):481−483. https://doi.org/10.1056/NEJMc2033369.
97. Engelhardt B, Coisne C. Fluids and barriers of the CNS establish immune privilege by confining immune surveillance to a two-walled castle moat surrounding the CNS castle. *Fluids Barriers CNS*. 2011;8(1):4. https://doi.org/10.1186/2045-8118-8-4.
98. Hallmann R, Zhang X, Di Russo J, et al. The regulation of immune cell trafficking by the extracellular matrix. *Curr Opin Cell Biol*. 2015;36:54−61. https://doi.org/10.1016/j.ceb.2015.06.006.
99. Cardot-Leccia N, Hubiche T, Dellamonica J, Burel-Vandenbos F, Passeron T. Pericyte alteration sheds light on micro-vasculopathy in COVID-19 infection. *Intensive Care Med*. 2020;46(9):1777−1778. https://doi.org/10.1007/s00134-020-06147-7.
100. Najjar S, Pahlajani S, De Sanctis V, Stern JNH, Najjar A, Chong D. Neurovascular unit dysfunction and blood-brain barrier hyperpermeability contribute to schizophrenia neurobiology: a theoretical integration of clinical and experimental evidence. *Front Psychiatry*. 2017;8:83. https://doi.org/10.3389/fpsyt.2017.00083.
101. Shigemoto-Mogami Y, Hoshikawa K, Sato K. Activated microglia disrupt the blood-brain barrier and induce chemokines and cytokines in a rat in vitro model. *Front Cell Neurosci*. 2018;12:494. https://doi.org/10.3389/fncel.2018.00494.
102. Cao X. COVID-19: immunopathology and its implications for therapy. *Nat Rev Immunol*. 2020;20(5):269−270. https://doi.org/10.1038/s41577-020-0308-3.
103. Moon C. Fighting COVID-19 exhausts T cells. *Nat Rev Immunol*. 2020;20(5):277. https://doi.org/10.1038/s41577-020-0304-7.
104. Grant MC, Geoghegan L, Arbyn M, et al. The prevalence of symptoms in 24,410 adults infected by the novel coronavirus (SARS-CoV-2; COVID-19): a systematic review and meta-analysis of 148 studies from 9 countries. *PLoS One*. 2020;15(6):e0234765. https://doi.org/10.1371/journal.pone.0234765.
105. Engstrom L, Ruud J, Eskilsson A, et al. Lipopolysaccharide-induced fever depends on prostaglandin E2 production specifically in brain endothelial cells. *Endocrinology*. 2012; 153(10):4849−4861. https://doi.org/10.1210/en.2012-1375.
106. Fritz M, Klawonn AM, Nilsson A, et al. Prostaglandin-dependent modulation of dopaminergic neurotransmission elicits inflammation-induced aversion in mice. *J Clin Invest*. 2016;126(2):695−705. https://doi.org/10.1172/JCI83844.
107. Banks WA, Farr SA, Morley JE. Entry of blood-borne cytokines into the central nervous system: effects on cognitive processes. *Neuroimmunomodulation*. 2002;10(6):319−327. https://doi.org/10.1159/000071472.
108. Dantzer R. Cytokine, sickness behavior, and depression. *Neurol Clin*. 2006;24(3): 441−460. https://doi.org/10.1016/j.ncl.2006.03.003.
109. Tobin MJ, Laghi F, Jubran A. Why COVID-19 silent hypoxemia is baffling to physicians. *Am J Respir Crit Care Med*. 2020;202(3):356−360. https://doi.org/10.1164/rccm.202006-2157CP.
110. Kashani KB. Hypoxia in COVID-19: sign of severity or cause for poor outcomes. *Mayo Clin Proc*. 2020;95(6):1094−1096. https://doi.org/10.1016/j.mayocp.2020.04.021.

111. Xie J, Covassin N, Fan Z, et al. Association between hypoxemia and mortality in patients with COVID-19. *Mayo Clin Proc*. 2020;95(6):1138−1147. https://doi.org/10.1016/j.mayocp.2020.04.006.
112. Yang Y, Rosenberg GA. Blood-brain barrier breakdown in acute and chronic cerebrovascular disease. *Stroke*. 2011;42(11):3323−3328. https://doi.org/10.1161/STROKEAHA.110.608257.
113. Mark KS, Davis TP. Cerebral microvascular changes in permeability and tight junctions induced by hypoxia-reoxygenation. *Am J Physiol Heart Circ Physiol*. 2002;282(4): H1485−H1494. https://doi.org/10.1152/ajpheart.00645.2001.
114. Josko J, Knefel K. The role of vascular endothelial growth factor in cerebral oedema formation. *Folia Neuropathol*. 2003;41(3):161−166.
115. Candelario-Jalil E, Yang Y, Rosenberg GA. Diverse roles of matrix metalloproteinases and tissue inhibitors of metalloproteinases in neuroinflammation and cerebral ischemia. *Neuroscience*. 2009;158(3):983−994. https://doi.org/10.1016/j.neuroscience.2008.06.025.
116. Plateel M, Teissier E, Cecchelli R. Hypoxia dramatically increases the nonspecific transport of blood-borne proteins to the brain. *J Neurochem*. 1997;68(2):874−877. https://doi.org/10.1046/j.1471-4159.1997.68020874.x.
117. Nzou G, Wicks RT, VanOstrand NR, et al. Multicellular 3D neurovascular unit model for assessing hypoxia and neuroinflammation induced blood-brain barrier dysfunction. *Sci Rep*. 2020;10(1):9766. https://doi.org/10.1038/s41598-020-66487-8.
118. Chen X, Sadowska GB, Zhang J, et al. Neutralizing anti-interleukin-1beta antibodies modulate fetal blood-brain barrier function after ischemia. *Neurobiol Dis*. 2015;73: 118−129. https://doi.org/10.1016/j.nbd.2014.09.007.
119. Patra A, Chen X, Sadowska GB, et al. Neutralizing anti-interleukin-1beta antibodies reduce ischemia-related interleukin-1beta transport across the blood-brain barrier in fetal sheep. *Neuroscience*. 2017;346:113−125. https://doi.org/10.1016/j.neuroscience.2016.12.051.
120. Chen X, Hovanesian V, Naqvi S, et al. Systemic infusions of anti-interleukin-1beta neutralizing antibodies reduce short-term brain injury after cerebral ischemia in the ovine fetus. *Brain Behav Immun*. 2018;67:24−35. https://doi.org/10.1016/j.bbi.2017.08.002.
121. Imperio GE, Lye P, Mughis H, et al. Hypoxia alters the expression of ACE2 and TMPRSS2 SARS-CoV-2 cell entry mediators in hCMEC/D3 brain endothelial cells. *Microvasc Res*. 2021;138:104232. https://doi.org/10.1016/j.mvr.2021.104232.
122. Oxley TJ, Mocco J, Majidi S, et al. Large-vessel stroke as a presenting feature of Covid-19 in the young. *N Engl J Med*. 2020;382(20):e60. https://doi.org/10.1056/NEJMc2009787.
123. Solomon IH, Normandin E, Bhattacharyya S, et al. Neuropathological features of Covid-19. *N Engl J Med*. 2020;383(10):989−992. https://doi.org/10.1056/NEJMc2019373.
124. Nicholson P, Alshafai L, Krings T. Neuroimaging findings in patients with COVID-19. *AJNR Am J Neuroradiol*. 2020;41(8):1380−1383. https://doi.org/10.3174/ajnr.A6630.
125. Batra A, Clark JR, LaHaye K, et al. Transcranial Doppler ultrasound evidence of active cerebral embolization in COVID-19. *J Stroke Cerebrovasc Dis*. 2021;30(3):105542. https://doi.org/10.1016/j.jstrokecerebrovasdis.2020.105542.
126. Merrill JT, Erkan D, Winakur J, James JA. Emerging evidence of a COVID-19 thrombotic syndrome has treatment implications. *Nat Rev Rheumatol*. 2020;16(10):581−589. https://doi.org/10.1038/s41584-020-0474-5.
127. Shatzel JJ, Taylor JA. Syndromes of thrombotic microangiopathy. *Med Clin North Am*. 2017;101(2):395−415. https://doi.org/10.1016/j.mcna.2016.09.010.
128. Java A, Apicelli AJ, Liszewski MK, et al. The complement system in COVID-19: friend and foe? *JCI Insight*. 2020;(15):5. https://doi.org/10.1172/jci.insight.140711.

129. Lindsberg PJ, Ohman J, Lehto T, et al. Complement activation in the central nervous system following blood-brain barrier damage in man. *Ann Neurol.* 1996;40(4):587–596. https://doi.org/10.1002/ana.410400408.
130. Bellander BM, Olafsson IH, Ghatan PH, et al. Secondary insults following traumatic brain injury enhance complement activation in the human brain and release of the tissue damage marker S100B. *Acta Neurochir (Wien).* 2011;153(1):90–100. https://doi.org/10.1007/s00701-010-0737-z.
131. Tyagi N, Roberts AM, Dean WL, Tyagi SC, Lominadze D. Fibrinogen induces endothelial cell permeability. *Mol Cell Biochem.* 2008;307(1–2):13–22. https://doi.org/10.1007/s11010-007-9579-2.
132. Patibandla PK, Tyagi N, Dean WL, Tyagi SC, Roberts AM, Lominadze D. Fibrinogen induces alterations of endothelial cell tight junction proteins. *J Cell Physiol.* 2009;221(1):195–203. https://doi.org/10.1002/jcp.21845.
133. Lee KR, Kawai N, Kim S, Sagher O, Hoff JT. Mechanisms of edema formation after intracerebral hemorrhage: effects of thrombin on cerebral blood flow, blood-brain barrier permeability, and cell survival in a rat model. *J Neurosurg.* 1997;86(2):272–278. https://doi.org/10.3171/jns.1997.86.2.0272.
134. Su EJ, Fredriksson L, Geyer M, et al. Activation of PDGF-CC by tissue plasminogen activator impairs blood-brain barrier integrity during ischemic stroke. *Nat Med.* 2008;14(7):731–737. https://doi.org/10.1038/nm1787.
135. Tang N, Li D, Wang X, Sun Z. Abnormal coagulation parameters are associated with poor prognosis in patients with novel coronavirus pneumonia. *J Thromb Haemost.* 2020;18(4):844–847. https://doi.org/10.1111/jth.14768.
136. Han H, Yang L, Liu R, et al. Prominent changes in blood coagulation of patients with SARS-CoV-2 infection. *Clin Chem Lab Med.* 2020;58(7):1116–1120. https://doi.org/10.1515/cclm-2020-0188.
137. Miners S, Kehoe PG, Love S. Cognitive impact of COVID-19: looking beyond the short term. *Alzheimer's Res Ther.* 2020;12(1):170. https://doi.org/10.1186/s13195-020-00744-w.
138. Montagne A, Barnes SR, Sweeney MD, et al. Blood-brain barrier breakdown in the aging human hippocampus. *Neuron.* 2015;85(2):296–302. https://doi.org/10.1016/j.neuron.2014.12.032.
139. Agnello L, Piccoli T, Vidali M, et al. Diagnostic accuracy of cerebrospinal fluid biomarkers measured by chemiluminescent enzyme immunoassay for Alzheimer disease diagnosis. *Scand J Clin Lab Invest.* 2020;80(4):313–317. https://doi.org/10.1080/00365513.2020.1740939.
140. Cairns DM, Rouleau N, Parker RN, Walsh KG, Gehrke L, Kaplan DL. A 3D human brain-like tissue model of herpes-induced Alzheimer's disease. *Sci Adv.* 2020;6(19):eaay8828. https://doi.org/10.1126/sciadv.aay8828.
141. Guo P, Benito Ballesteros A, Yeung SP, et al. Covcog 2: cognitive and memory deficits in long COVID: a second publication from the COVID and cognition study. *Front Aging Neurosci.* 2022;14:804937. https://doi.org/10.3389/fnagi.2022.804937.
142. Ciaccio M, Lo Sasso B, Scazzone C, et al. COVID-19 and Alzheimer's disease. *Brain Sci.* 2021;(3):11. https://doi.org/10.3390/brainsci11030305.
143. Shen WB, Logue J, Yang P, et al. SARS-CoV-2 invades cognitive centers of the brain and induces Alzheimer's-like neuropathology. *bioRxiv.* 2022. https://doi.org/10.1101/2022.01.31.478476.

144. Villa C, Rivellini E, Lavitrano M, Combi R. Can SARS-CoV-2 infection exacerbate Alzheimer's disease? An overview of shared risk factors and pathogenetic mechanisms. *J Pers Med.* 2022;12(1). https://doi.org/10.3390/jpm12010029.
145. Ramani A, Muller L, Ostermann PN, et al. SARS-CoV-2 targets neurons of 3D human brain organoids. *EMBO J.* 2020;39(20):e106230. https://doi.org/10.15252/embj.2020106230.
146. Kuo CL, Pilling LC, Atkins JL, et al. APOE e4 genotype predicts Severe COVID-19 in the UK Biobank community cohort. *J Gerontol A Biol Sci Med Sci.* 2020;75(11):2231−2232. https://doi.org/10.1093/gerona/glaa131.
147. Zhou Y, Xu J, Hou Y, et al. Network medicine links SARS-CoV-2/COVID-19 infection to brain microvascular injury and neuroinflammation in dementia-like cognitive impairment. *Alzheimer's Res Ther.* 2021;13(1):110. https://doi.org/10.1186/s13195-021-00850-3.
148. Akiyama H, Barger S, Barnum S, et al. Inflammation and Alzheimer's disease. *Neurobiol Aging.* 2000;21(3):383−421. https://doi.org/10.1016/s0197-4580(00)00124-x.
149. De Felice FG, Tovar-Moll F, Moll J, Munoz DP, Ferreira ST. Severe acute respiratory syndrome coronavirus 2 (SARS-CoV-2) and the central nervous system. *Trends Neurosci.* 2020;43(6):355−357. https://doi.org/10.1016/j.tins.2020.04.004.
150. Lahiri S, Regis GC, Koronyo Y, et al. Acute neuropathological consequences of short-term mechanical ventilation in wild-type and Alzheimer's disease mice. *Crit Care.* 2019;23(1): 63. https://doi.org/10.1186/s13054-019-2356-2.
151. Serrano-Castro PJ, Estivill-Torrus G, Cabezudo-Garcia P, et al. Impact of SARS-CoV-2 infection on neurodegenerative and neuropsychiatric diseases: a delayed pandemic? *Neurologia (Engl Ed).* 2020;35(4):245−251. https://doi.org/10.1016/j.nrl.2020.04.002. Influencia de la infeccion SARS-CoV-2 sobre enfermedades neurodegenerativas y neuropsiquiatricas: inverted question markuna pandemia demorada?
152. Fearon C, Fasano A. Parkinson's disease and the COVID-19 pandemic. *J Parkinsons Dis.* 2021;11(2):431−444. https://doi.org/10.3233/JPD-202320.
153. Park JH, Kim DH, Park YG, et al. Association of Parkinson disease with risk of cardiovascular disease and all-cause mortality: a nationwide, population-based cohort study. *Circulation.* 2020;141(14):1205−1207. https://doi.org/10.1161/CIRCULATIONAHA.119.044948.
154. Jang H, Boltz DA, Webster RG, Smeyne RJ. Viral parkinsonism. *Biochim Biophys Acta.* 2009;1792(7):714−721. https://doi.org/10.1016/j.bbadis.2008.08.001.
155. Tansey MG, Romero-Ramos M. Immune system responses in Parkinson's disease: early and dynamic. *Eur J Neurosci.* 2019;49(3):364−383. https://doi.org/10.1111/ejn.14290.
156. Prasad S, Holla VV, Neeraja K, et al. Parkinson's disease and COVID-19: perceptions and implications in patients and caregivers. *Mov Disord.* 2020;35(6):912−914. https://doi.org/10.1002/mds.28088.
157. Kwon M, Lee JH. Oro-pharyngeal dysphagia in Parkinson's disease and related movement disorders. *J Mov Disord.* 2019;12(3):152−160. https://doi.org/10.14802/jmd.19048.
158. Bhidayasiri R, Virameteekul S, Kim JM, Pal PK, Chung SJ. COVID-19: an early review of its global impact and considerations for Parkinson's disease patient care. *J Mov Disord.* 2020;13(2):105−114. https://doi.org/10.14802/jmd.20042.
159. Chao YX, He BP, Tay SS. Mesenchymal stem cell transplantation attenuates blood brain barrier damage and neuroinflammation and protects dopaminergic neurons against MPTP toxicity in the substantia nigra in a model of Parkinson's disease. *J Neuroimmunol.* 2009; 216(1−2):39−50. https://doi.org/10.1016/j.jneuroim.2009.09.003.
160. Kortekaas R, Leenders KL, van Oostrom JC, et al. Blood-brain barrier dysfunction in parkinsonian midbrain in vivo. *Ann Neurol.* 2005;57(2):176−179. https://doi.org/10.1002/ana.20369.

161. Zhai P, Ding Y, Li Y. The impact of COVID-19 on ischemic stroke. *Diagn Pathol*. 2020;15(1):78. https://doi.org/10.1186/s13000-020-00994-0.
162. Avula A, Nalleballe K, Narula N, et al. COVID-19 presenting as stroke. *Brain Behav Immun*. 2020;87:115–119. https://doi.org/10.1016/j.bbi.2020.04.077.
163. Beyrouti R, Adams ME, Benjamin L, et al. Characteristics of ischaemic stroke associated with COVID-19. *J Neurol Neurosurg Psychiatry*. 2020;91(8):889–891. https://doi.org/10.1136/jnnp-2020-323586.
164. Jain R, Young M, Dogra S, et al. COVID-19 related neuroimaging findings: a signal of thromboembolic complications and a strong prognostic marker of poor patient outcome. *J Neurol Sci*. 2020;414:116923. https://doi.org/10.1016/j.jns.2020.116923.
165. Bhatia R, Srivastava MVP. COVID-19 and stroke: incidental, triggered or causative. *Ann Indian Acad Neurol*. 2020;23(3):318–324. https://doi.org/10.4103/aian.AIAN_380_20.
166. Nian K, Harding IC, Herman IM, Ebong EE. Blood-brain barrier damage in ischemic stroke and its regulation by endothelial mechanotransduction. *Front Physiol*. 2020;11:605398. https://doi.org/10.3389/fphys.2020.605398.
167. Crescenzo F, Marastoni D, Bovo C, Calabrese M. Frequency and severity of COVID-19 in multiple sclerosis: a short single-site report from northern Italy. *Mult Scler Relat Disord*. 2020;44:102372. https://doi.org/10.1016/j.msard.2020.102372.
168. Ferini-Strambi L, Salsone M. COVID-19 and neurological disorders: are neurodegenerative or neuroimmunological diseases more vulnerable? *J Neurol*. 2021;268(2):409–419. https://doi.org/10.1007/s00415-020-10070-8.
169. Israelow B, Song E, Mao T, et al. Mouse model of SARS-CoV-2 reveals inflammatory role of type I interferon signaling. *J Exp Med*. 2020;(12):217. https://doi.org/10.1084/jem.20201241.
170. Takata F, Dohgu S, Matsumoto J, et al. Metformin induces up-regulation of blood-brain barrier functions by activating AMP-activated protein kinase in rat brain microvascular endothelial cells. *Biochem Biophys Res Commun*. 2013;433(4):586–590. https://doi.org/10.1016/j.bbrc.2013.03.036.
171. Wang X, Xue GX, Liu WC, et al. Melatonin alleviates lipopolysaccharide-compromised integrity of blood-brain barrier through activating AMP-activated protein kinase in old mice. *Aging Cell*. 2017;16(2):414–421. https://doi.org/10.1111/acel.12572.
172. Zhang R, Wang X, Ni L, et al. COVID-19: melatonin as a potential adjuvant treatment. *Life Sci*. 2020;250:117583. https://doi.org/10.1016/j.lfs.2020.117583.
173. El-Missiry MA, El-Missiry ZMA, Othman AI. Melatonin is a potential adjuvant to improve clinical outcomes in individuals with obesity and diabetes with coexistence of Covid-19. *Eur J Pharmacol*. 2020;882:173329. https://doi.org/10.1016/j.ejphar.2020.173329.
174. Castle RD, Williams MA, Bushell WC, et al. Implications for systemic approaches to COVID-19: effect sizes of remdesivir, tocilizumab, melatonin, vitamin D3, and meditation. *J Inflamm Res*. 2021;14:4859–4876. https://doi.org/10.2147/JIR.S323356.
175. Laksitorini MD, Yathindranath V, Xiong W, Hombach-Klonisch S, Miller DW. Modulation of Wnt/beta-catenin signaling promotes blood-brain barrier phenotype in cultured brain endothelial cells. *Sci Rep*. 2019;9(1):19718. https://doi.org/10.1038/s41598-019-56075-w.
176. Paolinelli R, Corada M, Ferrarini L, et al. Wnt activation of immortalized brain endothelial cells as a tool for generating a standardized model of the blood brain barrier in vitro. *PLoS One*. 2013;8(8):e70233. https://doi.org/10.1371/journal.pone.0070233.

The effects of lifestyle in Alzheimer's disease during the COVID-19 pandemic

12

Sparsh Ray[1], *Sonia Y. Khan*[1], *Shazma Khan*[1], *Kiran Ali*[1], *Zachery C. Gray*[2], *Pulak R. Manna*[2] *and P. Hemachandra Reddy*[3,4,5,6,7]
[1]School of Medicine, Texas University Health Sciences Center, Lubbock, TX, United States; [2]Department of Internal Medicine, School of Medicine, Texas Tech University Health Sciences Center, Lubbock, TX, United States; [3]Department of Pharmacology and Neuroscience, School of Medicine, Texas Tech University Health Sciences Center, Lubbock, TX, United States; [4]Neurology, Departments of School of Medicine, School of Medicine, Texas Tech University Health Sciences Center, Lubbock, TX, United States; [5]Public Health Department of Graduate School of Biomedical Sciences, School of Medicine, Texas Tech University Health Sciences Center, Lubbock, TX, United States; [6]Department of Speech, Language and Hearing Sciences, School of Health Professions, School of Medicine, Texas Tech University Health Sciences Center, Lubbock, TX, United States; [7]Department of Internal Medicine, Texas Tech University Health Sciences Center, Lubbock, TX, United States

Abstract

Alzheimer's disease (AD) is a multifactorial neurodegenerative disease affected by multiple elements such as exercise, food, and social stimulation. Research has demonstrated the positive effects of exercise such as community-based programs and aerobic activities in reducing rates of decline in cognition. Another protective measure is avoiding red meat and alcohol and instead incorporating a Mediterranean diet to reduce inflammation and inhibit free radicals. Finally, social stimulation can serve to reduce the progression of the disease by increasing a sense of connection and meaningful purpose. COVID-19 has made it difficult for AD patients, especially those living in nursing homes or advanced facilities, to participate in exercise classes due to restrictions, to eat a fresh diet due to resource shortages, and to see friends and family due to social distancing. This chapter delves into the effects of COVID-19 on elements such as physical activity, diet, and social interaction on the disease progression of AD.

Keywords: Aerobic; Alzheimer's disease; Dementia; Diet; Lifestyle; Social

1. Introduction

Lifestyle is a critical factor in slowing down the progression of AD in many patients. With such a multifactorial disease such as Alzheimer's disease, it is important to address several different components that contribute to overall neurocognitive

COVID-19 in Alzheimer's Disease and Dementia. https://doi.org/10.1016/B978-0-443-15256-6.00006-4

degeneration. Examples of these components include diet, exercise, and social interaction. Oxidative stress plays a major role in the progression of AD because of the high oxygen consumption of the brain.[1] Antioxidants can remove or decrease oxidants that cause physiologic changes in the brain in aging. Selenium is an example of an antioxidant, and low levels of this element have been associated with dementia.[1] Additionally, vitamin C and E prevent hyperphosphorylated tau protein dysfunction, while vitamin E specifically has been linked to reduced rates of neuronal death by beta-amyloid plaques within the hippocampus.[1] Understanding the progressive neurodegeneration in the pathogenesis of Alzheimer's disease, patients need to make active lifestyle choices to worsen the progression. The Mediterranean and dietary approaches to stop hypertension diets have been closely associated with improvement in cognitive performance and reduced risk of progression to AD.[1]

2. Exercise

Research has shown that there are significant benefits of exercise for Alzheimer's patients. Recent studies have linked physical exercise to spatial learning and memory, which involve the hippocampus.[2] This study conducted by Cassil has studied the positive impacts of physical activity in both rodents and humans. The hippocampus is made up of the cornus ammonis and the dentate gyrus.[2] The dentate gyrus is the only portion with the ability to generate new neurons, and it was shown this area grew two to three times more following physical activity in rodents.[2] Moreover, following aerobic exercise in adults, an enlarged volume of the left hippocampus proportional to cognitive findings was observed.[2] Neurogenesis is a well-known phenomenon that occurs in the adult hippocampal dentate gyrus. Although aging slows down this process, physical exercise seems to potentiate neuronal generation.[2] Specifically, within elderly people, physical conditioning had a positive correlation with the number of small vessels, indicating angiogenesis.[2] These processes serve as a clear model for the importance of exercise in all populations. Particularly in a population with such intricate pathophysiology, it is important for those suffering from AD to engage in some sort of physical activity.

In the study published by Dr. Anthea Vreugdenhil in the Scandinavian Journal of Caring Sciences, a randomized controlled trial was conducted to assess the effectiveness of a community-based exercise program in people with Alzheimer's.[3] The program consisted of daily exercises and walking under the supervision of a caretaker.[3] The results of the study demonstrated that the exercise group had an increased Mini-Mental Status Exam (MMSE) score by 2.6 points,[3] when compared to the control group that did not do any exercise suggesting improved cognition. These results illustrate the importance of neurocognitive improvement through routine activity, and how these daily exercise programs can improve cognition compared to simply preventing it from deteriorating further.

In an additional study conducted by Dr. Yu, the cognitive effects of aerobic exercise in Alzheimer's disease were studied. Within this study, patients with Alzheimer's performed moderate-intensity cycling for 20–50 min, 3 times per week for 6 months, while the control within the study engaged in light intensity stretching.[4] Following this activity, the participants were assessed with an ADAS-Cog score. This score is utilized as a neuropsychological assessment to determine the severity of symptoms in dementia patients and the higher the number, the greater the degree of cognitive impairment. The ADAS-Cog [AJ4] score of 1 ± 4.6 for cycling and 0.1 ± 4.1 for stretching were lower than the natural 3.2 ± 6.3 increase with natural disease progression.[4] These results suggest that exercise may reduce the decline in global cognition in adults with Alzheimer's disease.[4]

During the COVID-19 pandemic, many individuals were restricted in their activities due to social distancing and quarantine rules. A recent study published in September 2021 done by Bailey looked at the impact of "cocooning" during the COVID-19 pandemic. This study assessed the overall mental and physical health of 150 Irish citizens older than 70 years old during the pandemic. To decrease the risk of being infected with COVID-19, in late March 2020, all people aged 70 years or more were advised to stay home and limit seeing others.[5] Over 40% of participants reported a decline in physical health while "cocooning," while 57% reported feeling loneliness during this time.[5] The study highlights the physical and mental impact that the pandemic had on the elderly population as a whole. As a result, those with Alzheimer's disease were prevented from engaging in physical activity, and this may have contributed to the accelerating AD's neurodegenerative process. Furthermore, many AD patients reside in facilities where they can get a higher level of care. The pandemic disrupted the consistency of activities or exercise that residents typically have as a result of staffing shortages and illnesses.

3. Diet

A 2016 systematic review analyzing publications from 1996 to 2015 found that dietary factors have a strong correlation with AD.[6] Diet can affect AD outcomes by either causing or preventing an increased inflammatory response in the body. Fatty acids, fish oils, and antioxidants are all dietary factors that play a role in preserving cognitive function.[7] Antioxidants in particular help prevent damage from reactive oxygen species and aid in neuronal membrane formation.[8] Neuronal membrane stability is important for preventing synaptic loss in AD pathology.[9] This issue is further complicated by AD patients' decreased desire for food due to damage to the mesial temporal cortex, the center for food intake control. This makes diet an important modifiable factor when approaching the management of Alzheimer's. Unfortunately, economic factors during the COVID-19 pandemic made accessing certain foods more difficult for the general population, including those with AD.

3.1 The Mediterranean diet

The ideal diet for patients with AD is the Mediterranean diet. The Mediterranean diet emphasizes carbohydrates with a low glycemic index (fruits, legumes, vegetables, etc.), proteins with lower fat percentages (turkey, fish, etc.), and utilizes olive oil as the primary fat source.[10,11] Several observational studies show that following the Mediterranean diet has been associated with improved cognitive performance, reduced risk of cognitive impairment, and a slower rate of cognitive decline.[12] The antioxidants in the fruits and vegetables that are part of this diet suppress inflammatory processes via inhibition of free radicals and cytokine production.[13,14] Fish, a major component of this diet, also provides long-chain omega-3 fatty acids that decrease proinflammatory cytokines in microglia to assist in reducing brain inflammation.[15]

3.2 COVID-19 and diet

The pandemic led to significant supply shortages that increased food insecurity at risk and accessibility to good nutrition.[16–21] A survey conducted near the beginning of the pandemic found a ⅓ (32.3%) increase in food insecurity since the pandemic began with over ⅓ (35.5%) of food-insecure households recently food insecure.[22] This is particularly evident in poorer countries where agricultural production is more labor-intensive thus more susceptible to socioeconomic crises.[18,21] This can lead to people consuming more unhealthy diets. Unhealthy diets involve high levels of sugars, fats, and low-level lipoproteins (LDLs) and are low in vitamins, nutrients, and antioxidants, all of which are harmful to immune system function, and result in many health complications, including higher risk of COVID-19 infections.[23,24] Noteworthy, unhealthy foods show a significant correlation to weight gain and obesity which can lead to proinflammatory states, which, in turn, lead to many diseases, including diabetes and CVDs. Increased consumption of junk foods such as French fries, potato chips, candies, and refined grains shows an increase in body mass index, leading to many underlying complications and diseases as well. Diabetes, CVDs, and other medical conditions have been considered as higher risk factors with regards to severe COVID-19 infections and outcomes.[25,26]

Access to food particularly affects the health of the older population as they are often reliant on a care facility, meal delivery services, or family for their meals.[27] As a result of a lack of fresh food options, consumers, in general, turned toward unhealthy food consumption, as more processed foods are easier to access, alongside increased eating frequency and alcohol bingeing.[28] Processed meats, in particular, can lead to higher levels of IL-6 and c-reactive protein (CRP), thus contributing to systemic inflammation and decreased cognitive function.[12] A diet high in saturated fats induces a decrease in both T and B cell maturation and proliferation while also increasing B cell apoptosis thus impairing the adaptive immune system. This causes chronic inflammation against viruses such as COVID-19 and may lead to long-term health consequences such as cognitive decline.[29] Excessive alcohol consumption, drug usage, and smoking are deleterious to the immune system. Alcohol also affects the lungs and their ability to fight off respiratory infections due to a dysfunction of the ciliary movements to remove foreign particles as well as increased oxidative stress. Chronic alcohol intake has been shown to increase a proinflammatory state, hinders the function of immune cells (e.g., macrophages), and affects toll-like receptors by inhibiting ligand binding, which lowers immune responses in combating invaders.[30–32] The proinflammatory state found during a COVID-19 infection is only exacerbated due to alcoholism and further disrupts one's immune responses. Furthermore, substance abuse, involving immunosuppression, has been associated with many serious complications such as CVDs, asthma, cancers, and brain injury. Smoking is another key factor that is detrimental to the immune system and, thus, numerous complications, including lung cancers and CVDs. Of note, ~15% of global death is in some way due to first- or second-hand smoking of tobacco products.[33] Dysregulation of the immune system results in inadequate responses to pathogens through regulatory T cell production, increases pulmonary inflammation through T-helper cell promotion and pathogenic infection severity, and limits overall immune responses to combat infections. The proinflammatory state induced by smoking could increase inflammation due to the cytokine storm, and smoking has been recognized to be detrimental to the lung's health, resulting COVID-19 infections, severity, and fatal outcomes.[34,35] All of these situations lower the immune system to effectively combat COVID-19 or other relevant infections.

4. Social interaction

Alzheimer's disease can have a devastating impact on a person's ability to socialize and remain socially engaged. Impairments in the level of social interaction and emotion recognition have been traced to mild cognitive impairment. Most tangibly, memory loss and decline in cognitive ability in AD patients can impair activities of daily living. The inability to facilitate social interactions, such as driving and using the telephone, make it difficult for patients with AD to connect and maintain meaningful social interactions. Changes in individual behaviors related to social engagement such as personality attributes (apathy, disinhibition, and agitation) are also contributing factors.[36] Patients may also have difficulty expressing themselves, resulting in frustration that may result in further alienation from friends and family. Increased levels of anxiety may also contribute to a decreased inclination for social interaction and may even cause patients to dread interacting with people. One study found that social avoidance was reported in around 33% of AD patients, while a loss of interest was reported in 50%.[37]

While remaining socially connected can be challenging for patients with AD, it is important for their well-being. The level of social interaction is intimately tied to the development and progression of the disease. Social isolation is associated with a 50% increased risk of dementia.[38] Studies have shown that we can prevent cognitive decline and delay the onset of AD by staying mentally active and frequently participating in social activities.[39,40] Social engagement is associated with reduced rates of disability and mortality and may also reduce the risk for depression.[41] Maintaining and building social connections with trusted friends and family increases self-confidence, motivation, and a sense of purpose, which can further provide benefits of better eating habits, mood, and sleep.[42] The absence of such interaction is considered a major source of psychological stress. The positive impact of social interaction is seen on a physiologic level as well, as studies have shown that social interaction rescues AD patients' memory deficit by increasing brain-derived neurotrophic factor (BDNF) expression, increasing synaptic plasticity, and enhancing neurogenesis and cognitive functions.[43,44] Due to the importance of social connection for wellness, many people experiencing symptoms of memory loss move to a skilled care community where they can receive full-time support and the opportunity to maintain social interaction through a variety of social events and gatherings daily.

The COVID-19 pandemic, however, has brought about new concerns for the well-being of patients with AD in regards to social interaction. Within the span of a few weeks, what was once viewed as positive and encouraged, transformed into potentially life-threatening and dangerous. Around the world, the advice given by medical experts to stop the spread of the virus was: social isolation. Especially in long-term care facilities, which provide care for the most vulnerable populations of society, such advice was not taken lightly. Recognizing the high risk associated with these such institutions, the Centers for Disease Control in the United States interim guidelines.[45] Many facilities enforced strict measures to prevent outbreaks by canceling group meetings, halting communal dining and activities, and even prohibiting outside visitors. While social isolation was crucial to protect nursing home residents by decreasing the spread of COVID-19 infection, new concerns about the mental health of patients with AD have begun to develop. The major potential risk of social isolation is the risk for the development of loneliness, which can be especially prominent in patients living in nursing homes and assisted living communities.

Social isolation and loneliness are serious and underappreciated public health concerns that put people at risk for many serious complications.[46] Social isolation can be especially detrimental for patients with AD who are already at increased risk of stress, anxiety, and depression.[47] This fact makes it imperative that careful attention be paid to clinical management of the mental well-being of this population, especially for those living in care homes, since 98% of them present with neuropsychiatric symptoms already.[48,49] Emergence and worsening of neuropsychiatric symptoms have been described in patients with AD dementia after the enforcement of such social isolation measures.[50] One study found that patients with AD caregivers reported having experienced worsening of neuropsychiatric symptoms and significantly lower global cognitive status before social isolation enforcement.[50] In the context of the lockdown policy in long-term care facilities during the COVID-19 pandemic, many of the services and programs crucial for maintaining the social environment are no longer available. For such patients, family visits are a crucial component of their social connection, allowing a link to the outside world. Many family members of these residents visit often, sometimes every day. With limits on visitation during the pandemic, such visits are more limited. If the family and friends of patients with AD cannot visit, the fear that the resident will no longer recognize them may become an uncomfortable reality. The presence of neuropsychiatric symptoms is associated with a more severe progression of cognitive decline in adults with cognitive impairments.[51] It is important to be cognizant of the fact that social isolation to decrease the spread of COVID-19 may come at the cost of potentially profound negative impact on the well-being of patients with Alzheimer's disease.

5. Nursing homes

The COVID-19 pandemic has had devastating effects on nursing homes and other long-term care settings. Nursing homes especially, characterized by many immunosuppressed and elderly people in close quarters, became a "breeding ground" for the virus resulting in increased mortality.[52] A recent study has shown that long-term care residents represent approximately 6% of cases but 40% of deaths. As of December 2020, almost 729,000 residents had gotten the disease and over 100,000 had died.[53] But the impact on nursing home residents is beyond what can be quantified by cases and deaths, as the pandemic adversely has also affected the physical and emotional well-being of residents with loneliness and isolation playing an important role.[54]

While the pandemic has had a negative impact of COVID-19 on residents of nursing homes, it has also disrupted the lives of the professionals caring for this vulnerable population. Delays with COVID-19 testing of residents and staff impaired the overall function of nursing homes during the pandemic. Estimates show that as of late November 2020, at least 322,000 nursing home staff had been infected and more than 1100 had died.[55] Many staff left their jobs at care facilities due to the high viral spread rate in shared living. As a result, 16%−18% of nursing homes had shortages in both nursing staff and aides during the pandemic.[56] The pandemic puts considerable strain on this already vulnerable population, which historically already experiences high levels of turnover, chronic short staffing, and high burnout.[57−59].

Fortunately, the implementation of vaccination efforts in nursing homes continues to increase, which has led to a decrease in morbidity and mortality. As of mid-November, about 87% of residents and 78% of staff were fully vaccinated, up from 86% to 74% in mid-October.[60] Following the vaccine rollout in winter 2020−21, weekly cases and deaths in long-term care facilities dropped to an all-time low in June 2021, after which the Delta variant caused a rapid increase in cases and deaths.[60] Since then, the CDC endorsed an extra dose of the COVID-19 vaccine, which became available to residents of long-term care facilities.[45] In addition, the Biden administration recently announced a vaccination mandate for staff of nursing homes.[61] There does, however, remain a considerable number of vulnerable nursing home residents and unvaccinated staff who could put them at risk.

6. Conclusion

This chapter has touched upon optimal exercise, diet, and social interaction for those with AD. To slow the progression of the disease, steps must be taken to prevent increased neurodegeneration during the pandemic. Examples of this include decreasing sedentary lifestyles and increasing intake of healthy vitamins and minerals.[62] Other preventative measures include daily exercise and keeping in contact with loved ones, even through zoom. These tactics can reduce the rate of decline in patients with AD and have positive long-term implications on their health.

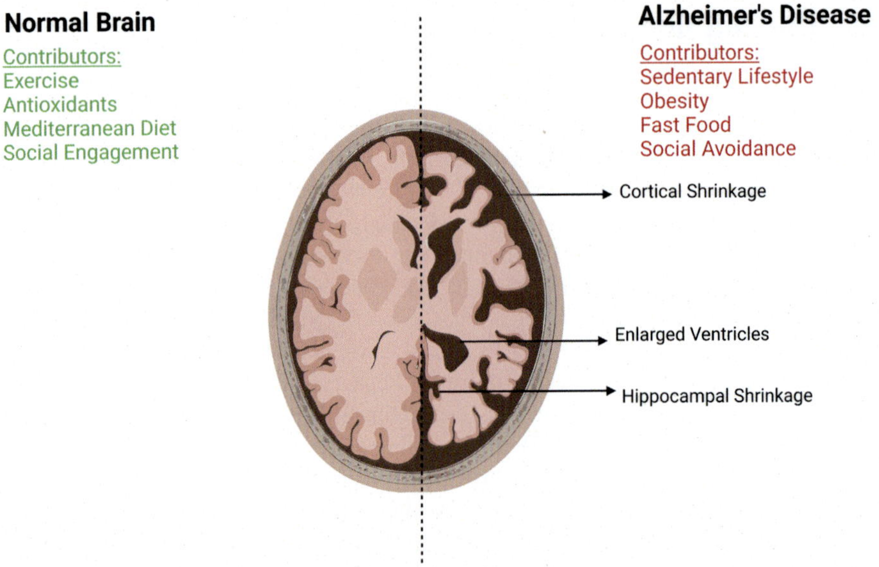

References

1. Cardoso BR, Cominetti C, Cozzolino SM. Importance and management of micronutrient deficiencies in patients with Alzheimer's disease. *Clin Interv Aging*. 2013;8:531–542. https://doi.org/10.2147/CIA.S27983.
2. Cassilhas RC, Tufik S, de Mello MT. Physical exercise, neuroplasticity, spatial learning and memory. *Cell Mol Life Sci*. Mar, 2016;73(5):975–983. https://doi.org/10.1007/s00018-015-2102-0.
3. Vreugdenhil A, Cannell J, Davies A, Razay G. A community-based exercise programme to improve functional ability in people with Alzheimer's disease: a randomized controlled trial. *Scand J Caring Sci*. Mar, 2012;26(1):12–19. https://doi.org/10.1111/j.1471-6712.2011.00895.x.
4. Yu F, Vock DM, Zhang L, et al. Cognitive effects of aerobic exercise in alzheimer's disease: a pilot randomized controlled trial. *J Alzheimers Dis*. 2021;80(1):233–244. https://doi.org/10.3233/JAD-201100.
5. Bailey L, Ward M, DiCosimo A, et al. Physical and mental health of older people while cocooning during the COVID-19 pandemic. *QJM*. Nov 13, 2021;114(9):648–653. https://doi.org/10.1093/qjmed/hcab015.
6. Yusufov M, Weyandt LL, Piryatinsky I. Alzheimer's disease and diet: a systematic review. *Int J Neurosci*. Feb, 2017;127(2):161–175. https://doi.org/10.3109/00207454.2016.1155572.
7. Smith PJ, Blumenthal JA. Diet and neurocognition: review of evidence and methodological considerations. *Curr Aging Sci*. Feb, 2010;3(1):57–66. https://doi.org/10.2174/1874609811003010057.

8. de Wilde MC, Vellas B, Girault E, Yavuz AC, Sijben JW. Lower brain and blood nutrient status in Alzheimer's disease: results from meta-analyses. *Alzheimers Dement (N Y)*. Sep, 2017;3(3):416–431. https://doi.org/10.1016/j.trci.2017.06.002.
9. Kosicek M, Hecimovic S. Phospholipids and Alzheimer's disease: alterations, mechanisms and potential biomarkers. *Int J Mol Sci*. Jan 10, 2013;14(1):1310–1322. https://doi.org/10.3390/ijms14011310.
10. Bach-Faig A, Berry EM, Lairon D, et al. Mediterranean diet pyramid today. Science and cultural updates. *Publ Health Nutr*. 2011;14(12A):2274–2284. https://doi.org/10.1017/S1368980011002515.
11. Davis C, Bryan J, Hodgson J, Murphy K. Definition of the mediterranean diet; a literature review. *Nutrients*. 2015;7(11):9139–9153. https://doi.org/10.3390/nu7115459.
12. McGrattan AM, McGuinness B, McKinley MC, et al. Diet and inflammation in cognitive ageing and alzheimer's disease. *Curr Nutr Rep*. 2019;8(2):53–65. https://doi.org/10.1007/s13668-019-0271-4.
13. Lau FC, Shukitt-Hale B, Joseph JA. Nutritional intervention in brain aging: reducing the effects of inflammation and oxidative stress. *Subcell Biochem*. 2007;42:299–318.
14. Mohammadzadeh Honarvar N, Saedisomeolia A, Abdolahi M, et al. Molecular anti-inflammatory mechanisms of retinoids and carotenoids in alzheimer's disease: a review of current evidence. *J Mol Neurosci*. 2017;61(3):289–304. https://doi.org/10.1007/s12031-016-0857-x.
15. Devassy JG, Leng S, Gabbs M, Monirujjaman M, Aukema HM. Omega-3 polyunsaturated fatty acids and oxylipins in neuroinflammation and management of alzheimer disease. *Adv Nutr*. 2016;7(5):905–916. https://doi.org/10.3945/an.116.012187.
16. R.B.. *The COVID-19 Pandemic: Anticipating its Effects on Canada's Agricultural Trade*. Canadian Journal of Agricultural Economics/Revue canadienne d'agroeconomie; 2020.
17. FAO. *How Is COVID-19 Affecting the Fisheries and Aquaculture Food Systems*. 2020.
18. *A F. COVID-19 Pandemic-Impact on Food and Agriculture Q1: Will Covid-19 Have Negative Impacts on Global Food Security*. Rome, Italy: FAO; 2020.
19. *V V. Coronavirus Could Disrupt Poultry Production*. Poultry World; 2020.
20. *S M. COVID-19: Effects on the Fertilizer Industry*. IHS Market; 2020.
21. Mumtaz M, Hussain N, Baqar Z, Anwar S, Bilal M. Deciphering the impact of novel coronavirus pandemic on agricultural sustainability, food security, and socio-economic sectors-a review. *Environ Sci Pollut Res Int*. 2021;28(36):49410–49424. https://doi.org/10.1007/s11356-021-15728-y.
22. Niles MT, Bertmann F, Belarmino EH, Wentworth T, Biehl E, Neff R. The early food insecurity impacts of COVID-19. *Nutrients*. 2020;(7):12. https://doi.org/10.3390/nu12072096.
23. Chung A, Zorbas C, Riesenberg D, et al. Policies to restrict unhealthy food and beverage advertising in outdoor spaces and on publicly owned assets: a scoping review of the literature. *Obes Rev*. 2022;23(2):e13386. https://doi.org/10.1111/obr.13386.
24. Clemente-Suárez VJ, Ramos-Campo DJ, Mielgo-Ayuso J, et al. Nutrition in the actual COVID-19 pandemic. *A Narrat Rev Nutr*. 2021;13(6). https://doi.org/10.3390/nu13061924.
25. Mozaffarian D, Hao T, Rimm EB, Willett WC, Hu FB. Changes in diet and lifestyle and long-term weight gain in women and men. *N Engl J Med*. 2011;364(25):2392–2404. https://doi.org/10.1056/NEJMoa1014296.
26. Elizabeth L, Machado P, Zinöcker M, Baker P, Lawrence M. Ultra-processed foods and health outcomes: a narrative review. *Nutrients.*. 2020;(7):12 https://doi.org/10.3390/nu12071955.

27. Schrack JA, Wanigatunga AA, Juraschek SP. After the COVID-19 pandemic: the next wave of health challenges for older adults. *J Gerontol A Biol Sci Med Sci.* 2020;75(9):e121−e122. https://doi.org/10.1093/gerona/glaa102.
28. Ammar A, Brach M, Trabelsi K, et al. Effects of COVID-19 home confinement on eating behaviour and physical activity: results of the ECLB-COVID19 international online survey. *Nutrients.* 2020;(6):12. https://doi.org/10.3390/nu12061583.
29. Butler MJ, Barrientos RM. The impact of nutrition on COVID-19 susceptibility and long-term consequences. *Brain Behav Immun.* 2020;87:53−54. https://doi.org/10.1016/j.bbi.2020.04.040.
30. Szabo G, Saha B. Alcohol's effect on host defense. *Alcohol Res.* 2015;37(2):159−170.
31. Cannon AR, Morris NL, Hammer AM, et al. Alcohol and inflammatory responses: highlights of the 2015 alcohol and immunology research interest group (AIRIG) meeting. *Alcohol.* 2016;54:73−77. https://doi.org/10.1016/j.alcohol.2016.06.005.
32. Yeligar SM, Chen MM, Kovacs EJ, Sisson JH, Burnham EL, Brown LA. Alcohol and lung injury and immunity. *Alcohol.* 2016;55:51−59. https://doi.org/10.1016/j.alcohol.2016.08.005.
33. Vardavas CI, Nikitara K. COVID-19 and smoking: a systematic review of the evidence. *Tob Induc Dis.* 2020;18:20. https://doi.org/10.18332/tid/119324.
34. Hu B, Huang S, Yin L. The cytokine storm and COVID-19. *J Med Virol.* 2021;93(1):250−256. https://doi.org/10.1002/jmv.26232.
35. Komiyama M, Hasegawa K. Smoking cessation as a public health measure to limit the coronavirus disease 2019 pandemic. *Eur Cardiol.* 2020;15:e16. https://doi.org/10.15420/ecr.2020.11.
36. Cerejeira J, Lagarto L, Mukaetova-Ladinska EB. Behavioral and psychological symptoms of dementia. *Front Neurol.* 2012;3:73. https://doi.org/10.3389/fneur.2012.00073.
37. Hackett RA, Steptoe A, Cadar D, Fancourt D. Social engagement before and after dementia diagnosis in the English longitudinal study of ageing. *PLoS One.* 2019;14(8):e0220195. https://doi.org/10.1371/journal.pone.0220195.
38. Holt-Lunstad J. Loneliness and social isolation as risk factors: the power of social connection in prevention. *Am J Lifestyle Med.* 2021;15(5):567−573. https://doi.org/10.1177/15598276211009454.
39. Hsiao YH, Chang CH, Gean PW. Impact of social relationships on Alzheimer's memory impairment: mechanistic studies. *J Biomed Sci.* 2018;25(1):3. https://doi.org/10.1186/s12929-018-0404-x.
40. Berkman LF, Glass T, Brissette I, Seeman TE. From social integration to health: durkheim in the new millennium. *Soc Sci Med.* 2000;51(6):843−857. https://doi.org/10.1016/s0277-9536(00)00065-4.
41. Norton S, Matthews FE, Barnes DE, Yaffe K, Brayne C. Potential for primary prevention of Alzheimer's disease: an analysis of population-based data. *Lancet Neurol.* 2014;13(8):788−794. https://doi.org/10.1016/S1474-4422(14)70136-X.
42. Forstmeier S, Maercker A. Motivational processes in mild cognitive impairment and Alzheimer's disease: results from the Motivational Reserve in Alzheimer's (MoReA) study. *BMC Psychiatr.* 2015;15:293. https://doi.org/10.1186/s12888-015-0666-8.
43. Hsiao YH, Hung HC, Chen SH, Gean PW. Social interaction rescues memory deficit in an animal model of Alzheimer's disease by increasing BDNF-dependent hippocampal neurogenesis. *J Neurosci.* 2014;34(49):16207−16219. https://doi.org/10.1523/JNEUROSCI.0747-14.2014.

44. Nagahara AH, Merrill DA, Coppola G, et al. Neuroprotective effects of brain-derived neurotrophic factor in rodent and primate models of Alzheimer's disease. *Nat Med.* 2009; 15(3):331–337. https://doi.org/10.1038/nm.1912.
45. Interim Infection Prevention and Control Recommendations to Prevent SARS-CoV-2 Spread in Nursing Homes. U.S. Department of Health and Human Services.
46. National Academies of Sciences E, and Medicine. *Social Isolation and Loneliness in Older Adults: Opportunities for the Health Care System.* 2020.
47. Lara B, Carnes A, Dakterzada F, Benitez I, Pinol-Ripoll G. Neuropsychiatric symptoms and quality of life in Spanish patients with Alzheimer's disease during the COVID-19 lockdown. *Eur J Neurol.* 2020;27(9):1744–1747. https://doi.org/10.1111/ene.14339.
48. Devita M, Bordignon A, Sergi G, Coin A. The psychological and cognitive impact of Covid-19 on individuals with neurocognitive impairments: research topics and remote intervention proposals. *Aging Clin Exp Res.* 2021;33(3):733–736. https://doi.org/10.1007/s40520-020-01637-6.
49. Velayudhan L, Aarsland D, Ballard C. Mental health of people living with dementia in care homes during COVID-19 pandemic. *Int Psychogeriatr.* 2020;32(10):1253–1254. https://doi.org/10.1017/S1041610220001088.
50. Boutoleau-Bretonniere C, Pouclet-Courtemanche H, Gillet A, et al. The effects of confinement on neuropsychiatric symptoms in alzheimer's disease during the COVID-19 crisis. *J Alzheimers Dis.* 2020;76(1):41–47. https://doi.org/10.3233/JAD-200604.
51. Campbell NL, Unverzagt F, LaMantia MA, Khan BA, Boustani MA. Risk factors for the progression of mild cognitive impairment to dementia. *Clin Geriatr Med.* 2013;29(4): 873–893. https://doi.org/10.1016/j.cger.2013.07.009.
52. McGarry BE, Grabowski DC, Barnett ML. Severe staffing and personal protective equipment shortages faced by nursing homes during the COVID-19 pandemic. *Health Aff.* 2020; 39(10):1812–1821. https://doi.org/10.1377/hlthaff.2020.01269.
53. Kaiser family foundation. Kaiser family foundation state reports of long-term care facility cases and deaths related to COVID-19. Accessed 12/18/2021,
54. Levere M, Rowan P, Wysocki A. The adverse effects of the COVID-19 pandemic on nursing home resident well-being. *J Am Med Dir Assoc.* 2021;22(5):948–954 e2. https://doi.org/10.1016/j.jamda.2021.03.010.
55. Nursing Home COVID-19 Public File. Centers for Disease Control and Prevention National Healthcare Safety Network. Accessed 11/2/2021,
56. Xu H, Intrator O, Bowblis JR. Shortages of staff in nursing homes during the COVID-19 pandemic: what are the driving factors? *J Am Med Dir Assoc.* 2020;21(10):1371–1377. https://doi.org/10.1016/j.jamda.2020.08.002.
57. Kash BA, Castle NG, Naufal GS, Hawes C. Effect of staff turnover on staffing: a closer look at registered nurses, licensed vocational nurses, and certified nursing assistants. *Gerontol.* 2006;46(5):609–619. https://doi.org/10.1093/geront/46.5.609.
58. White EM, Aiken LH, McHugh MD. Registered nurse burnout, job dissatisfaction, and missed care in nursing homes. *J Am Geriatr Soc.* 2019;67(10):2065–2071. https://doi.org/10.1111/jgs.16051.
59. Geng F, Stevenson DG, Grabowski DC. Daily nursing home staffing levels highly variable, often below CMS expectations. *Health Aff (Millwood).* 2019;38(7):1095–1100. https://doi.org/10.1377/hlthaff.2018.05322.
60. Priya Chidambaram RG. COVID-19 cases and deaths in long-term care facilities through June 2021. KFF. Accessed 11/20/2021,

61. *Biden to Require COVID Vaccines for Nursing Home Staff.* AP News; 2021. https://apnews.com/article/business-health-coronavirus-pandemic-nursing-homes-2e6189cd41068b1e0-f643ee7e4bfbb92. Accessed September 1, 2021.
62. John A, Ali K, Marsh H, Reddy PH. Can healthy lifestyle reduce disease progression of Alzheimer's during a global pandemic of COVID-19? *Age Res Rev*. 2021;70:101406. https://doi.org/10.1016/j.arr.2021.101406.

Dementia and COVID-19: An African American focused study

13

Shyam Sheladia [1], *Shivam Sheladia* [1], *Rishi Virani* [1,2] *and*
P. Hemachandra Reddy [1,3,4,5]
[1]Department of Internal Medicine, Texas Tech University Health Sciences Center, Lubbock, TX, United States; [2]Department of Biomedical Engineering, The University of Texas at Austin, Austin, TX, United States; [3]Department of Pharmacology and Neuroscience, School of Medicine, Texas Tech University Health Sciences Center, Lubbock, TX, United States; [4]Neurology, Departments of School of Medicine, School of Medicine, Texas Tech University Health Sciences Center, Lubbock, TX, United States; [5]Public Health Department of Graduate School of Biomedical Sciences, School of Medicine, Texas Tech University Health Sciences Center, Lubbock, TX, United States

Abstract

Dementia and coronavirus disease 2019 (COVID-19) are two separate illnesses responsible for high levels of morbidity and mortality within the general US population. Both diseases have differing origins, symptoms, and pathological processes within the human body as dementia is primarily a neurological disease, while COVID-19 is a disease of the respiratory system. Nevertheless, minority racial/ethnic groups within the United States, such as African Americans, face one of the highest burdens of both dementia and COVID-19 due to four specific risk factor categories: (1) unmodifiable risk factors, (2) modifiable risk factors, (3) age-related chronic diseases, and (4) environmental risk factors. Unmodifiable risk factors include increasing age and predisposing genetics. The major modifiable risk factors of concern are low income/socioeconomic status, low educational attainment, lack of exercise, poor diet, and smoking alongside the usage of tobacco products. Additionally, the higher prevalence of age-related chronic diseases such as diabetes, kidney disease, hypercholesterolemia, cardiovascular disease, and chronic lung diseases within the African American community places them at a higher risk for the future development of dementia as well as a fatal COVID-19 infection. Lastly, the African American population within the United States faces additional environmental risk factors, such as social inequalities and lack of access to healthcare, due to pre-existing systematic biases. With African Americans being one of the largest racial/ethnic minority groups with in the United States, this particular chapter will only focus upon the research and statistics associated with dementia and COVID-19 within the African American population of the United States. This chapter will also explore each of the four aforementioned risk factor categories in further detail as they greatly contribute to the development of dementia and COVID-19 within the African American population.

Keywords: African Americans; Age-related chronic diseases; Alzheimer's disease; COVID-19; Dementia; Risk factors; Social inequalities.

COVID-19 in Alzheimer's Disease and Dementia. https://doi.org/10.1016/B978-0-443-15256-6.00010-6

1. Introduction

Dementia and coronavirus disease 2019 (COVID-19) are both unique diseases with different effects on the human body and health. Dementia is primarily a neurological disease resulting in the loss of neurological function over a long period of time. COVID-19 is a viral disease that primarily debilitates the respiratory system with a more acute timeline and set of symptoms.

Despite the seemingly different nature of the two diseases, there are a number of key similarities among the risk factors that contribute to their development. This chapter will primarily focus on the four types of risk factor categories that lead to dementia and COVID-19: (1) unmodifiable risk factors, (2) modifiable risk factors, (3) age-related chronic diseases, and (4) environmental risk factors.

Furthermore, this chapter will narrow in on the aforementioned concepts in the context of the African American subpopulation as they constitute roughly 13.4% of the total US population with an expected steady increase as seen in Fig. 13.1. The most important lesson of this chapter is that African Americans carry a high degree of risk from all four of the aforementioned risk factor categories, which place them in a rather precarious position of being more likely to develop dementia, to contract a COVID-19 infection, or to be a victim of both.

2. Dementia/COVID-19

2.1 What is dementia?

Generally speaking, dementia is primarily characterized by a loss of neurological functions which may include an inhibited ability to speak, solve problems, and recall past events or information.[1] Alzheimer's disease (AD) is the primary causative agent behind the development of dementia in most humans and is actually responsible for

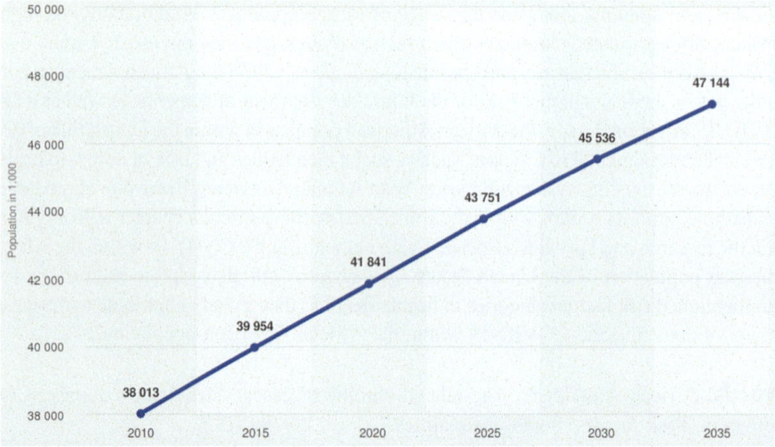

Figure 13.1 Projection of the African American population in the United States from 2010 to 2035. Chart created by Statista.

roughly 60%–80% of all dementia cases worldwide.[2] Other causes of dementia include cerebrovascular disease, traumatic brain injury, Lewy body disease, Parkinson's disease, hippocampal sclerosis, and fronto-temporal lobar degeneration.[2]

Dementia is an incredibly debilitating neurological disease that can greatly decrease the standard of living for an individual. Individuals with dementia often cannot remember their own family members, childhood, and recent events.[3] Furthermore, individuals with dementia are at a significantly higher risk of developing depression, anxiety, agitation, hallucinations, debilitating paranoia, and stark personality changes.[3] It is precisely due to this incredibly debilitating nature of dementia that it warrants a special discussion.

2.2 Prevalence of dementia in the US African American population

Currently, there are roughly 6.2 million Americans over the age of 65 suffering from AD, a statistic that is expected to increase to 7.2 million by 2025, to 12.7 million by 2050, and to 13.8 million by 2060.[4] The number of AD cases among the African American population is estimated to be 14%–100% greater than that among the non-Hispanic white American population.[5] Furthermore, researchers project that there will be 2.2 million African Americans with AD just by the year 2060.[6]

2.3 What is COVID-19?

Coronaviruses are a family of viruses named based on their crown-like form when viewed under an electron microscope. As shown in Fig. 13.2, the characteristic feature of viruses belonging to this family is spike glycoproteins that are used for attachment and binding to alveolar epithelial cells in humans. This is the primary method by which coronaviruses propagate and cause the onset of viral symptoms. While many patients experience moderate symptoms such as cough, increased mucous production, headache, loss of taste, diarrhea, sore throat, fatigue, and body aches, others are not as

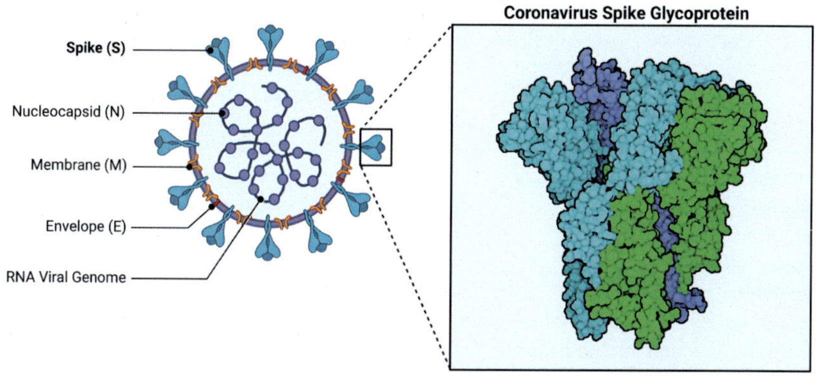

Figure 13.2 An in-depth look into the structure of the coronavirus spike glycoprotein.

fortunate.[7] In fact, roughly 17% of all individuals diagnosed with COVID-19 suffer life-threatening complications.[8] Such complications may encompass several or all of the following conditions: acute respiratory failure, pneumonia, acute kidney injury (AKI), acute cardiac injury, chronic fatigue, septic shock, blood clots, acute liver injury, rhabdomyolysis, and disseminated intravascular coagulation.[8] In many cases, patients are often afflicted with long-term COVID-19 symptoms, coined the term "long-haul COVID."[9] These conditions may include lasting damage to critical organs (such as the lungs, heart, and brain), chronic cough, fever, mood changes, and difficulty breathing.[9]

2.4 Prevalence of COVID-19 in the US African American population

There have been roughly 46 million cases and over 750,000 total deaths due to COVID-19 just within the United States as of November 2021.[10]

Narrowing in on the African American population, it is evident from Table 13.1 that African Americans alone account for 11.8% of all COVID-19 cases in the United States and 13.9% of all COVID-19-related deaths in the United States.[10] Upon closer examination of Table 13.2, it is clear from the two aforementioned pieces of data that African Americans are disproportionately affected by COVID-19 as they make up only 13.4% of the total population.[11]

2.5 Risk factors causing dementia and COVID-19

The major unmodifiable risk factors contributing to the development of AD are increasing age and predisposing genetics.[12] Modifiable risk factors for AD generally constitute income/socioeconomic status, education, diet, and exercise.[12] However, it

Table 13.1 Data as of November 01, 2021. COVID-19 information from the United States Centers for Disease Control and Prevention.[10]

COVID-19 cases and deaths by race/ethnicity for the United States		
Race/ethnicity	**Percent of cases (%)**	**Percent of deaths (%)**
White alone (not Hispanic or Latino)	52.5	59.7
Hispanic or Latino	26.1	17.8
Black or African American alone	11.8	13.9
Asian alone	3.0	3.5
American Indian and Alaska Native alone	1.1	1.1
Native Hawaiian and Pacific Islander alone	0.3	0.2
Multiple races and other races	5.2	3.8

Table 13.2 Most recent data available. Population information from the United States Census Bureau, 2019 data.[11]

Population breakdown by race/ethnicity for the United States (2019)		
Race/ethnicity	**Population percentage (%)**	**Population count**
White alone (not Hispanic or Latino)	60.1	197,271,953
Hispanic or Latino	18.5	60,724,311
Black or African American alone	13.4	43,984,096
Asian alone	5.9	19,366,132
American Indian and Alaska Native alone	1.3	4,267,114
Native Hawaiian and Pacific Islander alone	0.2	656,479
Multiple races and other races	2.8	9,190,706

is also important to note that age-related chronic diseases such as diabetes, kidney disease, hypercholesterolemia, cardiovascular disease (CVD), and chronic lung diseases also contribute to the development of AD.[12] Finally, it is important to note that out of all of the aforementioned risk factors, an advancing age is the single greatest risk factor that contributes to the development of AD.[12]

COVID-19 has many similarities to AD with regards to risk factors. In terms of unmodifiable risk factors, the main factors are increasing age and race/ethnicity.[13] Furthermore, the major modifiable risk factors that contribute to the contraction of COVID-19 include income/socioeconomic status, education, exercise, diet, and smoking/tobacco usage.[13] Similar to AD, age-related chronic diseases such as diabetes, kidney disease, hypercholesterolemia, CVD, chronic lung diseases, and dementia all contribute to an increasing likelihood of a COVID-19 infection.[13] Finally, similar to AD, out of all of the aforementioned risk factors, an advancing age is the single greatest risk factor that contributes to a severe illness, death, or long-term health issues from a COVID-19 infection.[13]

Here, it is crucial to understand that there is great overlap between the risk factors for AD and COVID-19. In other words, the vast majority of unmodifiable risk factors, modifiable risk factors, and age-related chronic diseases that contribute to the development and progression of AD also contribute to a COVID-19 infection and severe health complications from it.

However, the risk factors for both of these diseases are not just limited to the three aforementioned categories. A fourth category of risk factors can be defined here: environmental risk factors. Environmental risk factors for these diseases can be defined as the vast variety of systematic inequalities that assign minority groups, such as African Americans, to health and social inequalities, which force them to face several additional barriers in accessing healthcare resources.[13] These inequalities are a large reason

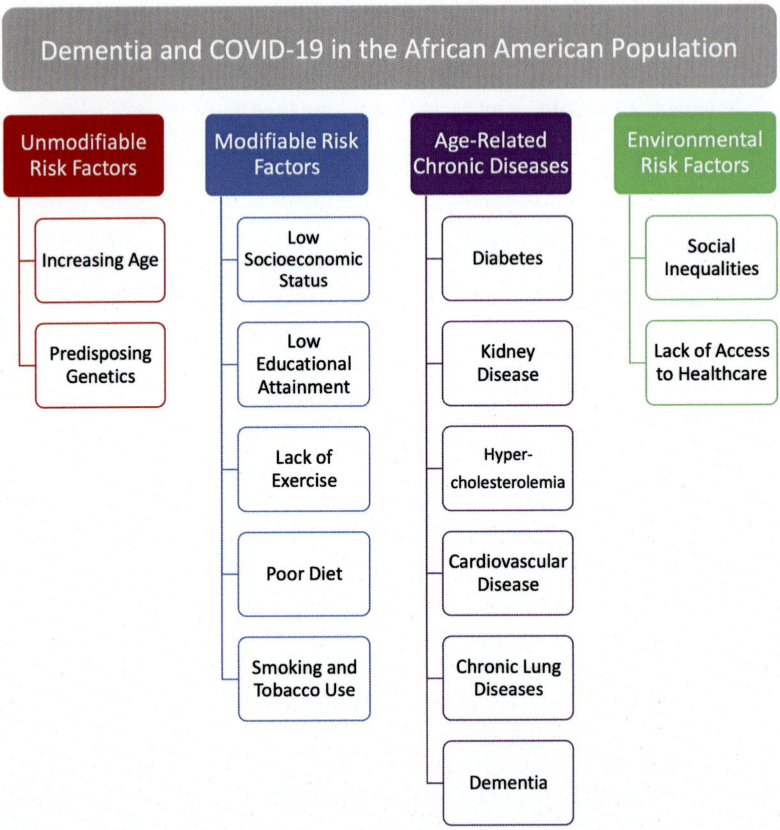

Figure 13.3 Summary of the disadvantages faced by African Americans in overcoming the burden of Alzheimer's disease, dementia, as well as a COVID-19 infection.

as to why an explicit discussion on the relationship between the African American population and the aforementioned risk factors is strongly warranted.

As can be seen in the hierarchy map shown in Fig. 13.3, this chapter will discuss each of the four major categories of risk factors: unmodifiable risk factors, modifiable risk factors, age-related chronic diseases, and environmental risk factors with greater detail in regards to AD, dementia, COVID-19, and the African American demographic of the United States.

3. Unmodifiable risk factors

3.1 Increasing age

As mentioned previously, elderly age is the single greatest risk factor out of all four major risk factor categories for the development of AD, dementia, or a severe COVID-19 infection.

In 2020, the United States Census Bureau reported that approximately 5.21 million African Americans within the United States were age 65 years and above.[14] This number is expected to rise to 8.97 million by the year 2040 and up to 12.15 million by the year 2060.[14]

3.2 Predisposing genetics

It has been well established within the medical literature that apolipoprotein E (APOE) ε4 allele inheritance carries the strongest association for the future development of AD and dementia.[5] In fact, APOE ε4 allele inheritance increases an individual's risk of developing AD by a factor of two.[12] Nevertheless, the association between the future risk of developing AD and APOE ε4 allele inheritance is rather inconsistent within the African American population.[15] To be specific, the presence of APOE ε4 allele inheritance is a strong determinant of future AD risk in non-Hispanic white individuals.[16]

However, research conducted within the past decade has revealed that the ABCA7 gene is a definitive genetic risk factor for the development of AD among the African American community.[17] ABCA7 is a gene, which regulates the homeostasis of cholesterol and lipids within the body.[17,18] This indicates that lipid metabolism may play an important role in the development of AD within the African American population.[17,18] Extensive studies conducted by neurologists from Columbia University Medical Center revealed that African Americans with a variant of the ABCA7 gene have a twofold risk of developing AD as compared to African Americans who lack the ABCA7 gene variant.[17]

As of November 2021, a total of 13 genome-wide significant loci have been reported to be associated with a COVID-19 infection.[19] To be specific, four of the genome-wide significant loci have been linked to an individual's risk of contracting a COVID-19 infection, while the remaining nine genome-wide significant loci have been linked to the severity of the disease a COVID-19 infection will cause within an individual's body.[19] Nevertheless, no specific genetic links have been made between a COVID-19 infection and the African American population at large. The research exploring the specific genes associated with a COVID-19 infection is being conducted on an international scale as of November 2021.[19] The collective international effort is known as *The COVID-19 Host Genetics Initiative*.[19]

4. Modifiable risk factors

4.1 Income/socioeconomic status

The United States Census Bureau reported in 2019 that the median household income for all race/ethnicities was approximately $68,703.[20] However, the median household income for African Americans in 2019 was the lowest among any other racial/ethnic group at only $46,073.[21] As depicted in Fig. 13.4, African Americans within the United States have historically been at a disadvantage as they always fall at the bottom of the charts with regards to median household income.[21]

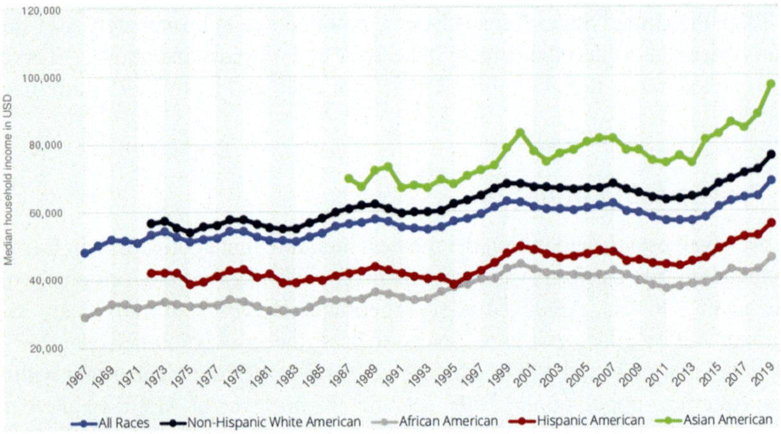

Figure 13.4 Median household income by race/ethnicity from 1967 to 2019.
Chart created by Statista.[21]

The effects of low median household incomes within the African American community on a yearly basis directly lead to an increase within the poverty level of African Americans. To be specific, the United States Census Bureau reported a poverty percentage of 10.5% in 2019 for the general US population.[22] As shown in Table 13.3, the African American population registered a poverty percentage of 21.2%, which is greater than two times that of the general United States population.[23] As a result,

Table 13.3 Most recent data available. Poverty information from the United States Census Bureau, 2019 ACS 1-year estimates data.[23] Education information from the United States Census Bureau, 2019 ACS 1-year estimates data.[25]

Poverty level and educational attainment by race/ethnicity for the United States (2019)

Race/ethnicity	Percent individuals below poverty level (%)	Percent individuals holding bachelor's degree or higher (%)
White alone (not Hispanic or Latino)	10.3	34.4
Hispanic or Latino	17.2	17.6
Black or African American alone	21.2	22.5
Asian alone	9.6	55.6
American Indian and Alaska Native alone	23.0	16.1
Native Hawaiian and Pacific Islander alone	16.5	18.1
Multiple races and other races	15.2	33.4

the data show that over one in five African Americans in 2019 lived in poverty within the United States. On the other hand, non-Hispanic white Americans accounted for 60.1% of the total US population as shown in Table 13.2 but registered a poverty percentage level of 10.3%, which is lower than the national average.[11,23] The yearly trend of extremely low median household incomes combined with an increase in the overall African American population will only continue to increase the poverty gap between African Americans and other racial/ethnic groups within the United States.

Here, it is important to discuss the link between poverty and increased rates of COVID-19 and AD. Individuals from disadvantaged backgrounds are less likely to have access to adequate healthcare, knowledge about healthy habits (especially in the context of COVID-19), healthy and nutritive foods, and have a lower quality of physical and mental health.[24] As such, it has been found in several studies that individuals from less financially stable backgrounds are more prone to COVID-19.[24] Furthermore, the same aforementioned consequences of a lack of financial stability causes disadvantaged individuals to be at a higher risk for developing AD.

4.2 Education

With regards to educational attainment, a similar negative outcome is seen when African Americans are compared to the US general population. To be specific, the percentage of individuals who held a bachelor's degree or higher within the United States in 2019 was 32.1%, while African Americans drastically fell below the US national average at only 22.5% as seen in Table 13.3.[22,25] Furthermore, the data from Table 13.3 reveal that African Americans registered a significantly lower level of educational attainment than Asian Americans and non-Hispanic white Americans.[25]

The demographic data presented previously, combined with the data presented in this section, reveal that African Americans are disproportionately disadvantaged on *both* ends of the spectrum with regards to income/socioeconomic status and educational attainment. As a result, both disadvantages can work to exacerbate each other where (1) a low income/socioeconomic status leads to a lower amount of educational enrichment opportunities and (2) a poor education leads to a lower income/socioeconomic status. Therefore, members of the African American community are often stuck within this cycle that carries on from one generation to the next. Furthermore, it is important to discuss the relevance of education in the context of AD. Several studies have found that individuals who have not graduated high school consistently exhibit unhealthier lifestyles and diets, which greatly contribute to the development of AD.[26] Furthermore, similar studies explain that individuals with higher levels of education have greater *cognitive reserve*, which is simply the brain's ability to function despite damage.[26]

4.3 Exercise and diet

Exercise and diet are both key modifiable risk factors that play a crucial role in the development of several age-related chronic diseases, which ultimately leads to the future development of AD, dementia, or a severe COVID-19 infection. Data collected by various national agencies within the United States revealed alarming patterns of inactivity and poor diet throughout the overall African American community.

With regards to exercise, the Centers for Disease Control and Prevention (CDC) revealed that African Americans had one of the highest prevalences of physical inactivity at 30.3% in 2020.[27] Furthermore, the American Heart Association reported in 2018 that only 19.9% of African Americans over the age of 18 regularly met the aerobic and muscle-strengthening guidelines set by the U.S. Department of Health and Human Services.[28]

With regards to diet, the American Heart Association reported that in 2016, approximately 58.3% of African Americans were classified as having an unhealthy overall diet.[28] The alarming rates of unhealthy overall diets throughout the African American community can be attributed to what is known as the "Southern diet."[29] A study conducted by the University of Alabama at Birmingham revealed that African Americans are more likely to adhere to a Southern diet, which is known to heavily rely upon unhealthy products such as fried foods, organ meats, processed meats, dairy, sugar-sweetened beverages, and bread.[29] Furthermore, the argument can be made that a lower income/socioeconomic status combined with a poor education gives African Americans a greater access and need to rely upon unhealthy products.

An unhealthy diet combined with the lack of physical activity directly leads to African Americans having the highest prevalence of obesity among all racial/ethnic groups within the United States at 49.6% as reported by the CDC in 2018.[30] Furthermore, the CDC reported that 75.7% of African Americans were overweight in 2018.[30] The extremely high rates of obesity and overweight individuals throughout the African American community place them at a greater risk of developing several age-related chronic diseases, which directly contribute to the future development of AD, dementia, or a severe COVID-19 infection.

In terms of the specific link between obesity and COVID-19, obesity decreases the capacity of the lungs to hold oxygen and also decreases lung reserves, which makes ventilation more difficult and subsequently increases the risk of COVID-19-related hospitalizations.[31] Furthermore, there is an intimate link between obesity and AD: obesity results in high levels of inflammation, toxicity to brain cells, and decreased blood flow to the brain.[32] When the aforementioned conditions persist for long periods of time, the brain naturally degrades, resulting in irreversible AD.[32]

4.4 Smoking and tobacco use

Smoking and the overall usage of tobacco products are one of the leading preventable modifiable risk factors within the African American community. Furthermore, smoking and the usage of tobacco products place African Americans at a higher risk to the development of age-related chronic diseases such as diabetes, kidney disease, CVD, and chronic lung diseases. These age-related chronic diseases in turn increase the risk for African Americans to develop AD, dementia, or a severe COVID-19 infection in the future.

The CDC reports that African Americans smoke or use tobacco products at higher rates than the general US population.[33] To be specific, 14.9% of African American adults above the age of 18 years reported smoking or using tobacco products, while the US national average was 14.0%.[33] Furthermore, African Americans have higher smoking and tobacco-related death rates on a yearly basis than non-Hispanic white Americans due to an increased risk of developing age-related chronic diseases.[34]

One of the most significant factors responsible for the increased rates of smoking and tobacco usage within the African American community is the fact that tobacco companies aggressively target young African Americans through advertising and promotions within urban communities.[33] The CDC reports that major tobacco companies have historically attempted to maintain a positive image within the African American community by sponsoring community and cultural events.[33] Furthermore, the CDC reports that a greater number of tobacco retailers are located within regions where the African American population density is high.[33] Additionally, data show that African Americans are less successful at quitting the usage of tobacco products due to the fact that they have less access to cessation treatments such as professional counseling and medication.[33] The brutal combination of increased usage of tobacco products with decreased access to proper healthcare and cessation treatments leads to smoking being one of the leading preventable causes of morbidity and mortality within the African American population in the United States.[33]

As one can reasonably assume, there is an intimate connection between COVID-19 and smoking. More specifically, smokers have weaker lungs than individuals who do not smoke.[35] Therefore, when they are exposed to infections that primarily target the airways, they fare much worse than nonsmokers; this is especially true for individuals with COVID-19 infections.[35] Smoking also increases the risk of AD and associated dementia. More specifically, toxins that are inhaled when one smokes greatly increase inflammation and oxidative stress, which are intimately linked to the development of AD.[36]

5. Age-related chronic diseases

5.1 Diabetes

The proportion of African Americans with diabetes, or at risk of developing diabetes, is significantly higher than non-Hispanic white individuals. More specifically, African Americans are 60% more likely than non-Hispanic white individuals to be given a diabetes diagnosis within the United States.[37] To further illustrate the issue in the context of each demographic taken separately, according to the CDC's National Health Interview Survey conducted in 2018, 13% of African Americans have diabetes while just 8% of non-Hispanic white Americans have diabetes.[38]

The issue of diabetes disproportionately affecting the African American population takes on a more serious and urgent significance when one considers the ramifications it has on the frequency of AD and the subsequent occurrence of dementia within the African American population of the United States as the two diseases have been shown to display a mutually comorbid association.[39] This is further supported by the conclusion that individuals with type 2 diabetes are twice as likely to develop AD than normally functioning nondiabetic individuals.[40] It is important to note that although research is currently ongoing in efforts to better understand the link between diabetes and AD, it is suspected that the link between these two conditions lies in the way that type 2 diabetes affects the body's ability to utilize glucose and effectively respond to insulin.[41]

However, the collateral effects of diabetes within the African American population are not simply limited to an increased risk of developing AD. Increased frequency of diabetes among individuals of the African American population provides impetus for concern regarding a COVID-19 infection as well. Given the contemporary nature of the COVID-19 pandemic, as of November 2021, not enough data has been provided to support the notion that individuals with diabetes are more likely to *contract* COVID-19 than those not suffering from diabetes. However, there is significant evidence from recent studies showing that diabetic individuals are more prone to a severe infection, long-term complications (commonly known as "long-haul COVID"), or death from a COVID-19 infection as compared to the general population.[42] As a clear reference, data published in December 2020 showed that roughly 25% of severe COVID-19 cases involved diabetic patients.[42]

Thus, from the fact that the African American population is disproportionately affected by diabetes comes the clear implication that it is also at a greater risk of developing AD, dementia, and severe COVID-19 infections.

5.2 *Kidney disease*

African Americans are about four times more likely than non-Hispanic white Americans to develop kidney disease and subsequent kidney failure according to the National Kidney Foundation.[43] Hypertension, obesity, heart disease, and diabetes (all diseases that African Americans are more likely to develop as compared to non-Hispanic white Americans) can be implicated as the agents causing a disproportionate number of kidney disease cases among the African American population.[43]

Studies have shown that patients of chronic kidney disease (CKD) are more likely to develop AD and that the cognitive function of CKD patients significantly improves upon receiving a kidney transplant.[44] Both of the aforementioned findings have provided sufficient evidence for the notion that there is a possible link between CKD, AD, and general cognitive decline.[44] Furthermore, when comparing the risk factors that contribute to AD and dementia, it has been statistically proven that approximately 10% of all dementia cases are due to an estimated glomerular filtration rate (eGFR) that is less than 60 mL/min/1.73 m^2, which shows that a higher proportion of dementia cases can be attributed to kidney disease as compared to other dementia risk factors (mainly diabetes and cardiovascular complications).[45] It is important to note however that current research only points to an associative relationship between eGFR and dementia as no causative mechanisms/relationships have been identified yet.[46]

Similar implications are present for the link between CKD and COVID-19 infections. This is due to the fact that many CKD patients with kidney transplants have to take antirejection medications, which function to decrease overall immune system activity.[47] As a result, they are more susceptible to both contracting COVID-19 and suffering a more serious long-term infection.[47] Furthermore, early reports have suggested that approximately 57% of hospitalized patients with COVID-19 develop AKI.[48] However, irrespective of whether patients develop a case of AKI or CKD over the course of their COVID-19 hospitalization, it has been shown that COVID-19 has the potential to worsen pre-existing CKD or cause novel AKI within a patient.[48]

Thus, given the fact that African Americans are approximately four times more likely than non-Hispanic white Americans to develop kidney disease, they are also at a greater risk of developing AD and dementia and are more likely to suffer from severe COVID-19 infections that have consequences such as long-term complications, amplification of pre-existing kidney disease, or the development of novel kidney disease.

5.3 Hypercholesterolemia

As per reports from the American Heart Association, African Americans have slightly lower cholesterol counts than non-Hispanic white Americans (44.8% of African American males have borderline/high cholesterol as compared to 47.9% of non-Hispanic white males; 42.1% of African American females have borderline/high cholesterol levels as compared to 49.7% of non-Hispanic white females).[49] However, African American males *do* have slightly higher counts of low density lipoprotein cholesterol (LDL-C) than non-Hispanic white males (32.4% of African American males have borderline/high levels of LDL-C; 31.7% non-Hispanic white males have borderline/high levels of LDL-C).[49] However, despite these very *slight* differences in total cholesterol and LDL-C counts between the African American and non-Hispanic white American populations, there are *still* discernible consequences related to hypercholesterolemia that disproportionately affect African Americans. More specifically, coronary heart disease (CHD) mortality among African Americans is much higher than among non-Hispanic white Americans.[50] Furthermore, the decline in CHD occurrence is much less within the African American population than in the non-Hispanic white American population.[50] It is important to note that there are three primary suspects that are responsible for the aforementioned disparity with regards to CHD within the African American population and the non-Hispanic white American population: (1) African Americans are significantly less likely than non-Hispanic white Americans to undergo serum cholesterol screening, (2) when diagnosed with high cholesterol levels, African Americans are less likely than non-Hispanic white Americans to take cholesterol-lowering medications, and (3) due to diagnostic biases among physicians, African Americans are often provided with differential care, oftentimes of relatively poor quality, for CHD.[50,51]

As with diabetes and kidney disease, the more pronounced effect of hypercholesterolemia among the African American population warrants concern in regards to the frequency of occurrence of AD and dementia among African Americans. Research has shown that there is a statistically significant effect of elevated cholesterol levels on AD development.[52] More specifically, results have shown that there is evidence suggesting that elevated LDL-C levels increase an individual's chances of developing

AD.[52] However, it is important to note that as of November 2021, further studies investigating the link between hypercholesterolemia and AD need to be conducted in order to definitively claim that hypercholesterolemia *is* indeed a causal agent for the future development of AD.[52] Current research only points to an associative relationship between elevated cholesterol levels and AD, but researchers are currently performing observational studies to determine the presence of a causative relationship and potential mechanism which link these two conditions.[53] Nonetheless, it can surely be asserted that the increased number of hypercholesterolemia presentations among African Americans holds significant implications for the occurrence of AD and dementia within this minority population.

However, the effects of hypercholesterolemia are not just limited to AD and dementia. Elevated serum cholesterol levels also hold implications for the African American demographic in terms of COVID-19. Research has shown that familial hypercholesterolemia generates many complications in the both the short-term phase and long-term phase of a COVID-19 infection.[54] This is largely due to the fact that in individuals with high cholesterol levels from birth, endothelial cells are prone to dysfunction which in turn makes these individuals more susceptible to a viral attack and inflammatory responses that are characteristic of COVID-19 infections.[54] However, the risk associated with hypercholesterolemia in the context of a COVID-19 infection also extends to those with developed hypercholesterolemia (as opposed to familial hypercholesterolemia). More specifically, elevated cholesterol levels facilitate viral entry, as the SARS-CoV-2 virus enters cells via GM-1 lipid rafts which are composed of cholesterol and monosialotetrahexosylganglioside1.[55] Thus, elevated cholesterol levels lead to members of the African American population being disproportionately affected by severe and long-lasting COVID-19 infections.

5.4 Cardiovascular disease

Several of the aforementioned risk factors such as diabetes, kidney disease, and hypercholesterolemia are also risk factors for the development of CVD.[56] As such, the fact that these risk factors are disproportionately prevalent within the African American population leads to the implication that CVD disproportionately affects the African American demographic.[56] More specifically, as of 2018, African American individuals are 30% more likely to die of CVD than non-Hispanic white Americans and are 40% more likely to develop hypertension over the course of their lives.[57] Just as with diabetes, kidney disease, and hypercholesterolemia, greater frequency of occurrence of CVD among African Americans is cause for concern with regards to their susceptibility to AD, dementia, and a severe COVID-19 infection.

Although not yet thoroughly proven as of November 2021, there are a great number of emergent studies suggesting CVD to be intimately connected with AD and the subsequent development of dementia. It is important to understand that the brain is incredibly sensitive to changes in oxygen levels.[58] This is largely due to the dependence of neurons on oxygen for proper and healthy function and that CVD greatly reduces the heart's efficiency in circulating oxygen via red blood cells throughout the body.[58] As a result, there is a stark reduction in the amount of oxygen that is delivered to the brain.[58]

Vasculature damage and abnormalities due to CVD can greatly diminish blood flow to the brain thereby causing damage to brain cells, which significantly catalyzes the development of AD.[58] However, it is important to note that as of November 2021, there is no clear and rigorously proven mechanism of causation between CVD and AD, and that the aforementioned phenomenon regarding vasculature damage and abnormalities is merely one posited hypothesis.[58,59]

A recent study has shown that there are similar implications of CVD with regards to COVID-19 infections by demonstrating that the risk of developing a severe or fatal COVID-19 infection was considerably higher in individuals who presented with risk factors for CVD and CVD itself.[60] There are several mechanisms by which the link between these two conditions is made, but the most important and pressing ones are as follows. First, COVID-19 often causes myocarditis (inflammation of the heart muscle), which can exacerbate the pre-existing effects of CVD on a patient, thus leading to a significantly worsened and more consequential COVID-19 infection.[61] Second, a viral attack by SARS-CoV-2 virions can damage the respiratory system, which subsequently leads to a diminished ability to acquire proper amounts of oxygen from the environment. A direct consequence of this will be the requirement for the heart to pump with more force to deliver increasingly spare oxygen-rich blood to major body systems. This is of particular danger for an individual who already has suboptimal cardiovascular health as their heart may not be able to pump effectively enough to compensate for the decreased oxygen levels due to a damaged respiratory system.[62] Finally, a COVID-19 infection can induce the formation of blood clots more frequently in patients with CVD by way of causing inflammation to occur within the blood vessel lining.[63] This has dire effects for the patient as blood clot formation in the blood vessels greatly increases the risk of a stroke or a myocardial infarction.[63]

5.5 Chronic lung diseases

Several studies have shown that the African American demographic is more likely to develop chronic lung diseases (CLD), which have implications for their susceptibility to a COVID-19 infection and AD development.[64] For example, studies have shown that African Americans are three times more likely to die of asthma-related complications, 50% more likely to develop lung cancer, and three times more likely to suffer from sarcoidosis than non-Hispanic white Americans.[64] Several hypotheses for why these disparities are present have been posited, but the most convincing and evidence-supported hypotheses come from the American Lung Association. As per their findings, a large part of why African Americans suffer CLD at a much greater rate than non-Hispanic white Americans is due to demographic distribution.[64] More specifically, African Americans are more likely to be located near transportation corridors where there is a poor quality of air.[64] Furthermore, there are many systematic biases within the current US healthcare system that contribute to African Americans being diagnosed much later on along a CLD timeline, at which point the disease is likely to have progressed enough to be significantly difficult to treat.[64] Along the same lines, pulmonology specialists are much less accessible to African Americans, which as one may reason, leads to this minority demographic being disproportionately affected by CLD.

CLD, more specifically chronic obstructive pulmonary disease (COPD), is a proven risk factor for the development of AD and dementia. The low amounts of oxygen and the greater accumulation of carbon dioxide that is an immediate result of COPD can cause harm to the brain.[65] This makes one more susceptible to cognitive decline and accelerates AD development.[65] To further illustrate this point, a recent population-based study by the Mayo Clinic has shown that patients presenting with COPD are approximately two times more likely than their healthy counterparts to develop mild cognitive impairments including memory loss.[66] For full understanding of the relationship between COPD and AD, it is important to note that several meta-analyses of cohort studies have shown that there only exists an *associative* relationship between COPD and AD. However, research is currently ongoing in terms of identifying a causative link between COPD and AD. These results are indeed alarming, especially in the context of the matter at hand, as they indicate that the African American population is more likely to develop cognitive impairments and AD simply due to being at a greater risk for CLD, which is primarily due to systematic biases within the US healthcare system as mentioned above.

Given the mechanism behind a COVID-19 infection, the link between CLD and a severe COVID-19 infection (in terms of intensive care unit time, fatal symptoms, and "long-haul COVID") is clear. A series of meta-analyses has demonstrated that COPD is a serious risk factor for COVID-19-related hospitalizations and mortality as COPD increases the risk of hospitalization and mortality by up to a factor of four.[67] It is important to note that these meta-analyses pertained to studies that derived results from patients who *already* had COVID-19 and thus indicate nothing about the relationship between CLD and increased likelihood of contracting COVID-19.[67] Just as CLD causes a disproportionate rate of occurrence of AD within the African American demographic, it causes a disproportionate frequency of COVID-19 infection among them as well. It is clear that the consequences of the systematic biases that have led to African Americans being more likely to develop CLD are not limited to just worsened respiratory health but also to cognitive decline and lethal COVID-19 infections among African Americans.

5.6 Dementia

Dementia is a key concern among the African American population as elderly African Americans are approximately twice as likely to develop dementia than their non-Hispanic white American counterparts.[68] Furthermore, it has been predicted by the CDC that by 2060, the number of African Americans who will have developed dementia will be four times greater than the current number.[68] The reason for this disproportionate affliction of the African American community is that African Americans are statistically more likely to suffer from stress, depression, and diabetes, which are all risk factors that *significantly* catalyze the development of dementia.[68] The greater likelihood that African Americans will develop dementia unfortunately has implications that lie far beyond just mental health, especially in the context of a COVID-19 infection.

In general, individuals with dementia are at an increased risk for contracting COVID-19, requiring subsequent hospitalization, and experiencing more severe and potentially even fatal infections.[69] Furthermore, African Americans with dementia are at an even *greater* risk than their non-Hispanic white American counterparts for the aforementioned health complications. Results from a study conducted in 2021 showed that 73% of African American individuals with dementia were admitted within 6 months of a COVID-19 diagnosis as compared to 54% of non-Hispanic white Americans with dementia.[69] It is also important to note that only 25% of COVID-19 patients *without* dementia had to undergo hospitalization.[69] Moreover, COVID-19 patients with dementia were found to be four times more likely to die from their infection than patients without dementia.[69]

As one can reason, the greater occurrence of dementia within the African American population is seeing unprecedented consequences in the context of the COVID-19 pandemic as of November 2021 leading to tremendous and unnecessary loss of life and emotional strain. Issues regarding systemic disadvantages conferred unto minority populations, more specifically in terms of healthcare, economic status, and class mobility need to be addressed as the consequences of these disadvantages are not only limited to social inequality but also to human health and quality of living.

6. Environmental risk factors

6.1 Social inequalities

As alluded to previously, the African American demographic of the United States experience great disadvantages when it comes to social equality and income/socioeconomic status. In fact, a study conducted by the University of Chicago found that when African American and non-Hispanic white American students were given a similar quality of instruction, they performed equally well in their academics.[70] However, the same study found that African American children are often given a much lower quality of education than non-Hispanic white American students which is contributing to the racial gap within educational attainment.[70] What this indicates is that genetic or biological phenomena are not the main cause leading to this racial gap in educational achievement. Rather, the root cause behind this disparity lies within *extrinsic* factors, namely, a lower quality of instruction for African American students. This lack in educational opportunity and equity affects their quality of life, socioeconomic status as adults, and subsequent access to quality healthcare.[71]

6.2 Lack of access to healthcare

Findings published by the CDC have revealed that (1) a lower percentage of African Americans have health insurance compared to non-Hispanic white Americans, (2) fewer African Americans are vaccinated against influenza, (3) fewer African American women have access to prenatal care during their first trimester, and (4) fewer African Americans participate in regular physical activity.[72] To explain some of the aforementioned findings, the CDC has found that fewer African Americans can see a physician due to cost and fewer African Americans are regularly exposed to health education and awareness.[73]

As one may reasonably predict, access to quality healthcare has a profound impact on health. More specifically, poor access to healthcare will lead to a poor quality of life and overall health, which is a key issue that the US' African American population faces today. This is due to the fact that African Americans simply do not have access to the same quality of healthcare that their non-Hispanic white American counterparts do due to many long-standing systematic biases against African Americans. As a result, they are at a much greater risk for contracting various illnesses, diseases, and chronic medical conditions than non-Hispanic white Americans.

7. Concluding remarks

The various unmodifiable risk factors, modifiable risk factors, age-related chronic diseases, and environmental risk factors that affect African Americans come together to place the African American demographic of the United States at a high risk for developing AD, dementia, a COVID-19 infection, and consequent medical complications.

With regards to unmodifiable risk factors, there are unfortunately no meaningful solutions that would make a considerable impact at this time and is an area of study that warrants a great amount of further research. Along the same lines, solutions to some of the aforementioned modifiable risk factors, namely, systematic income/socioeconomic status and educational disparities, are extremely difficult to implement in the short run. However, more *long-term* solutions can be effectively implemented. For example, campaigns to bring greater levels of health awareness and education to African American communities may prove to be significantly effective in improving their access to quality healthcare.

In combination with this, the US government and major hospital systems across the nation should make sincere and persistent efforts to bring quality healthcare *to* African Americans, instead of relying upon them to seek out healthcare on their own, which in today's economic climate, unfortunately proves to be a difficult task for many African Americans. Ways that such efforts could be implemented are as follows: (1) implementation of curriculums in all levels of schooling covering healthy lifestyles and habits, (2) development of community outreach centers within African American communities that make quality healthcare more accessible to them, and (3) allocation of federal funds to building the aforementioned outreach centers and medical facilities within underserved African American communities across the United States.

Acknowledgments

The research and relevant findings presented in this chapter were supported by the National Institutes of Health (NIH) grants AG042178, AG047812, NS105473, AG060767, AG069333 and AG066347.

References

1. Alzheimer's Association, "What is Dementia?," Alzheimer's Association. https://www.alz.org/alzheimers-dementia/what-is-dementia.
2. 2020 Alzheimer's disease facts and figures. *Alzheimer's Dementia*. Mar. 2020;16(3): 391−460. https://doi.org/10.1002/alz.12068.
3. Mayo Clinic Staff, "Dementia," Mayo Clinic Disease & Conditions. https://www.mayoclinic.org/diseases-conditions/dementia/symptoms-causes/syc-20352013.
4. Alzheimer's Association. *2021 Alzheimer's Disease Facts and Figures*; 2021 [Online]. Available: https://www.alz.org/media/documents/alzheimers-facts-and-figures.pdf.
5. Alzheimer's Association, "African-Americans and Alzheimer's Disease: The Silent Epidemic." [Online]. Available: https://www.alz.org/media/Documents/african-americans-silent-epidemic-r.pdf.
6. Centers for Disease Control and Prevention. *U.S. Burden of Alzheimer's Disease, Related Dementias to Double by 2060*; 2018 [Online]. Available: https://www.cdc.gov/media/releases/2018/p0920-alzheimers-burden-double-2060.html.
7. Centers for Disease Control and Prevention. *Symptoms of COVID-19*. CDC Your Health; 2021. https://www.cdc.gov/coronavirus/2019-ncov/symptoms-testing/symptoms.html.
8. M. W. Smith, "Complications Coronavirus Can Cause," WebMD Coronavirus Reference. https://www.webmd.com/lung/coronavirus-complications#3.
9. Centers for Disease Control and Prevention. *Post-COVID Conditions*. 2CDC Your Health; 2021. https://www.cdc.gov/coronavirus/2019-ncov/long-term-effects/index.html.
10. Centers for Disease Control and Prevention, "COVID Data Tracker," CDC COVID Data Tracker. https://covid.cdc.gov/covid-data-tracker/#datatracker-home.
11. United States Census Bureau, "United States Census Bureau QuickFacts," United States Census Bureau QuickFacts. https://www.census.gov/quickfacts/fact/table/US/PST045219.
12. Sheladia S, Reddy PH. Age-related chronic diseases and Alzheimer's disease in Texas: a Hispanic focused study. *J Alzheimer's Dis Rep*. Feb. 2021;5(1):121−133. https://doi.org/10.3233/ADR-200277.
13. Centers for Disease Control and Prevention. *Assessing Risk Factors for Severe COVID-19 Illness*; 2021. https://www.cdc.gov/coronavirus/2019-ncov/covid-data/investigations-discovery/assessing-risk-factors.html.
14. Profile of African Americans Age 65 and Over. U.S. Department of Health and Human Services. Published October 2019. Accessed October 28, 2021. https://acl.gov/sites/default/files/Aging%20and%20Disability%20in%20America/2018AA_OAProfile.pdf.
15. Alzheimer Disease in African Americans. Boston University Biomedical Genetics. Accessed October 13, 2021. https://www.bumc.bu.edu/genetics/research/alzheimers-disease/alzheimer-disease-in-african-americans/.
16. Tang M, Stern Y, Marder K, et al. The *APOE*-ε4 allele and the risk of Alzheimer disease among African Americans, Whites, and Hispanics. *JAMA*. 1998;279(10):751−755. https://doi.org/10.1001/jama.279.10.751.
17. Reitz C, Jun G, Naj A, et al. Variants in the ATP-binding cassette transporter (*ABCA7*), apolipoprotein E ε4, and the risk of late-onset Alzheimer disease in African Americans. *JAMA*. 2013;309(14):1483−1492. https://doi.org/10.1001/jama.2013.2973.
18. Berg CN, Sinha N, Gluck MA. The effects of APOE and ABCA7 on cognitive function and Alzheimer's disease risk in African Americans: a focused mini review. Front Hum Neurosci. 2019;13:387. Published 2019 Nov 5. https://doi.org/10.3389/fnhum.2019.00387.

19. COVID-19 Host Genetics Initiative. Mapping the human genetic architecture of COVID-19. *Nature*. 2021. https://doi.org/10.1038/s41586-021-03767-x.
20. Median Income by Race. Peter G. Peterson Foundation. https://www.pgpf.org/chart-archive/0257_income_by_race. Published September 28, 2020. Accessed October 22, 2021.
21. O'Neill A. Median Household Income in the United States 1967−2019, by Race. Statista. Accessed November 1, 2021. https://www.statista.com/statistics/1086359/median-household-income-race-us/.
22. U.S. Census Bureau QuickFacts: United States. U.S. Census Bureau. https://www.census.gov/quickfacts/fact/table/US/PST045219. Published 2021. Accessed October 25, 2021.
23. Poverty Status In The Past 12 Months. U.S. Census Bureau. https://data.census.gov/cedsci/table?q=poverty&tid=ACSST1Y2019.S1701. Published 2020. Accessed August 16, 2021.
24. Finch WH, Hernández Finch ME. Poverty and covid-19: rates of incidence and deaths in the United States during the first 10 Weeks of the pandemic. *Front Sociol*. 2020;5. https://doi.org/10.3389/fsoc.2020.00047.
25. Explore Census Data. U.S. Census Bureau. https://data.census.gov/cedsci/table?q=education&tid=ACSST1Y2019.S1501. Published 2020. Accessed August 16, 2021.
26. ScienceDaily. *Low Education Level Linked to Alzheimer's, Study Shows*. ScienceDaily; 2007. Published Online https://www.sciencedaily.com/releases/2007/10/071001172855.htm.
27. Fulton J. Physical Inactivity is More Common among Racial and Ethnic Minorities in Most States | | Blogs | CDC. Centers for Disease Control and Prevention. Published April 1, 2020. Accessed October 24, 2021. https://blogs.cdc.gov/healthequity/2020/04/01/physical-inactivity/.
28. Virani SS, Alonso A, Aparicio HJ, et al. American heart association council on epidemiology and prevention statistics committee and stroke statistics subcommittee. Heart disease and stroke statistics-2021 update: a report from the American heart association. *Circulation*. February 23, 2021;143(8):e254−e743. https://doi.org/10.1161/CIR.0000000000000950. Epub 2021 Jan 27. PMID: 33501848.
29. Howard G, Cushman M, Moy CS, et al. Association of clinical and social factors with excess hypertension risk in black compared with white US adults. *JAMA*. 2018;320(13):1338−1348. https://doi.org/10.1001/jama.2018.13467.
30. Hales C, Carroll M, Fryar C, Ogden C. Prevalence of Obesity and Severe Obesity Among Adults: United States, 2017−2018. Centers for Disease Control and Prevention. Published February 2020. Accessed October 24, 2021. https://www.cdc.gov/nchs/products/databriefs/db360.htm.
31. Centers for Disease Control and Prevention. Obesity, Race/Ethnicity, and COVID-19. CDC Overweight & Obesity. https://www.cdc.gov/obesity/data/obesity-and-covid-19.html#:∼:text=Obesity decreases lung capacity and reserve and can make ventilation more difficult.&text=A study of COVID-19,are higher with increasing BMI.
32. The Conversation. Alzheimer's Disease: Obesity May Worsen its Effects—New Research. Conversation Health. Published online 2021. https://theconversation.com/alzheimers-disease-obesity-may-worsen-its-effects-new-research-154214.
33. Burden of Tobacco Use in the U.S. Centers for Disease Control and Prevention. https://www.cdc.gov/tobacco/campaign/tips/resources/data/cigarette-smoking-in-united-states.html. Published 2021. Accessed October 22, 2021.
34. Tobacco Use in the African American Community. Truth Initiative. Published May 28, 2020. Accessed November 3, 2021. https://truthinitiative.org/research-resources/targeted-communities/tobacco-use-african-american-community.

35. WebMD. Coronavirus and Smoking. WebMD. https://www.webmd.com/lung/covid-19-smoking-vaping#1.
36. Alzheimer's Society. Smoking and Dementia. Alzheimer's Society. https://www.alzheimers.org.uk/about-dementia/risk-factors-and-prevention/smoking-and-dementia#:~:text=It is known that smoking,to developing of Alzheimer's disease.
37. U.S. Department of Health and Human Services Office of Minority Health, "Diabetes and African Americans." [Online]. Available: https://minorityhealth.hhs.gov/omh/browse.aspx?lvl=4&lvlid=18.
38. National Center for Health Statistics. *Selected Diseases and Conditions Among Adults Aged 18 and Over, By Selected Characteristics: United States, 2018*. vol. 8. 2018:1–9 [Online]. Available: https://ftp.cdc.gov/pub/Health_Statistics/NCHS/NHIS/SHS/2018_SHS_Table_A-4.pdf.
39. Karki R, Kodamullil AT, Hofmann-Apitius M. Comorbidity analysis between Alzheimer's disease and type 2 diabetes mellitus (T2DM) based on shared pathways and the role of T2DM drugs. *J Alzheimers Dis*. 2017;60(2):721–731. https://doi.org/10.3233/JAD-170440.
40. Reitz C, Brayne C, Mayeux R. Epidemiology of Alzheimer disease. *Nat Rev Neurol*. Mar. 2011;7(3):137–152. https://doi.org/10.1038/nrneurol.2011.2.
41. Mayo Clinic. Diabetes and Alzheimer's Linked. Published 2021. https://www.mayoclinic.org/diseases-conditions/alzheimers-disease/in-depth/diabetes-and-alzheimers/art-20046987.
42. Watson S. *COVID-19 and Diabetes*. WebMD; 2020. https://www.webmd.com/diabetes/diabetes-and-coronavirus.
43. National Kidney Foundation. *Race, Ethnicity, & Kidney Disease*. National Kidney Foundation; 2021. https://www.kidney.org/atoz/content/minorities-KD.
44. Zhang C-Y, He F-F, Su H, Zhang C, Meng X-F. Association between chronic kidney disease and Alzheimer's disease: an update. *Metab Brain Dis*. 2020;35(6):883–894. https://doi.org/10.1007/s11011-020-00561-y.
45. Xu H, Garcia-Ptacek S, Trevisan M, et al. Kidney function, kidney function decline, and the risk of dementia in older adults. *Neurology*. Jun. 2021;96(24):e2956–e2965. https://doi.org/10.1212/WNL.0000000000012113.
46. Kang MW, Park S, Lee S, et al. Glomerular hyperfiltration is associated with dementia: a nationwide population-based study. *PLoS One*. 2020;15(1):e0228361. https://doi.org/10.1371/journal.pone.0228361.
47. National Kidney Foundation, "Kidney Disease & COVID-19," National Kidney Foundation. https://www.kidney.org/coronavirus/kidney-disease-covid-19.
48. AJMC, "Study Illustrates Kidney Impact After COVID-19 Resolves," AJMC. https://www.ajmc.com/view/study-illustrates-kidney-impact-after-covid-19-resolves.
49. Lee-Frye B, Ali YS. *Cholesterol, Heart Disease, and African Americans*. VerywellHealth; Aug. 2021.
50. Nelson K, Norris K, Mangione CM. Disparities in the diagnosis and pharmacologic treatment of high serum cholesterol by race and ethnicity. *Arch Intern Med*. Apr. 2002;162(8):929. https://doi.org/10.1001/archinte.162.8.929.
51. Fincher C, Williams JE, MacLean V, Allison JJ, Kiefe CI, Canto J. Racial disparities in coronary heart disease: a sociological view of the medical literature on physician bias. *Ethn Dis*. 2004;14(3):360–371 [Online]. Available: http://www.ncbi.nlm.nih.gov/pubmed/15328937.

52. Sáiz-Vazquez O, Puente-Martínez A, Ubillos-Landa S, Pacheco-Bonrostro J, Santabárbara J. Cholesterol and Alzheimer's disease risk: a meta-meta-analysis. *Brain Sci.* Jun. 2020;10(6):386. https://doi.org/10.3390/brainsci10060386.
53. Alzheimer's Society. Cholesterol and Dementia. Alzheimer's Society. https://www.alzheimers.org.uk/about-dementia/risk-factors-and-prevention/cholesterol-and-dementia.
54. Vuorio A, Raal F, Kaste M, Kovanen PT. Familial hypercholesterolaemia and COVID-19: a two-hit scenario for endothelial dysfunction amenable to treatment. *Atherosclerosis.* 2021; 320:53−60. https://doi.org/10.1016/j.atherosclerosis.2021.01.021.
55. Thomas L. *Does Cholesterol Play a Role in COVID-19?* News Medical Life Sciences; 2020. https://www.news-medical.net/news/20200512/Does-cholesterol-play-a-role-in-COVID-19.aspx.
56. Carnethon MR, Pu J, Howard G, et al. Cardiovascular health in African Americans: a scientific statement from the American heart association. *Circulation.* Nov. 2017;136(21). https://doi.org/10.1161/CIR.0000000000000534.
57. *U.S. Department of Health and Human Services Office of Minority Health.* Heart Disease and African Americans; 2021 [Online]. Available: https://minorityhealth.hhs.gov/omh/browse.aspx?lvl=4&lvlid=19.
58. Schein C. *The Link Between Alzheimer's and Cardiovascular Disease.* Aegis Living; 2019.
59. Stewart R. Cardiovascular factors in Alzheimer's disease. *J Neurol Neurosurg Psychiatry.* Aug. 1998;65(2):143−147. https://doi.org/10.1136/jnnp.65.2.143.
60. Bae S, Kim SR, Kim M-N, Shim WJ, Park S-M. Impact of cardiovascular disease and risk factors on fatal outcomes in patients with COVID-19 according to age: a systematic review and meta-analysis. *Heart.* Mar. 2021;107(5):373 LP−380. https://doi.org/10.1136/heartjnl-2020-317901.
61. Lee Lewis DK. *How Does Cardiovascular Disease Increase the Risk of Severe Illness and Death from COVID-19.* Harvard Health Publishing; 2020.
62. Scripps. *Why is Coronavirus Dangerous to People with Heart Disease?* Scripps; 2020. https://www.scripps.org/news_items/6989-why-is-coronavirus-dangerous-to-people-with-heart-disease.
63. McCallum K. *How Does COVID-19 Affect the Heart?* Houston Methodist; 2021. https://www.houstonmethodist.org/blog/articles/2021/mar/how-does-covid-19-affect-the-heart/.
64. DeNoon DJ. *Why 7 Deadly Diseases Strike Blacks Most.* WebMD; 2005. https://www.webmd.com/hypertension-high-blood-pressure/features/why-7-deadly-diseases-strike-blacks-most.
65. Healthline, Sampson S. *Recognizing Serious COPD Complications.* Healthline; 2020. https://www.healthline.com/health/copd/serious-complications. Accessed June 10, 2021.
66. Singh B, Parsaik AK, Mielke MM, et al. Chronic obstructive pulmonary disease and association with mild cognitive impairment: the Mayo clinic study of aging. *Mayo Clin Proc.* 2013;88(11):1222−1230, Nov. https://doi.org/10.1016/j.mayocp.2013.08.012.
67. Gerayeli FV, Milne S, Cheung C, et al. COPD and the risk of poor outcomes in COVID-19: a systematic review and meta-analysis. *EClinicalMedicine.* Mar. 2021;33:100789. https://doi.org/10.1016/j.eclinm.2021.100789.
68. Medical News Today, Biggers A. *Dementia in the Black Community.* Medical News Today; 2021. https://www.medicalnewstoday.com/articles/dementia-in-the-black-community. Accessed August 10, 2021.
69. Wang Q, Davis PB, Gurney ME, Xu R. COVID-19 and dementia: analyses of risk, disparity, and outcomes from electronic health records in the US. *Alzheimers Dement.* 2021; 17(8):1297−1306. https://doi.org/10.1002/alz.12296.

70. Darling-Hammond L. *Unequal Opportunity: Race and Education*. Brookings; Mar. 01, 1998.
71. Taylor J. *Racism, Inequality, and Health Care for African Americans*. The Century Foundation; 2019. https://tcf.org/content/report/racism-inequality-health-care-african-americans/?agreed=1.
72. Centers for Disease Control and Prevention. *Health Disparities Experienced by Black or African Americans—United States*. CDC MMWR; 2005.
73. Centers for Disease Control and Prevention, "African American Health," CDC Vital Signs.

Further reading

1. United States Census Bureau. *United States Census Bureau QuickFacts*; 2019 [Online]. Available: https://www.census.gov/quickfacts/fact/table/US/PST045219.
2. Statista. *Projected African-American Population in the United States from 2010 to 2035*; 2021 [Online]. Available: https://www.statista.com/statistics/549348/projected-african-american-population-in-the-us/.
3. World Health Organization, "Coronavirus Disease (COVID-19)," World Health Organization Health Topics. https://www.who.int/health-topics/coronavirus#tab=tab_1.

Dementia and COVID-19: A Hispanic focused study

14

Shyam Sheladia [1], *Shivam Sheladia* [1], *Rishi Virani* [1,2] *and*
P. Hemachandra Reddy [1,3,4,5,6]
[1]Department of Internal Medicine, Texas Tech University Health Sciences Center, Lubbock, TX, United States; [2]Department of Biomedical Engineering, The University of Texas at Austin, Austin, TX, United States; [3]Department of Pharmacology and Neuroscience, School of Medicine, Texas Tech University Health Sciences Center, Lubbock, TX, United States; [4]Neurology, Departments of School of Medicine, School of Medicine, Texas Tech University Health Sciences Center, Lubbock, TX, United States; [5]Public Health Department of Graduate School of Biomedical Sciences, School of Medicine, Texas Tech University Health Sciences Center, Lubbock, TX, United States; [6]Department of Speech, Language and Hearing Sciences, School of Health Professions, School of Medicine, Texas Tech University Health Sciences Center, Lubbock, TX, United States

Abstract

While both dementia and coronavirus disease 2019 (COVID-19) have differing etiology, there is a complex interplay between the two, especially when looking into their effects on certain sub-populations. Hispanic Americans face a higher burden of dementia and COVID-19 due to both modifiable and unmodifiable risk factors, age-related chronic diseases, and environmental factors. The major unmodifiable risk factors include increasing age and predisposing genetics, while the major modifiable risk factors include income/socioeconomic status, educational attainment, exercise, diet, and smoking/tobacco use. Furthermore, specific age-related chronic diseases such as diabetes, kidney disease, hypercholesterolemia, cardiovascular disease, and chronic lung diseases place Hispanic Americans at high risk for dementia and COVID-19. Lastly, Hispanic Americans face the additional disadvantage of environmental factors, such as social inequalities and lack of access to adequate healthcare resources. Given that Hispanic Americans are the largest racial/ethnic minority group within the United States, this chapter will focus upon the research associated with dementia and COVID-19 within the Hispanic American population of the United States. Furthermore, this chapter will explore the four major risk factor categories (unmodifiable risk factors, modifiable risk factors, age-related chronic diseases, and environmental factors), which contribute to the development of dementia and COVID-19 within the Hispanic American population of the United States.

Keywords: Age-related chronic diseases; Alzheimer's disease; COVID-19; Dementia; Hispanic Americans; Risk factors; Social inequalities

COVID-19 in Alzheimer's Disease and Dementia. https://doi.org/10.1016/B978-0-443-15256-6.00007-6

1. Introduction

Dementia is a disease of the brain characterized by a loss of memory and deteriorating mental functions, which often takes years to develop. On the other hand, COVID-19 is a viral illness, which primarily attacks the respiratory system and can be fatal within a matter of days to weeks. Although both of these illnesses cause harm to the human body through differing biological mechanisms, there is a great deal of overlap among the key risk factors for developing either one of these debilitating illnesses. This chapter will focus upon the four main categories of risk factors that are relevant to the development of both dementia and COVID-19. The four risk factor categories of focus are (1) unmodifiable risk factors, (2) modifiable risk factors, (3) age-related chronic diseases, and (4) environmental factors.

Additionally, this chapter will only focus on a subset of the US population, the Hispanic population that makes up approximately 18.5% of the total United States population and is growing as seen in Fig. 14.1.[1] However, the Hispanic population carries a high degree of burden from all four risk factor categories previously mentioned, which places them at a higher risk than other racial/ethnic groups for the development of dementia, COVID-19, or both.

2. Dementia/COVID-19

2.1 What is dementia?

Dementia is a generic term used to describe the loss of memory, language, and problem-solving abilities that interferes with activities of daily life.[2] Alzheimer's disease (AD) is the leading cause of dementia, accounting for between 60% and 80% of cases worldwide.[3] Other common causes of dementia are Lewy body disease, cerebrovascular disease, fronto-temporal lobar degeneration, Parkinson's disease, hippocampal sclerosis, traumatic brain injury, and other mixed pathologies.[3]

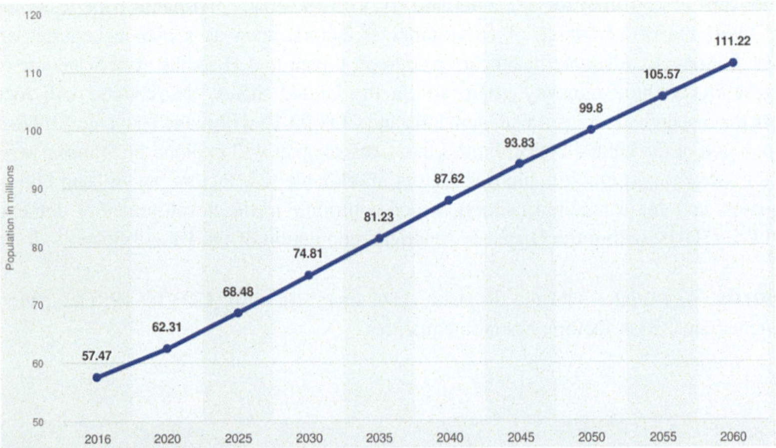

Figure 14.1 Projection of the Hispanic American Population of the United States Through 2060. Population Information from The United States Census Bureau, 2017 National Population Projections. Chart Created by Statista.[1]

Patients suffering from dementia often begin to display noticeable symptoms such as memory loss, confusion, and difficulties in communication.[3] These patients also struggle with daily tasks such as tying their shoes, are prone to forgetting recent events, and have difficulty naming close family relatives among other deficits.[3]

Extensive research has been conducted on AD as it is the leading cause of dementia in the overall population. Therefore, this chapter will utilize the published statistics, facts, and figures regarding AD to generalize the prevalence of dementia in the population at large.

2.2 Prevalence of dementia in the US Hispanic population

According to the *2021 Alzheimer's Disease Facts and Figures* report, approximately 6.2 million Americans age 65 years and older are living with the burden of AD in 2021.[4] Furthermore, this number is projected to increase within the United States to reach 7.2 million Americans by the year 2025, 12.7 million Americans by the year 2050, and 13.8 million Americans by the year 2060, given that no significant medical developments occur with regards to curing or delaying the progression of AD.[3,4] To put this into perspective, an individual is currently diagnosed with AD every 60 s within the United States, and by the year 2050, this number is projected to be around one individual every 33 s on average.[5]

With regards to Hispanic Americans in particular, data from the Chicago Health and Aging Project reveal that 14% of Hispanics age 65 years and above suffer from Alzheimer's dementia compared to only 10% of the non-Hispanic white population age 65 years and above.[4] Furthermore, studies have indicated that elderly Hispanics are one-and-a-half times more likely to suffer from Alzheimer's dementia than are elderly non-Hispanic whites.[3,4] Recent statistics as of the year 2020 show that Hispanic Americans account for approximately 628,000 AD cases within the United States.[6] Unfortunately, it is estimated that by the year 2060, the number of Hispanic Americans with AD may increase approximately nine-fold reaching up to 3.5 million (Fig. 14.2).[6]

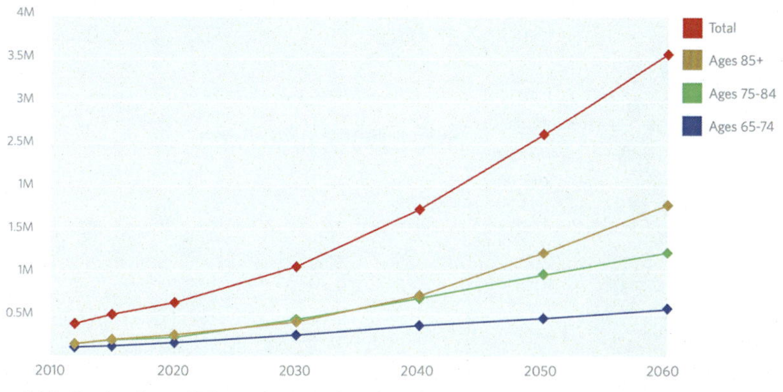

Figure 14.2 Projection of Hispanic Americans with Alzheimer's Disease Through 2060 by Age Category (In Millions). Chart Created by UsAgainstAlzheimer's.[6]

2.3 What is COVID-19?

Coronaviruses are a group of viruses, which are named due to their spherical and crown-like appearance when viewed with electron microscopy.[7] As seen in Fig. 14.3, their unique appearance is mainly due to spike proteins located on their envelopes, which can be used to attach and bind to human lung alveolar epithelial cells for the purposes of survival and replication.[7,8] Coronaviruses are usually prominent among various animals such as bats, cats, camels, cows, pigs, and chickens; however, they possess the ability to evolve and infect humans and human coronaviruses were first discovered in the mid-1960s.[9] Human coronaviruses are commonly transmitted via respiratory droplets.[9]

The majority of individuals infected with COVID-19, a novel coronavirus that emerged in late 2019 from the city of Wuhan, China, experience mild to moderate respiratory symptoms and recover without the need for hospitalization or special treatment.[10] These symptoms include fever, chills, shortness of breath, fatigue, headache, sore throat, congestion, nausea, diarrhea, and new loss of taste or smell, which usually appears between 2 and 14 days after exposure to the virus.[10] However, severe symptoms and often life-threatening complications will occur in approximately one out of every six individuals diagnosed with a COVID-19 infection.[11] These complications include acute respiratory distress syndrome, acute respiratory failure, pneumonia, acute liver injury, acute cardiac injury, acute kidney injury, septic shock, secondary infection, disseminated intravascular coagulation, blood clots, rhabdomyolysis, and chronic fatigue.[11]

Furthermore, long-term effects and complications after recovery from the initial COVID-19 infection include permanent organ damage to the heart, lungs, and brain.[12] With regards to a COVID-19 infection, the severe symptoms and long-term effects are

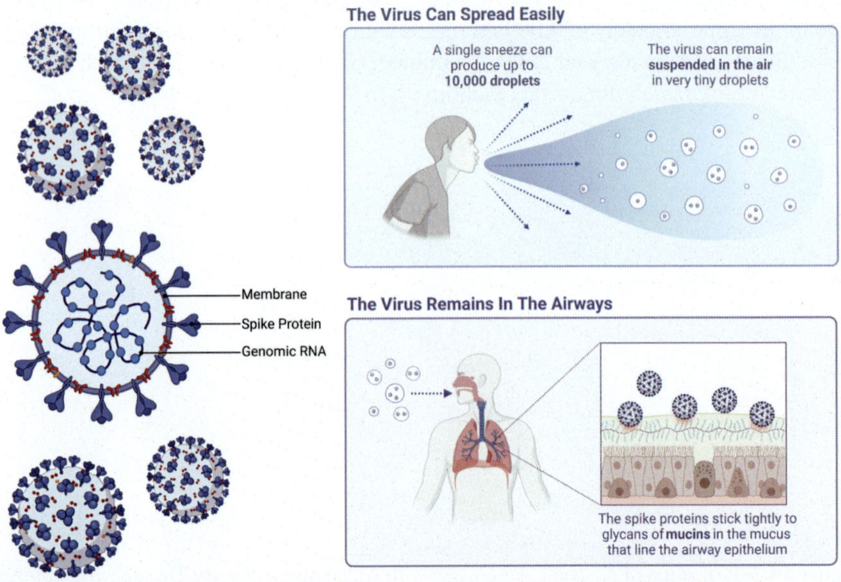

Figure 14.3 Structure of Coronaviruses Depicting Spike Proteins and Mechanism of Transmission.

Table 14.1 Data as of October 16, 2021. COVID-19 information from the United States Centers for Disease Control and Prevention.[13]

COVID-19 cases and deaths by race/ethnicity for the United States		
Race/ethnicity	**Percent of cases**	**Percent of deaths**
White alone (not Hispanic or Latino)	52.0%	59.3%
Hispanic or Latino	26.4%	18.0%
Black or African American alone	11.9%	13.9%
Asian alone	3.0%	3.6%
American Indian and Alaska Native alone	1.1%	1.1%
Native Hawaiian and Pacific Islander alone	0.3%	0.2%
Multiple races and other races	5.3%	3.8%

of greatest concern as they lead to permanent damage and future complications for an individual.

2.4 Prevalence of COVID-19 in the US Hispanic population

As of October 2021, there have been approximately 45 million total cases and over 725,000 total deaths recorded within the United States due to the COVID-19 pandemic.[13]

Furthermore, as seen in Table 14.1, Hispanic Americans account for approximately 26.4% of total COVID-19 cases as well as 18.0% of total COVID-19 deaths within the United States.[13] Although the non-Hispanic white population accounts for 52.0% of total COVID-19 cases and 59.3% of total COVID-19 deaths within the United States, as seen in Table 14.2, the non-Hispanic white population accounts for 60.1% of the total US general population.[13,14] Therefore, according to the Centers for Disease

Table 14.2 Most recent data available. Population information from the United States Census Bureau, 2019 data.[14]

Population breakdown by race/ethnicity for the United States (2019)		
Race/ethnicity	**Population percentage**	**Population count**
White alone (not Hispanic or Latino)	60.1%	197,271,953
Hispanic or Latino	18.5%	60,724,311
Black or African American alone	13.4%	43,984,096
Asian alone	5.9%	19,366,132
American Indian and Alaska Native alone	1.3%	4,267,114
Native Hawaiian and Pacific Islander alone	0.2%	656,479
Multiple races and other races	2.8%	9,190,706

Control and Prevention (CDC), Hispanic Americans are 1.9 times more likely to be infected with COVID-19, 2.8 times more likely to be hospitalized due to a COVID-19 infection, and 2.3 times more likely to die due to a COVID-19 infection than non-Hispanic white individuals within the United States.[15]

2.5 Risk factors causing dementia and COVID-19

With regards to Alzheimer's dementia and the development of AD in general, the major unmodifiable risk factors are age and predisposing genetics.[3] The major modifiable risk factors include income/socioeconomic status, educational attainment, exercise, and diet.[3] Furthermore, age-related chronic diseases contributing to the advancement and rapid progression of Alzheimer's dementia and AD include diabetes, kidney disease, hypercholesterolemia, cardiovascular disease, and chronic lung diseases.[3] With regards to unmodifiable risk factors, modifiable risk factors, and age-related chronic diseases, advancing age is the most prominent and greatest risk factor in the development of Alzheimer's dementia and AD.[3]

With regards to COVID-19 and the development of a severe illness from a COVID-19 infection, the major unmodifiable risk factors are age and race/ethnicity.[16] The major modifiable risk factors include income/socioeconomic status, educational attainment, exercise, and diet, as well as smoking and tobacco use.[16] Furthermore, certain age-related chronic diseases and medical conditions such as diabetes, kidney disease, hypercholesterolemia, cardiovascular disease, chronic lung diseases, and dementia place individuals at high risk for the development of a severe illness or death from a COVID-19 infection.[16] With regards to unmodifiable risk factors, modifiable risk factors, and age-related chronic diseases, advancing and elderly age is the greatest risk factor for severe illness, long-term complications, or death from a COVID-19 infection as over 80% of COVID-19-related deaths occur in individuals over the age of 65 within the United States.[16]

It should be noted that a majority of the unmodifiable risk factors, modifiable risk factors, and age-related chronic diseases, which contribute to the development of Alzheimer's dementia and AD, as well as a severe illness or death from a COVID-19 infection are the same. There is a great amount of overlap in all three categories with respect to the two separate conditions at hand.

Furthermore, the fourth risk factor category of environmental factors can be established as long-standing systemic health and social inequalities place various racial and ethnic minority groups such as Hispanic Americans at a disproportionate disadvantage as they face a greater number of barriers in accessing adequate healthcare.[16] Such inequalities within the United States healthcare system greatly increases the risk for Hispanic Americans to develop AD and Alzheimer's dementia, a COVID-19 infection leading to a severe illness or death, as well as any number of other illnesses, diseases, or medical conditions, which plague the population of the United States today.

As depicted in the relationship map shown in Fig. 14.4, this chapter will now continue to explore each one of the four major risk factor categories (unmodifiable risk factors, modifiable risk factors, age-related chronic diseases, and environmental

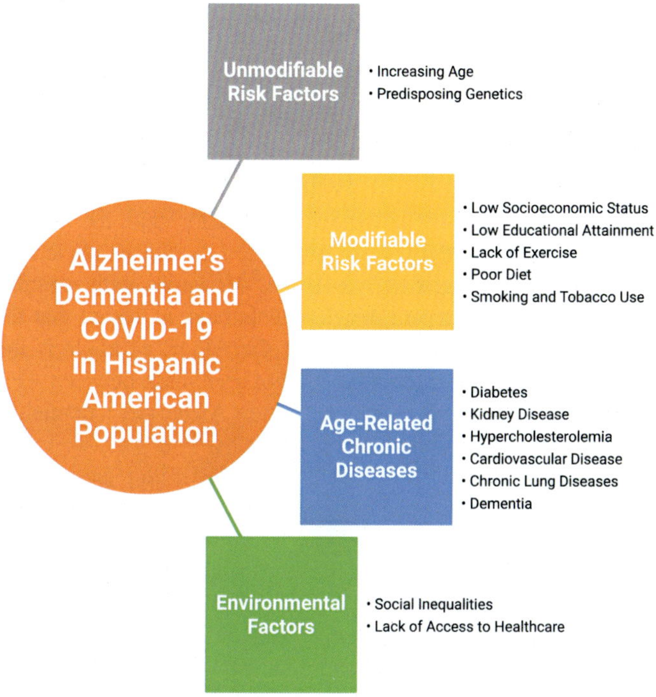

Figure 14.4 Summary of the Disadvantages Faced by Hispanic Americans in Overcoming the Burden of AD and Alzheimer's Dementia as well as a COVID-19 Infection Leading to Severe Illness, Long-Term Complications, or Death.

factors) in further detail with respect to AD and Alzheimer's dementia as well as COVID-19 infection within the Hispanic American population of the United States.

3. Unmodifiable risk factors

3.1 Increasing age

As mentioned previously, advancing and elderly age is the single greatest risk factor out of all four major risk factor categories for the development of AD and Alzheimer's dementia, as well as severe infection, long-term complications, or death from a COVID-19 infection.

The U.S. Department of Health and Human Services reported that approximately 4.75 million Hispanic Americans within the United States were age 65 years and above in the year 2020.[17] This number is expected to double by the year 2030 and quadruple by the year 2060.[17]

The cumulative effects of an increasing overall Hispanic population as well as an increasing elderly Hispanic population within the United States means that a greater number of Hispanic Americans will face both dementia and COVID-19 infection–related complications in the future years to come.[3]

3.2 Predisposing genetics

Extensive prior research has indicated that apolipoprotein E (APOE) ε4 allele inheritance carries the strongest association for the development of AD and future Alzheimer's dementia than any other genetic risk factor that has been identified.[3] Studies have indicated that in the general population, APOE ε4 allele inheritance increases one's risk of developing AD by twofold.[3] However, with regards to the Hispanic population specifically, recent developments have indicated that the increased risk of AD associated with the inheritance of the APOE ε4 allele is inconsistent.[3] The reasoning as to why a clear association has not been made between APOE ε4 allele inheritance, and the development of AD and future Alzheimer's dementia within the Hispanic population is not understood at this time.[3]

Nevertheless, specific mutations have been identified with regards to familial cases of AD.[3] Mutations within genes such as presenilin-1, presenilin-2, amyloid precursor protein, bridging integrator 1, ephrin type-A receptor 1, progranulin, microtubule-associated protein tau, and mitochondrial cytochrome c-oxidase gene II have been strongly linked to familial cases of AD within the Hispanic population.[3] Therefore, these predisposing genetic factors ultimately indicate a higher risk of AD and Alzheimer's dementia within the Hispanic community.[3]

As of October 2021, four genome-wide significant loci have been linked to an individual's risk of being infected with COVID-19, while nine other genome-wide significant loci have been linked to the severity of the disease caused by a COVID-19 infection.[18] However, no specific genetic links have been made between COVID-19 and the general Hispanic population. Extensive research regarding specific genes associated with COVID-19 is being led by an international effort known as *The COVID-19 Host Genetics Initiative* and is still in its early stages as of October 2021.[18]

4. Modifiable risk factors

4.1 Income/socioeconomic status

According to the United States Census Bureau, the median household income for all races/ethnicities in 2019 was $68,703.[19] On the other hand, the median household income for Hispanic Americans in 2019 fell well below the national median at only $56,113.[19] As seen in Fig. 14.5, Hispanic Americans have historically been at a disadvantage with regards to household income as they fall well below the national median household income on a yearly basis.[19]

Furthermore, a significant difference is also noted within the poverty level of Hispanic Americans when compared to the overall United States. In 2019, the US poverty percentage was 10.5%.[14] As shown in Table 14.3, Hispanic Americans registered a significantly higher poverty percentage of 17.2%.[20] Therefore, over one in six

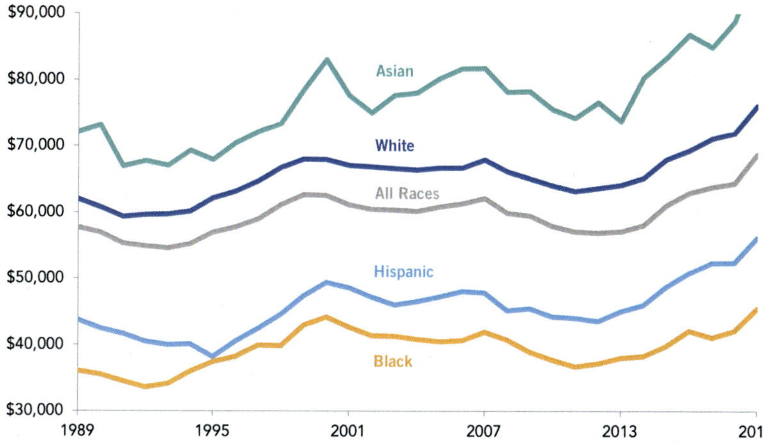

Figure 14.5 Median Household Income by Race/Ethnicity in 2019. Chart Created by Peter G. Peterson Foundation.[19]

Table 14.3 Most recent data available. Poverty Information from The United States Census Bureau, 2019 ACS 1-Year Estimates data.[20] Education Information from The United States Census Bureau, 2019 ACS 1-Year Estimates data.[22]

Poverty level and educational attainment by race/ethnicity for the United States (2019)		
Race/ethnicity	**Percent individuals below poverty level**	**Percent individuals holding Bachelor's degree or higher**
White alone (not Hispanic or Latino)	10.3%	34.4%
Hispanic or Latino	17.2%	17.6%
Black or African American alone	21.2%	22.5%
Asian alone	9.6%	55.6%
American Indian and Alaska Native alone	23.0%	16.1%
Native Hawaiian and Pacific Islander alone	16.5%	18.1%
Multiple races and other races	15.2%	33.4%

Hispanic Americans lived in poverty within the United States in 2019. This is in stark contrast to other groups, such as non-Hispanic white Americans who accounted for over 60% of the total population of the United States but still had a lower poverty percentage level than the national average at 10.3%.[14,20] Given the steadily rapid increase of the Hispanic American population within the United States, the groups poverty percentage level is also projected to increase, thereby further increasing the poverty gap between Hispanic Americans and other racial/ethnic groups.

This poverty gap between Hispanic Americans and other racial/ethnic groups has significant consequences in terms of access to healthcare and living conditions. For example, reports show that Hispanic Americans have an inferior access to healthcare as compared to non-Hispanic white Americans and, as a result of this, are at a greater risk for a COVID-19 infection, diabetes, and other long-term health complications such as AD.[21] Furthermore, lack of financial resources often forces minority demographics such as Hispanic Americans to purchase lower quality food items (in terms of nutritive value) and as such, greatly contributes to poorer overall Hispanic American health.

4.2 Education

As with the income and poverty level, a similar pattern is seen with regards to education when Hispanic Americans are compared to the United States national average. In 2019, the percentage of individuals aged 25 years and older who held a bachelor's degree or higher in the United States was 32.1%, while Hispanic Americans fell significantly below the national average.[14,22]

As shown in Table 14.3, Hispanic Americans had one of the lowest levels of educational attainment when directly compared to other racial/ethnic groups in the United States.[22] Although they accounted for 18.5% of the total population of the United States in 2019, only a mere 17.6% of Hispanic Americans held a bachelor's degree or higher.[14,22]

From the demographic data presented previously and within this section, it is clear that Hispanic Americans are disproportionately disadvantaged with regards to income/socioeconomic status and educational attainment. Furthermore, both disadvantages can further feed into each other to create a vicious cycle, where the argument can be made that a low income/socioeconomic status leads to a lower educational attainment and vice versa.[3]

This educational disparity, similar to the financial disparity, is yet another causative agent leading to Hispanic Americans being at a greater risk for a COVID-19 infection and AD. Given that Hispanic Americans do not have the same level of educational attainment as other demographics in the United States, there is often at times a lack of awareness for various health issues and methods of preventing them or obtaining care; this includes conditions such as COVID-19 and AD. As such, a lack of educational attainment has proven to be a significant and substantial risk factor for a COVID-19 infection and AD rates among the Hispanic American demographic.

4.3 Exercise and diet

Modifiable risk factors such as exercise and diet play a critical role in the future development of various age-related chronic diseases, which contribute to development of dementia or a COVID-19 infection.[3] National data collected on the Hispanic American youth and adult populations of the United States reveal interesting patterns regarding exercise and diet.[3]

With regards to exercise, national data reveal that Hispanics regularly do not meet the recommendations for physical activity.[3] Furthermore, the data reveal that such a lack of physical activity often begins from a young age, and this poor habit is often carried into adulthood.[3] In 2017, the American Heart Association reported that only 33.3% of Hispanic boys and 18.1% of Hispanic girls in grades 9–12 were active for at least 60 min a day on all 7 days of the week.[3] Additionally, with regards to adults over the age of 18 years, only 48.7% of Hispanic males and 41.0% of Hispanic females met aerobic guidelines set by the U.S. Department of Health and Human Services in 2017.[3] According to the CDC, Hispanic Americans had the highest prevalence of self-reported physical inactivity than any other racial/ethnic group in the year 2020 at 31.7%.[23]

It is important to note that the primary reason why many Hispanic Americans do not regularly meet recommendations for physical activity is because they feel a "lack of time," "very tired," and feel that they have a "lack of self-discipline" to exercise.[24] The most likely reason for these issues stems from the fact that many Hispanic Americans are forced to work long hours and multiple jobs (this relates to the poverty gap between Hispanic Americans and non-Hispanic white Americans) and do not have the same level of educational attainment as other racial/ethnic groups (this contributes to the "lack of self-discipline" that many Hispanic Americans feel they have).[24]

With regards to diet, similar patterns are seen where poor dietary habits were often developed from a young age within the Hispanic community.[3] The CDC Youth Risk Behavior Survey revealed that Hispanic youths had lower vegetable, fruit, milk, and water consumption on a daily basis.[3] Furthermore, the CDC reported that Hispanic students ate or drank approximately 47 more low-nutrient calories per day than their peers.[3] The issue of poor diet stems from the combination of a lower income/socioeconomic status and a poor dietary and overall education, which leads to Hispanic Americans having more access than non-Hispanic whites and Asians to unhealthy foods and drinks.[3] It is also important to note that from 2011 to 2016, a study was conducted concerning the dietary habits and typical food consumption of Hispanic Americans.[25] This study found that the overall traditional diet of Hispanic Americans as evaluated using the Healthy Eating Index (HEI) is in need of improvement and is a major causative agent leading to health issues (such as obesity and subsequent diseases/conditions) among the Hispanic American population.[25]

The combination of poor diet and lack of physical activity directly leads to Hispanic Americans having one of the highest rates of obesity within the United States.[3] According to the CDC, approximately 38.0% of Hispanic children are overweight or obese.[3] Furthermore, the American Heart Association reveals that 80.8% of Hispanic males and 77.8% of Hispanic females are either overweight or obese.[3] Such alarming rates of overweight and obesity directly places Hispanic Americans at a higher risk of developing age-related chronic diseases, which ultimately increases their risk of developing AD and Alzheimer's dementia as well as a COVID-19 infection in the future.[3]

4.4 Smoking and tobacco use

A key modifiable risk factor within the Hispanic American community is smoking and the usage of tobacco products. Smoking and the usage of tobacco products directly lead to the development of various severe age-related chronic diseases such as diabetes, kidney disease, cardiovascular disease, and chronic lung diseases. In turn, these age-related chronic diseases can contribute to the future development of AD and Alzheimer's dementia, as well as severe infection, long-term complications, or death from a COVID-19 infection.

According to the CDC, the prevalence of smoking within the Hispanic American community is generally lower than that of other racial/ethnic groups.[26] In 2019, 8.8% of Hispanic American adults above the age of 18 years smoked, while the national average for all adults above the age of 18 years in the United States was 14.0%.[26] However, research indicates that smoking prevalence is higher among Hispanic Americans born within the United States due to acculturation.[26]

Furthermore, tobacco companies have historically targeted the Hispanic community via various marketing techniques since the early 1970s.[27] Internal documents from the largest global tobacco companies reveal that they deem the Hispanic population to be "lucrative," "easy to reach," and "undermarketed."[27] Major tobacco companies also routinely sponsor various Hispanic cultural events, provide scholarships, as well as fund Hispanic political action committees in order to maintain their presence within the Hispanic American community at large.[27] Additionally, data show that Hispanic Americans are less likely to receive quitting advice from a healthcare professional than non-Hispanic white individuals as they face systemic health and social inequalities.[27] As a result, smoking is the leading preventable cause of morbidity and mortality within the Hispanic American community in the United States.[27]

5. Age-related chronic diseases

5.1 Diabetes

The proportion of Hispanic Americans with diabetes or at risk of developing diabetes is significantly higher than that of non-Hispanic whites. Hispanic Americans are twice as likely to develop diabetes during their lifetime than are non-Hispanic whites.[3] Furthermore, the CDC's National Health Interview Survey in 2018 found that 13.2% of Hispanic adults have diabetes compared to 8.0% of non-Hispanic white adults.[28]

The significantly higher percentage of Hispanic Americans with diabetes compared to non-Hispanic whites is a contributing factor for the higher proportion of Hispanic Americans diagnosed with AD and Alzheimer's dementia. Furthermore, observational studies have concluded that individuals with type 2 diabetes are twice as likely to develop AD than nondiabetic individuals.[29]

In addition, the higher percentage of diabetic Hispanic Americans is a cause for concern with regards to the COVID-19 pandemic. As of October 2021, there is not enough data to determine whether or not a diabetic individual is more likely to contract a COVID-19 infection than the general population.[30] However, studies have shown that diabetic individuals are more likely to develop a severe infection, long-term complications, or death from a COVID-19 infection when compared to the general public.[30] Data published in December 2020 have shown that approximately 25% of severe COVID-19 cases involved diabetic individuals.[30]

5.2 Kidney disease

According to the National Kidney Foundation, Hispanics are one-and-a-half times more likely to have kidney disease and kidney failure than other racial/ethnic groups.[3] This is due to the fact that the leading cause of kidney disease is diabetes.[3]

Furthermore, studies have shown that chronic kidney disease (CKD) not only allows for the development of AD and Alzheimer's dementia but promotes it.[31] Although seemingly unrelated, there exists an intimate mechanism linking CKD and AD.[32] More specifically, the cognitive impairments associated with AD is often associated with the buildup of uremic toxins and neurotoxicity in the brain after renal failure commonly caused by CKD.[32] Clinical studies have revealed that the degree of cognitive impairment is linked to the progression of CKD and renal failure.[31] The spillage of protein in the urine, known as albuminuria, is one of the earliest signs of kidney disease in patients.[31] According to a research study conducted by the Emory University School of Medicine, patients with albuminuria were approximately 50% more likely to have dementia than patients without albuminuria.[31]

With regards to a COVID-19 infection, research indicates that the virus causes severe damage to the kidneys in those who are hospitalized due to a serious infection. A COVID-19 infection causes acute kidney injury in up to 57% of COVID-19 infection–related hospitalizations whether or not they have prior established CKD.[33] Nevertheless, acute kidney injury from a COVID-19 infection increases an individual's risk of developing CKD, worsening existing CKD, or renal failure.[33]

Given that Hispanic Americans are more likely to have pre-existing kidney disease and kidney failure than any other racial/ethnic groups, they are at a greater risk for the development of AD and Alzheimer's dementia as well as severe infection, long-term complications, or death from a COVID-19 infection.

5.3 Hypercholesterolemia

The American Heart Association reported in 2020 that 39.9% of Hispanic males and 38.9% of Hispanic females had total cholesterol levels of 200 mg/dL or higher, while 13.0% of Hispanic males and 10.1% of Hispanic females had total cholesterol levels of 240 mg/dL or higher.[3] Furthermore, 33.5% of Hispanic males and 23.8% of Hispanic females had low–density-lipoprotein cholesterol (LDL-C) levels of at least 130 mg/dL.[34] It is important to note that high levels of LDL-C are immensely damaging to health and have been shown to consistently cause the development of cardiovascular disease, which is currently the leading cause of death in the United States.[35] According to the US National Library of Medicine, the normal total cholesterol level for both men and women is between 125 mg/dL and 200 mg/dL, while the normal LDL-C level for both men and women is less than 100 mg/dL.[36] It was also noted by the American Heart Association that the percentage of adults voluntarily getting screened for hypercholesterolemia between 2015–2020 was lower for Hispanic adults than non-Hispanic white, African American, and Asian adults.[3]

The large proportion of Hispanic males and females with hypercholesterolemia is a cause for concern regarding the future development of AD and Alzheimer's dementia.

Research has shown that hypercholesterolemia and AD are strongly linked such that hypercholesterolemia directly increases the risk of developing AD.[37] Moreover, some research also suggests that higher LDL-C levels could contribute to development of early-onset Alzheimer's disease, a rare form of AD that manifests before the age of 65 years.[37]

As of October 2021, ongoing and rapidly evolving research indicates that individuals with hypercholesterolemia are at an increased risk of getting a severe illness, long-term complications, or death from a COVID-19 infection.[38] In particular, high LDL-C levels contribute to various vasculopathies in patients with a COVID-19 infection.[38] The virus invades the endothelial cells, which line the lumen of the blood vessels thereby causing injury.[38] This triggers an inflammatory reaction, which leads to widespread blood clotting and coagulopathies.[38] Such widespread blood clotting can directly lead to life-threatening complications such as a stroke or heart attack.[38]

Furthermore, it should be noted that hypercholesterolemia and high LDL-C levels in particular directly lead to the exacerbation of several other age-related chronic diseases such as diabetes, kidney disease, and cardiovascular disease, which all promote the development of AD and Alzheimer's dementia, as well as severe infection, long-term complications, or death from a COVID-19 infection.

5.4 Cardiovascular disease

Cardiovascular disease is the leading cause of morbidity and mortality for most racial and ethnic groups within the United States.[39] A recent report released by the American Heart Association revealed that approximately 49.0% of Hispanic males and 42.6% of Hispanic females had cardiovascular disease between 2013 and 2016.[3] Furthermore, the CDC reported that cardiovascular disease was the leading cause of death within the Hispanic American population accounting for approximately 20.3% of Hispanic American deaths within the United States in 2015.[39]

Extensive studies and research conducted within the past decade have indicated that there is a strong association between cardiovascular disease and the development of AD and Alzheimer's dementia.[40] Cardiovascular disease starves the brain of blood, and the brain is particularly sensitive to changes in oxygen supply.[40] The gradual drop in oxygen supply will lead to the damage of brain cells over time ultimately increasing an individual's risk of developing AD and Alzheimer's dementia.

With regards to a COVID-19 infection, the CDC reports that individuals with pre-existing cardiovascular disease are more likely to experience severe illness or death due to a serious infection.[41] The severe viral illness places an increased demand on the heart during a time when oxygen levels are decreased throughout the body as the virus primarily attacks the lungs.[42] Furthermore, individuals with underlying cardiovascular disease often develop acute cardiac injury as a COVID-19 infection triggers inflammation of the heart muscle.[42] Blood tests reveal elevated levels of troponin, which demonstrates that the heart muscle has been significantly damaged.[42] As of October 2021, data have shown that 10% of patients with an underlying cardiovascular disease who have contracted a COVID-19 infection will die.[42] This is significantly higher than the 1% of patients who are relatively healthy and will die due to a COVID-19 infection.[42]

5.5 Chronic lung diseases

According to the most recent data released by the CDC, Hispanic Americans had the lowest prevalence of chronic lung diseases at 2.6% than any other racial/ethnic groups.[43] Within the United States, non-Hispanic whites accounted for the highest prevalence of chronic lung diseases at 5.7%.[43] The lower prevalence of chronic lung diseases among the Hispanic American community is due to the generally lower prevalence of smoking and tobacco use as discussed previously. Nevertheless, factors such as acculturation as well as various marketing techniques used by major tobacco companies are increasing the smoking prevalence among Hispanic Americans born within the United States.[26]

Studies conducted within the past five years have indicated that individuals with chronic lung diseases have an increased risk of developing AD and Alzheimer's dementia in the future.[44] The link is due to the fact that individuals with underlying chronic lung diseases generally have low blood oxygen levels ultimately causing damage to brain cells. Observational studies conducted have revealed that compared to normal healthy individuals, individuals with chronic lung diseases are 33%–58% more likely to be diagnosed with dementia or cognitive impairment.[44]

It has been well established within the medical literature that a COVID-19 infection primarily attacks and causes severe damage to an individual's lungs. In particular, COVID-19 infects and kills cells within the alveolar epithelium leading to damage within the lung parenchyma as well as the capillaries, which are responsible for gas exchange within the lungs.[45] This directly leads to several clinical conditions which can often be fatal, such as bronchopneumonia, acute respiratory distress syndrome, as well as sepsis.[45] A population cohort study conducted between January 2020 and April 2020 analyzed 8,256,161 patients.[46] The study revealed that the risk for hospitalization due to a COVID-19 infection was increased by up to 50% in patients who had an underlying chronic lung disease.[46] Furthermore, individuals diagnosed with lung cancer were twice as more likely to be hospitalized due to a COVID-19 infection than individuals without lung cancer.[46]

5.6 Dementia

As discussed previously, AD as well as Alzheimer's dementia is rapidly increasing within the Hispanic American population. To be specific, current analytics project that approximately 3.5 million Hispanic Americans will be living with AD by the year 2060.

Individuals diagnosed with underlying dementia are at a higher risk of contracting a COVID-19 infection according to the CDC. Studies have shown that the blood–brain barrier of dementia patients is altered thereby predisposing them to various viral and bacterial infections.[47] Furthermore, underlying dementia may interfere with an individual's ability to follow the latest preventative measures as established by the CDC and WHO to protect themselves from a COVID-19 infection.[47]

As of October 2021, many questions regarding the long-term effects of a COVID-19 infection still remain unanswered. However, research reported at the Alzheimer's Association International Conference held in July 2021 found associations revealing that a COVID-19 infection led to the acceleration of AD and Alzheimer's dementia pathology and symptoms.[48] The prevalence of low blood oxygen levels during a COVID-19 infection revealed increased levels of biological markers, signifying brain injury and neuroinflammation, which accelerate the pathological processes responsible for the development of AD and Alzheimer's dementia.[48]

6. Environmental factors

6.1 Social inequalities

As discussed previously, Hispanic Americans face disproportionate disadvantages with regards to income/socioeconomic status as well as educational attainment. However, these issues are further worsened due to the fact that Hispanic Americans face various social inequalities and obstacles within the United States. A study conducted in 2019 by the Pew Research Center revealed that 51% of individuals believed that being Hispanic hindered a person's ability to "get ahead" and be successful within the United States.[49] Furthermore, in 2018, 40% of Hispanic Americans revealed that they experienced discrimination in some form.[50] The four most common discriminatory incidents experienced were (1) being called offensive names, (2) being told to go back to their home country, (3) being criticized for speaking Spanish, and (4) experiencing unfair treatment due to being Hispanic.[50]

As a result, 62% of Hispanic Americans revealed in 2018 that they are dissatisfied with the overall direction of the United States.[50] Such widespread discrimination and social inequalities within the United States also cause 55% of Hispanic Americans to worry about deportation regardless of legal status.[50]

6.2 Lack of access to healthcare

Long-standing social inequalities and discrimination also prevent Hispanic Americans from having access to adequate healthcare resources. Research conducted by the CDC has revealed that Hispanic Americans are three times more likely to lack a regular healthcare provider than non-Hispanic whites.[51] Furthermore, Hispanic Americans had the highest uninsured rate at 25% than any other racial/ethnic group in 2018.[52] As a result, up to 25% of Hispanic American adults claim that they receive no healthcare information from licensed medical personnel on a yearly basis.[51] However, over 80% of Hispanic Americans report that they resort to receiving health information from public media sources such as television and radio.[51]

An individual's overall health status and their ability to access healthcare resources are closely linked. Poor access to healthcare resources ultimately leads to poor overall health. An increasing Hispanic American population combined with a lack of access to adequate healthcare resources ultimately places Hispanic Americans at a high risk for various illnesses, diseases, and medical conditions, which plague the population of the United States.

7. Concluding remarks

The combining detrimental effects of various unmodifiable risk factors, modifiable risk factors, age-related chronic diseases, and environmental factors (Fig. 14.4) place the Hispanic American community at an extremely high risk for the future development of AD and Alzheimer's dementia as well as a COVID-19 infection leading to a severe illness, long-term complications, or death.

With regards to unmodifiable risk factors, no meaningful or impactful solutions can be provided at this time. Furthermore, it is also difficult to provide solutions to certain modifiable risk factors such as income/socioeconomic status and educational attainment. However, steps can be taken to provide education and guidance to the Hispanic American population regarding lifestyle changes (exercise, diet, and smoking/tobacco use) that can be made in order to significantly reduce the risk of developing various age-related chronic diseases, which contribute to the development of AD and Alzheimer's dementia as well as a severe COVID-19 infection.

Furthermore, a sincere effort spearheaded by the US government and major hospital systems throughout the country should be made in order to provide equitable and adequate healthcare resources to the underserved Hispanic American community.

These challenges can be addressed by (1) developing community outreach centers within underserved Hispanic American communities across the United States, which educate the population about the importance of a healthy lifestyle and the negative impacts of age-related chronic diseases, (2) the development and implementation of educational curriculums within middle schools, high schools, and colleges/universities, which focus specifically upon the biology of age-related chronic diseases, (3) creating a national database, which tracks Hispanic American residents of the United States who are diagnosed with or are at risk for the development of an age-related chronic disease, and (4) allocating federal funds for the construction of healthcare facilities within the underserved Hispanic American communities of the United States.

Acknowledgments

The research and relevant findings presented in this chapter were supported by the National Institutes of Health (NIH) grants AG042178, AG047812, NS105473, AG060767, AG069333 and AG066347.

References

1. U.S. Hispanic population 2016–2060. https://www.statista.com/statistics/251238/hispanic-population-of-the-us/. Published January 20, 2021. Accessed 16 June 2021.
2. What is dementia? Alzheimer's disease and dementia. https://www.alz.org/alzheimers-dementia/what-is-dementia. Accessed 22 June 2021.

3. Sheladia S, Reddy PH. Age-related chronic diseases and Alzheimer's disease in Texas: a Hispanic focused study. *J Alzheimers Dis Rep*. 2021;5(1):121−133. https://doi.org/10.3233/ADR-200277. Published 2021 Feb 24.
4. Alzheimer's Disease Facts and Figures. *Ebook*. Chicago: Alzheimer's Association; 2021. https://www.alz.org/media/documents/alzheimers-facts-and-figures.pdf.
5. The Alzheimer's disease crisis − by the numbers. UsAgainstAlzheimer's. https://www.usagainstalzheimers.org/learn/alzheimers-crisis. Accessed 16 July 2021.
6. Wu S, Vega WA, Resendez J, Jin H. Latinos and Alzheimer's disease: new numbers behind the crisis. Accessed 3 December 2016; Projection of the costs for US Latinos living with Alzheimer's disease through 2060. http://www.usagainstalzheimers.org/sites/default/files/Latinos-and-AD_USC_UsA2-Impact-Report.pdf.
7. Ganji R, Reddy PH. Impact of COVID-19 on mitochondrial-based immunity in aging and age-related diseases. *Front Aging Neurosci*. 2021;12:614650. https://doi.org/10.3389/fnagi.2020.614650. Published 2021 Jan 12.
8. Bortoletti M. Understanding SARS-CoV-2 and the drugs that might lessen its power. Economist. https://www.economist.com/briefing/2020/03/12/understanding-sars-cov-2-and-the-drugs-that-might-lessen-its-power. Published March 14, 2020. Accessed 16 October 2021.
9. C. National Foundation for Infectious Diseases. https://www.nfid.org/infectious-diseases/coronaviruses/. Published July 27, 2021. Accessed 16 August 2021.
10. Coronavirus. World Health Organization. https://www.who.int/health-topics/coronavirus#tab=tab_3. Accessed 16 October 2021.
11. Smith MW. What are the complications of coronavirus (COVID-19)? WebMD. https://www.webmd.com/lung/coronavirus-complications#3. Published August 9, 2021. Accessed 16 August 2021.
12. COVID-19 (coronavirus): long-term effects. Mayo Foundation for Medical Education and Research. https://www.mayoclinic.org/diseases-conditions/coronavirus/in-depth/coronavirus-long-term-effects/art-20490351. Published October 14, 2021. Accessed 16 October 2021.
13. CDC COVID Data Tracker. Centers for Disease Control and Prevention. https://covid.cdc.gov/covid-data-tracker/#demographics. Accessed 16 October 2021.
14. U.S. Census Bureau QuickFacts: United States. U.S. Census Bureau. https://www.census.gov/quickfacts/fact/table/US/PST045219. Published 2021. Accessed 16 August 2021.
15. Risk for COVID-19 infection, hospitalization, and death by race/ethnicity. Centers for Disease Control and Prevention. https://www.cdc.gov/coronavirus/2019-ncov/covid-data/investigations-discovery/hospitalization-death-by-race-ethnicity.html. Published 2021. Accessed 16 August 2021.
16. Assessing risk factors for severe COVID-19 illness. Centers for Disease Control and Prevention. https://www.cdc.gov/coronavirus/2019-ncov/covid-data/investigations-discovery/assessing-risk-factors.html. Published 2021. Accessed 20 August 2021.
17. Profile of Hispanic Americans age 65 and over. U.S. Department of Health and Human Services. https://acl.gov/sites/default/files/Aging%20and%20Disability%20in%20America/2018HA_OAProfile.pdf. Published 2019. Accessed 16 September 2021.
18. COVID-19 host genetics initiative. Mapping the human genetic architecture of COVID-19. Nature. 2021. https://doi.org/10.1038/s41586-021-03767-x.
19. Median income by race. Peter G. Peterson Foundation. https://www.pgpf.org/chart-archive/0257_income_by_race. Published September 28, 2020. Accessed 16 September 2021.
20. Poverty status in the past 12 months. U.S. Census Bureau. https://data.census.gov/cedsci/table?q=poverty&tid=ACSST1Y2019.S1701. Published 2020. Accessed 16 August 2021.
21. Shiro AG, Reeves R V. Latinos often lack access to healthcare and have poor health outcomes. Here's how we can change that. Brookings. Published 2020. https://www.

brookings.edu/blog/how-we-rise/2020/09/25/latinos-often-lack-access-to-healthcare-and-have-poor-health-outcomes-heres-how-we-can-change-that/.

22. Explore Census Data. U.S. Census Bureau. https://data.census.gov/cedsci/table?q=education&tid=ACSST1Y2019.S1501. Published 2020. Accessed 16 August 2021.
23. Fulton J. Physical inactivity is more common among racial and ethnic minorities in most states | Blogs | CDC. Centers for Disease Control and Prevention. https://blogs.cdc.gov/healthequity/2020/04/01/physical-inactivity/. Published April 01, 2020. Accessed 16 September 2021.
24. Bautista L, Reininger B, Gay JL, Barroso CS, McCormick JB. Perceived barriers to exercise in Hispanic adults by level of activity. *J Phys Activ Health*. 2011;8(7):916–925. https://doi.org/10.1123/jpah.8.7.916.
25. Overcash F, Reicks M. Diet quality and eating practices among Hispanic/Latino men and women: NHANES 2011–2016. *Int J Environ Res Publ Health*. 2021;18(3):1–11. https://doi.org/10.3390/ijerph18031302.
26. Burden of tobacco use in the U.S. Centers for Disease Control and Prevention. https://www.cdc.gov/tobacco/campaign/tips/resources/data/cigarette-smoking-in-united-states.html#hispanic. Published 2021. Accessed 16 September 2021.
27. Hispanic/Latino Americans. Truthinitiative.org. https://truthinitiative.org/sites/default/files/media/files/2020/05/Truth_Race-ethnicity%20Series%20Factsheets-Hispanic_final.pdf. Published 2020. Accessed 16 September 2021.
28. Diabetes and Hispanic Americans. U.S. Department of Health and Human Services Office of Minority Health. https://minorityhealth.hhs.gov/omh/browse.aspx?lvl=4&lvlid=63. Published March 01, 2021. Accessed 16 September 2021.
29. Reitz C, Brayne C, Mayeux R. Epidemiology of Alzheimer disease. *Nat Rev Neurol*. 2011;7(3):137–152. https://doi.org/10.1038/nrneurol.2011.2.
30. Watson S. Diabetes during the COVID-19 pandemic. WebMD. https://www.webmd.com/diabetes/diabetes-and-coronavirus. Published December 13, 2020. Accessed 16 September 2021.
31. 3. Kidney disease linked to dementia. National Kidney Foundation. https://www.kidney.org/news/ekidney/august08/Dementia_august08. Published August 12, 2014. Accessed 20 September 2021.
32. Shi Y, Liu Z, Shen Y, Zhu H. A novel perspective linkage between kidney function and Alzheimer's disease. *Front Cell Neurosci*. 2018;12:384. https://doi.org/10.3389/fncel.2018.00384.
33. Inserro A. Study illustrates kidney impact after COVID-19 resolves. AJMC. https://www.ajmc.com/view/study-illustrates-kidney-impact-after-covid-19-resolves. Published March 10, 2021. Accessed 8 August 2021.
34. Virani SS, Alonso A, Benjamin EJ, et al. Heart disease and stroke statistics-2020 update: a report from the American heart association. *Circulation*. 2020;141(9):e139–e596. https://doi.org/10.1161/CIR.0000000000000757.
35. Centers for Disease Control and Prevention. Heart Disease Facts. CDC Heart Disease. https://www.cdc.gov/heartdisease/facts.htm#:~:text=Heartdiseaseistheleading,groupsin theUnitedStates.&text=Onepersondiesevery36,UnitedStatesfromcardiovasculardisease.&text=About 659%2C000peopleinthe,1inevery4deaths.
36. Cholesterol levels: what you need to know: MedlinePlus. MedlinePlus. https://medlineplus.gov/cholesterollevelswhatyouneedtoknow.html. Published 2020. Accessed 8 September 2021.
37. Wingo TS, Cutler DJ, Wingo AP, et al. Association of early-onset Alzheimer disease with elevated low-density lipoprotein cholesterol levels and rare genetic coding variants of *APOB*. *JAMA Neurol*. 2019;76(7):809–817. https://doi.org/10.1001/jamaneurol.2019.0648.

38. Charles S. What to know about high cholesterol and COVID-19. Verywell Health. https://www.verywellhealth.com/high-cholesterol-and-covid-19-5118092. Published April 27, 2021. Accessed 8 October 2021.
39. Heart Disease Facts. Centers for Disease Control and Prevention. https://www.cdc.gov/heartdisease/facts.htm. Published September 27, 2021. Accessed 8 October 2021.
40. Schein C. Alzheimer's and Cardiovascular Disease | Aegis Living. Aegis Living. https://www.aegisliving.com/resource-center/the-link-between-alzheimers-and-cardiovascular-disease/. Published December 11, 2019. Accessed 12 October 2021.
41. COVID-19 and your health. Centers for Disease Control and Prevention. https://www.cdc.gov/coronavirus/2019-ncov/need-extra-precautions/people-with-medical-conditions.html. Published October 14, 2021. Accessed 16 October 2021.
42. Lewis D. How does cardiovascular disease increase the risk of severe illness and death from COVID-19? - Harvard Health. Harvard Health. https://www.health.harvard.edu/blog/how-does-cardiovascular-disease-increase-the-risk-of-severe-illness-and-death-from-covid-19-2020040219401. Published April 02, 2020. Accessed 16 October 2021.
43. Akinbami L, Liu X. Chronic obstructive pulmonary disease among adults aged 18 and over in the United States, 1998–2009. Centers for Disease Control and Prevention. https://wwwcdcgov/nchs/products/databriefs/db63htm#fig3. Published November 06, 2015. Accessed 16 October 2021.
44. Lutsey PL, Chen N, Mirabelli MC, et al. Impaired lung function, lung disease, and risk of incident dementia. *Am J Respir Crit Care Med.* 2019;199(11):1385–1396. https://doi.org/10.1164/rccm.201807-1220OC.
45. Radchenko C. Short & long-term effects of COVID-19 on the lungs. UC Health. https://www.uchealth.com/en/media-room/covid-19/short-and-long-term-lung-damage-from-covid-19. Published November 08, 2020. Accessed 16 October 2021.
46. Aveyard P, Gao M, Lindson N, et al. Association between pre-existing respiratory disease and its treatment, and severe COVID-19: a population cohort study. *Lancet Respir Med.* 2021;9(8):909–923. https://doi.org/10.1016/S2213-2600(21)00095-3.
47. Wood H. Elevated risk of COVID-19 in people with dementia. *Nat Rev Neurol.* 2021;17:194. https://doi.org/10.1038/s41582-021-00473-0.
48. COVID-19 Associated with long-term cognitive dysfunction, acceleration of Alzheimer's symptoms | AAIC 2021. Alzheimer's Association. https://www.alz.org/aaic/releases_2021/covid-19-cognitive-impact.asp. Published 2021. Accessed 13 October 2021.
49. Horowitz J, Brown A, Cox K. Views of racial inequality in America. Pew Research Center. https://www.pewresearch.org/social-trends/2019/04/09/views-of-racial-inequality/. Published April 09, 2019. Accessed 16 October 2021.
50. Lopez M, Gonzalez-Barrera A, Krogstad J. Latinos' experiences with discrimination. Pew Research Center. https://www.pewresearch.org/hispanic/2018/10/25/latinos-and-discrimination/. Published October 25, 2018. Accessed 16 October 2021.
51. Livingston G, Minushkin S, Cohn D. Hispanics and Health Care in the United States. Pew Research Center. https://www.pewresearch.org/hispanic/2008/08/13/hispanics-and-health-care-in-the-united-states-access-information-and-knowledge/. Published August 13, 2008. Accessed 16 October 2021.
52. Reeves A. Latinos often lack access to healthcare and have poor health outcomes. Here's how we can change that. Brookings. https://www.brookings.edu/blog/how-we-rise/2020/09/25/latinos-often-lack-access-to-healthcare-and-have-poor-health-outcomes-heres-how-we-can-change-that/. Accessed 16 October 2021.

Women and Alzheimer's disease risk: a focus on gender

15

Emma Schindler[1] *and P. Hemachandra Reddy*[2,3,4,5]
[1]University of Miami Miller School of Medicine, Miami, FL, United States; [2]Department of Internal Medicine, Texas Tech University Health Sciences Center, Lubbock, TX, United States; [3]Nutritional Sciences Department, Texas Tech University, Lubbock, TX, United States; [4]Department of Pharmacology and Neuroscience, School of Medicine, Texas Tech University Health Sciences Center, Lubbock, TX, United States; [5]Department of Neurology, Texas Tech University Health Sciences Center, Lubbock, TX, United States

Abstract

A previous chapter highlighted the biological mechanisms by which female sex contributes to Alzheimer's disease (AD) risk and outcomes. However, discussion of AD in women is incomplete without considering the impact of female gender on AD risk, as gender encompasses psychosocial and cultural differences between women and men that also modulate risk for cognitive decline. The current chapter discusses several main social determinants of health and explains how women, as a historically oppressed population, may be particularly vulnerable to the effect of each on cognition. This chapter also considers the disproportionate female burden of dementia caregiving, how associated stresses augment risk for later cognitive decline among caregivers themselves, and how the COVID-19 pandemic may add to this risk. Understanding the gender-specific factors that affect AD risk and disease progression is essential for developing targeted preventative interventions and treatments. Future research is necessary to better characterize how social determinants of health uniquely impact female cognition compared to males. Moreover, future studies focused on gender identities outside of the male—female binary are critical to developing a holistic understanding of how gender may impact late-life cognition.

Keywords: Alzheimer's disease; Caregiver burden; COVID-19; Gender; Identity; Race; Women.

1. Introduction

Sex is biological, whereas gender is how a person identifies.[1] Gender is a complex label that includes social and cultural attitudes, expectations, and opportunities that vary by self-identification male, female, or nonbinary.[2,3] Although gender is a continuum,[4] there is minimal literature discussing gender differences in Alzheimer's disease (AD) outside of the female—male binary. For this reason, this chapter will focus on female—male gender differences in AD, while acknowledging that there are significant gaps in

COVID-19 in Alzheimer's Disease and Dementia. https://doi.org/10.1016/B978-0-443-15256-6.00002-7

our understanding of how individuals with nonbinary gender identities are affected by AD. Awareness of gender differences in AD risk, presentation, progression, and treatment outcomes is critical to guide the clinical management of women as they age.

2. Education

Increased years of formal education is associated with a lower risk of developing AD.[5–10]

Education is one mechanism by which individuals may build up a "cognitive reserve" to protect against neurodegeneration. Cognitive reserve describes the ability to efficiently utilize cognitive networks made up of neuron-to-neuron connections within the brain that enables an individual to maintain cognitive function despite pathologic brain changes.[9,11]

Women have historically had fewer educational opportunities than men, and this remains true today across many cultures.[12,13] In one European study, low levels of education increased the risk of developing AD more significantly in females than in males.[14] Another study showed that lower levels of educational attainment were associated with a greater risk of dementia death in women but not men.[15] The gender disparity in education has shifted in the United States in the past few decades, such that women now have higher average educational attainment than men.[16] Increased educational opportunities for women may contribute to a recently observed decline in the incidence of dementia among women.[17] Existing works highlight the importance of expanding educational opportunities for women both to progress toward gender equality and build up cognitive reserve to protect against cognitive decline later in life. Future studies will be essential to uncover how changing trends in educational attainment impact late-life cognition and AD risk differentially in males and females.

3. Employment

In both men and women, employment is associated with slower rates of cognitive decline and better cognition later in life,[12,18,19] whereas retirement is associated with accelerated cognitive decline.[20,21] Similar to education, there is a significant gender discrepancy in this protective factor, as women have historically had fewer occupational opportunities than men.[22] A recent study found that average memory decline after age 60 years was 50% slower among women who participated in the paid workforce in early adulthood and midlife, regardless of family structure and educational attainment, compared to women who did not work at all.[23] Importantly, rates of memory decline were also similar regardless of whether or not mothers took time off from workforce participation to raise young children at home.[23] Similar to education, participation in the workforce may provide a means to bolster cognitive reserve through social and cognitive stimulation.[24–26]

Employment trends have shifted in recent decades, with more educational and occupational opportunities for women than ever before.[22] Currently studied populations of elderly women grew up during a period in which women had fewer educational and employment opportunities—their experiences differ from those of young and middle-aged women today. Future studies will be necessary to understand how changes in workforce participation will affect late-life cognition and AD risk in generations of women to come. Future research should evaluate unpaid volunteer work, as nonworking women may reap the cognitive benefit of social and intellectual stimulation through volunteering outside of the home.[24]

4. Race

Historic oppression of racial and ethnic minorities in the United States has contributed to the development of significant and well-documented health disparities that are hypothesized to contribute to a heightened risk for AD and other dementias.[27] Indeed, older black individuals have about twice the incidence of dementia and older Hispanic individuals have about 1.5 times the incidence of dementia compared to whites.[28] Specific drivers may include a higher burden of chronic disease, chronic stress due to oppression, limited availability of quality healthcare, and low socioeconomic status.[29,30] However, these studies have not included sex-stratified analyses.

Disproportionate risk of developing AD may also stem from disparities in educational attainment, as low education is a risk factor for AD.[8,28] Racial and ethnic minorities have historically had fewer educational and employment opportunities due to systemic racism. This disadvantage persists through a generation-to-generation cycle in which lower household income restricts access to higher education and subsequently, high-paying occupations.[31,32] Gender gaps in education and employment may be amplified by racial/ethnic minority status, compounding minority women's risk for cognitive decline.

Lastly, racial minorities—minority women in particular—are underrepresented in AD clinical trials; therefore, much AD research is not generalizable to these populations.[28,30,33] Concerted efforts need to be taken to recruit racially and ethnically diverse study populations to improve understanding, prevention, and treatment of AD in diverse populations.

Female gender increases the risk for cognitive decline later in life; this risk is likely exacerbated by racial and/or ethnic minority status. However, studies examining this interaction are limited. Future research focused on the intersection of gender and race/ethnicity is vital to fully understand a woman's dementia risk. Providers must also be aware of (1) the probable interaction between sex, gender, and race and dementia risk as they care for aging women and (2) how their own implicit biases may affect patient satisfaction and healthcare-seeking behaviors among historically marginalized groups.

5. Sexual and gender identity

The LGBT+ (lesbian, gay, bisexual, transgender, other) community faces significant health disparities that stem from historic societal and institutional discrimination.[34,35] The higher prevalence of certain risk factors and limited access to healthcare services suggest that LGBT + older adults have an elevated risk of cognitive decline and dementia. However, very few studies have explored dementia risk in this population, underscoring a need for future research in this area.

The stress related to societal oppression and discrimination in healthcare contributes to higher rates of hypertension,[36] cardiovascular disease,[37] diabetes,[36] smoking, and heavy alcohol use[37–40] among LGBT + individuals—all of which are risk factors for developing dementia. Discrimination and identity concealment also increase LGBT + individuals' risk of social isolation and loneliness,[41,42] which negatively affect mental and physical health.[43] Indeed, LGBT + individuals experience a higher prevalence of depression than cis-gendered, heterosexual counterparts, and one-third of LGBT + older adults report depression.[36,37,44,45] This is particularly concerning because depression is associated with a 2–3-fold increase in the risk of dementia.[46,47]

As LGBT + individuals age, compounding effects of stigma due to sexual and gender minority status, old age, and cognitive difficulties, contribute to obstacles accessing healthcare that may cause cognitive impairment to go unnoticed and untreated.[48] The National Longitudinal Aging with Pride study found that 13% of LGBT + older adults (and 40% of transgender participants) had at some point been denied healthcare or had received inferior care due to their sexual or gender identity.[49,50]

Among biologically female individuals who identify with a sexual and/or gender minority group, this risk may be compounded. One study reports that lesbian and bisexual women demonstrate higher rates of cardiovascular disease and obesity than heterosexual women, both of which increase the risk of cognitive decline and dementia.[37] However, whether LGBT + status interact to modulate AD risk has not been sufficiently examined. Future research must work toward better understanding the intersection of LGBT + adults' unique risk factors and cognitive impairment, AD, and dementia through continued longitudinal studies. Meanwhile, providers must be aware of social drivers of poor physical and mental health—including ageism, and discrimination by sexuality—to understand the unique risk profile of each aging patient. It is equally important that providers understand how prejudice against sexual/gender minorities infiltrates the healthcare system, discouraging access to healthcare services and contributing to health disparities that modulate dementia risk.

6. Exercise

Studies have demonstrated that regular physical activity is linked to a lower risk of AD compared to those who do not exercise.[51–55] In fact, regular aerobic exercise has been associated with reduced brain atrophy and improved cognition and memory[56] and has been shown to decrease Aß deposition in animal models.[57,58] Regular exercise also slows the loss of sex hormones with aging; this is particularly relevant in women, as loss of estrogen during menopause is a sex-specific predictor of cognitive decline.[52]

However, women tend to exercise less frequently than men. This is in part due to family structure and the demands of being a primary caregiver.[59,60] As a result, physical inactivity is a risk factor for cognitive decline that disproportionately impacts women. A community-wide cohort study also found that obesity—a consequence of poor diet and physical inactivity—was a risk factor for AD onset after 3-year follow-up in women over age 65, but not in men.[61] Interestingly, women with MCI may reap a greater cognitive benefit from introducing regular aerobic exercise than men. Baker et al. report that women with MCI randomized to regular aerobic exercise demonstrated improved executive function, glucose metabolism, and reduced cortisol compared to the control group.[62] This effect was not observed among men. A similar Canadian study supports these findings, as women who participated in moderate to heavy exercise had a reduced risk of cognitive impairment, while men did not.[63]

7. Depression

Women are twice as likely to develop depression than men[64]; this risk increases after menopause.[65] Depression in middle age may increase the risk of developing AD by 70%,[46] and the WHIMS study found that women with depressive symptoms had twice the risk of developing MCI and dementia.[66] Similar studies have shown that the risk of transitioning from MCI to AD is higher in females than males with depression.[67,68] Importantly, depression has been associated with hippocampal atrophy in women but not in men.[69,70]

8. Caregiver burden

Over 80% of Americans living with AD require at-home assistance from a caregiver.[71] Caregivers are often unpaid family members who help individuals with AD complete activities of daily living (including bathing, dressing, paying bills, transportation, and taking medications), as well as provide emotional support.[72,73] Women are more likely than men to become full-time caregivers for people living with AD.[74–76]

In 2020, there were an estimated 11.2 million unpaid caregivers for people with dementia in the United States—over two-thirds of which were women.[27] Caregiving is associated with significant financial, emotional, and physical burdens. Caregivers report higher rates of unemployment and financial strain,[77] as well as psychosocial (depression, social isolation, and sleep disturbances), behavioral (poor diet and lack of exercise), and physiological (metabolic syndrome and inflammation) manifestations of chronic stress that predispose caregivers to develop dementia themselves.[78–80] Multiple studies report higher levels of cortisol and impaired attention and executive function among caregivers, underscoring the impact that chronic stress has on cognition (Fig. 15.1).[81]

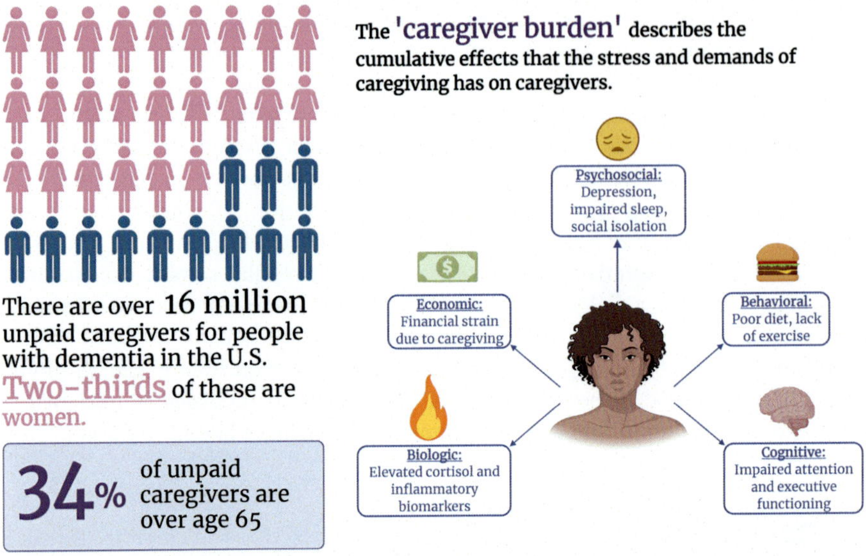

Figure 15.1 Summary of the cumulative financial, psychosocial, and physical effects of caregiving for someone with Alzheimer's disease on unpaid caregivers.

Because women comprise two-thirds of dementia caregivers, they are disproportionately affected by the listed risk factors. In addition, females spend more time caregiving, assume more caregiving tasks, and report higher rates of depression, impaired mood, and poor health compared to male caregivers.[78,79] For these reasons, female caregivers are estimated to have twice the caregiver burden of male caregivers, compounding their risk for adverse outcomes associated with chronic stress, such as dementia.[78,79]

Awareness of the disproportionate burden of caregiving on females is essential, as interventions to support the physical and emotional well-being of female caregivers may mitigate long-term adverse health outcomes. Understanding the challenges of unpaid caregiving also highlights the importance of AD prevention strategies, as slowing the progression of normal cognition or MCI to AD may help alleviate the caregiver burden.

9. COVID-19 pandemic

The consequences of the COVID-19 pandemic are extensive and include high levels of mortality, global economic hardship, poor mental health, and changes to quality of life.[82] Those with dementia are at a higher risk of severe morbidity and mortality from COVID-19,[83] and multiple studies report worsening independence and neuropsychiatric manifestations of dementia since the beginning of the pandemic.[84–86]

The COVID-19 pandemic has also placed new challenges on informal caregivers for people with dementia, most of which are women.[27] Multiple studies demonstrate an increase in the demands of caregiving since the beginning of the pandemic.[27,82,84,85,87,88] Caregivers report increasing difficulty accessing healthcare services for those they care for, as well as having to take on additional caregiving tasks due to limited access to the home health services available prepandemic.[84,85,89] Changes in caregiver workload also contribute to an increase in caregiver burden, reflected by increasing rates of anxiety, depression, and burnout among informal caregivers.[82,87–90]

These stresses may be amplified in women, who make up the majority of informal caregivers.[27,82] Though reports of the differential impact of the COVID-19 pandemic on male and female caregivers are sparse, one Italian study found that increases in anxiety and depression among informal caregivers during the COVID-19 pandemic were more pronounced in women than in men.[87] Among female caregivers, the COVID-19 pandemic may exacerbate pre-existing gender disparities in mental health, contributing to worse long-term psychologic outcomes.[87]

These findings highlight the importance of providing support to both people with dementia and their caregivers as the COVID-19 pandemic continues to evolve. People with dementia are a particularly vulnerable subgroup not only because of susceptibility to severe disease but also because of unmet care needs due to isolation. Developing ways to streamline access to healthcare services is also critical, as this may help alleviate the increase in caregiver burden.

10. Conclusion

With female gender identity comes a multitude of social determinants of health that can moderate risk for AD and other dementias. These include limited access to education, employment, or exercise, and a higher incidence of depression. It is important to pay special attention to certain subgroups in which multiple marginalized identities intersect (i.e., racial and/or ethnic minorities and LGBT women), as these women may have a particularly heightened risk for adverse health outcomes, including AD, necessitating targeted prevention and treatment strategies. In addition, the burden of caregiving for individuals with AD and other dementias falls predominantly on women. Understanding how the stresses of caregiving can impact late-life cognition is imperative to developing strategies to support caregivers and implementing prevention strategies in this group as well.

Providers must be aware of how gender impacts health, especially as it relates to cognitive decline in the aging population. However, due to the paucity of literature comparing how social determinants of health impact AD risk in women compared to men, our understanding of gender and AD is incomplete. Future studies are needed to uncover how social determinants of cognitive decline affect women differently from men. Finally, future research emphasizing gender identities outside of the male–female binary will be essential to generate a more comprehensive understanding of how gender effects cognitive decline.

References

1. Wizemann TM, Pardue ML, eds. *Exploring the Biological Contributions to Human Health: Does Sex Matter?* The National Academies Collection: Reports funded by National Institutes of Health; 2001.
2. Peterson A, Tom SE. A lifecourse perspective on female sex-specific risk factors for later life cognition. *Curr Neurol Neurosci Rep*. 2021;21(9):46. https://doi.org/10.1007/s11910-021-01133-y.
3. Mielke MM, Vemuri P, Rocca WA. Clinical epidemiology of Alzheimer's disease: assessing sex and gender differences. *Clin Epidemiol*. 2014;6:37–48. https://doi.org/10.2147/CLEP.S37929.
4. Hyde JS, Bigler RS, Joel D, Tate CC, van Anders SM. The future of sex and gender in psychology: five challenges to the gender binary. *Am Psychol*. February–March 2019; 74(2):171–193. https://doi.org/10.1037/amp0000307.
5. Evans DA, Bennett DA, Wilson RS, et al. Incidence of Alzheimer disease in a biracial urban community: relation to apolipoprotein E allele status. *Arch Neurol*. 2003;60(2):185–189.
6. Fitzpatrick AL, Kuller LH, Ives DG, et al. Incidence and prevalence of dementia in the cardiovascular health study. *J Am Geriatr Soc*. 2004;52(2):195–204.
7. Kukull WA, Higdon R, Bowen JD, et al. Dementia and Alzheimer disease incidence: a prospective cohort study. *Arch Neurol*. 2002;59(11):1737–1746.
8. Sando SB, Melquist S, Cannon A, et al. Risk-reducing effect of education in Alzheimer's disease. *Int J Geriatr Psychiatr J Psychiatry Late Life Allied Sci*. 2008;23(11):1156–1162.
9. Stern Y. Cognitive reserve in ageing and Alzheimer's disease. *Lancet Neurol*. 2012;11(11): 1006–1012.
10. Hendrie HC, Smith-Gamble V, Lane KA, Purnell C, Clark DO, Gao S. The association of early life factors and declining incidence rates of dementia in an elderly population of African Americans. *J Gerontol: Ser Bibliogr*. 2018;73(suppl_1):S82–S89.
11. Stern Y, Arenaza-Urquijo EM, Bartrés-Faz D, et al. Whitepaper: defining and investigating cognitive reserve, brain reserve, and brain maintenance. *Alzheimer's Dementia*. 2020;16(9): 1305–1311.
12. Vemuri P, Knopman DS, Lesnick TG, et al. Evaluation of amyloid protective factors and Alzheimer disease neurodegeneration protective factors in elderly individuals. *JAMA Neurol*. June 1, 2017;74(6):718–726. https://doi.org/10.1001/jamaneurol.2017.0244.
13. Pusswald G, Lehrner J, Hagmann M, et al. Gender-specific differences in cognitive profiles of patients with Alzheimer's disease: results of the prospective dementia registry Austria (PRODEM-Austria). *J Alzheimers Dis*. 2015;46(3):631–637. https://doi.org/10.3233/jad-150188.
14. Launer LJ, Andersen K, Lobo A, et al. Rates and risk factors for dementia and Alzheimer's disease : results from EURODEM pooled analyses. *Neurology*. 1999;52(1):78–84. https://doi.org/10.1212/WNL.52.1.78.
15. Russ TC, Stamatakis E, Hamer M, Starr JM, Kivimäki M, Batty GD. Socioeconomic status as a risk factor for dementia death: individual participant meta-analysis of 86 508 men and women from the UK. *Br J Psychiatry*. 2013;203(1):10–17. https://doi.org/10.1192/bjp.bp.112.119479.
16. Ryan CL, Siebens J. *Educational Attainment in the United States: 2009. Population Characteristics. Current Population Reports. P20-566*. US Census Bureau; 2012.

17. Langa KM, Larson EB, Crimmins EM, et al. A comparison of the prevalence of dementia in the United States in 2000 and 2012. *JAMA Intern Med.* 2017;177(1):51−58. https://doi.org/10.1001/jamainternmed.2016.6807.
18. Pool LR, Weuve J, Wilson RS, Bültmann U, Evans DA, Mendes de Leon CF. Occupational cognitive requirements and late-life cognitive aging. *Neurology.* April 12, 2016;86(15):1386−1392. https://doi.org/10.1212/wnl.0000000000002569.
19. Kobayashi LC, M Feldman J. Employment trajectories in mid-life and cognitive performance in later-life: longitudinal study of older AMERICAN men and women. *Innov Aging.* 2018;2(suppl 1):9. https://doi.org/10.1093/geroni/igy023.029.
20. Dufouil C, Pereira E, Chêne G, et al. Older age at retirement is associated with decreased risk of dementia. *Eur J Epidemiol.* May 2014;29(5):353−361. https://doi.org/10.1007/s10654-014-9906-3.
21. Xue B, Cadar D, Fleischmann M, et al. Effect of retirement on cognitive function: the Whitehall II cohort study. *Eur J Epidemiol.* October 2018;33(10):989−1001. https://doi.org/10.1007/s10654-017-0347-7.
22. Goldin C. The quiet revolution that transformed women's employment, education, and family. *Am Econ Rev.* 2006;96(2):1−21. https://doi.org/10.1257/000282806777212350.
23. Mayeda ER, Mobley TM, Weiss RE, Murchland AR, Berkman LF, Sabbath EL. Association of work-family experience with mid-and late-life memory decline in US women. *Neurology.* 2020;95(23):e3072−e3080.
24. Mielke MM, James BD. Women who participated in the paid labor force have lower rates of memory decline: working to remember. *Neurology.* 2020;95(23):1027−1028. https://doi.org/10.1212/WNL.0000000000010987.
25. James BD, Wilson RS, Barnes LL, Bennett DA. Late-life social activity and cognitive decline in old age. *J Int Neuropsychol Soc.* November 2011;17(6):998−1005. https://doi.org/10.1017/s1355617711000531.
26. Wilson RS, Segawa E, Boyle PA, Bennett DA. Influence of late-life cognitive activity on cognitive health. *Neurology.* 2012;78(15):1123−1129. https://doi.org/10.1212/WNL.0b013e31824f8c03.
27. 2021 Alzheimer's disease facts and figures. Alzheimer's Dementia. March 2021;17(3):327−406. https://doi.org/10.1002/alz.12328.
28. Quiñones AR, Kaye J, Allore HG, Botoseneanu A, Thielke SM. An agenda for addressing multimorbidity and racial and ethnic disparities in Alzheimer's disease and related dementia. *Am J Alzheimer's Dis Other Dementias.* 2020;35. https://doi.org/10.1177/1533317520960874.
29. Amjad H, Carmichael D, Austin AM, Chang C-H, Bynum JP. Continuity of care and health care utilization in older adults with dementia in fee-for-service Medicare. *JAMA Intern Med.* 2016;176(9):1371−1378.
30. Zuelsdorff M, Okonkwo OC, Norton D, et al. Stressful life events and racial disparities in cognition among middle-aged and older adults. *J Alzheim Dis.* 2020;73:671−682. https://doi.org/10.3233/JAD-190439.
31. Chetty R, Hendren N, Jones MR, Porter SR. Race and economic opportunity in the United States: an intergenerational perspective. *Q J Econ.* 2020;135(2):711−783. https://doi.org/10.1093/qje/qjz042.
32. Bloome D. Racial inequality trends and the intergenerational persistence of income and family structure. *Am Socio Rev.* 2014;79(6):1196−1225. https://doi.org/10.1177/0003122414554947.

33. Olson NL, Albensi BC. Race- and sex-based disparities in Alzheimer's disease clinical trial enrollment in the United States and Canada: an indigenous perspective. *J Alzheimers Dis Rep*. 2020;4(1):325–344. https://doi.org/10.3233/ADR-200214.
34. Flatt JD, Johnson JK, Karpiak SE, Seidel L, Larson B, Brennan-Ing M. Correlates of subjective cognitive decline in lesbian, gay, bisexual, and transgender older adults. *J Alzheimers Dis*. 2018;64(1):91–102. https://doi.org/10.3233/JAD-171061.
35. Meyer IH. Prejudice, social stress, and mental health in lesbian, gay, and bisexual populations: conceptual issues and research evidence. *Psychol Bull*. 2003;129(5):674–697. https://doi.org/10.1037/0033-2909.129.5.674.
36. Wallace SP, Cochran SD, Durazo EM, Ford CL. The health of aging lesbian, gay and bisexual adults in California. *Policy Brief UCLA Cent Health Policy Res*. March 2011; (Pb2011–2):1–8.
37. Fredriksen-Goldsen KI, Kim HJ, Barkan SE, Muraco A, Hoy-Ellis CP. Health disparities among lesbian, gay, and bisexual older adults: results from a population-based study. *Am J Publ Health*. October 2013;103(10):1802–1809. https://doi.org/10.2105/AJPH.2012.301110.
38. Fredriksen-Goldsen KI, Kim HJ, Barkan SE. Disability among lesbian, gay, and bisexual adults: disparities in prevalence and risk. *Am J Public Health*. January 2012;102(1): e16–e21. https://doi.org/10.2105/ajph.2011.300379.
39. Gonzales G, Przedworski J, Henning-Smith C. Comparison of health and health risk factors between lesbian, gay, and bisexual adults and heterosexual adults in the United States: results from the national health interview survey. *JAMA Intern Med*. September 1, 2016; 176(9):1344–1351. https://doi.org/10.1001/jamainternmed.2016.3432.
40. Max WB, Stark B, Sung H-Y, Offen N. Sexual identity disparities in smoking and secondhand smoke exposure in California: 2003–2013. *Am J Publ Health*. 2016;106(6): 1136–1142. https://doi.org/10.2105/AJPH.2016.303071.
41. Kim H-J, Fredriksen-Goldsen KI. Living arrangement and loneliness among lesbian, gay, and bisexual older adults. *Gerontol*. 2016;56(3):548–558. https://doi.org/10.1093/geront/gnu083.
42. Kuyper L, Fokkema T. Loneliness among older lesbian, gay, and bisexual adults: the role of minority stress. *Arch Sex Behav*. October 2010;39(5):1171–1180. https://doi.org/10.1007/s10508-009-9513-7.
43. Cornwell EY, Waite LJ. Social disconnectedness, perceived isolation, and health among older adults. *J Health Soc Behav*. 2009;50(1):31–48. https://doi.org/10.1177/002214650905000103.
44. Fredriksen-Goldsen KI, Emlet CA, Kim H-J, et al. The physical and mental health of lesbian, gay male, and bisexual (LGB) older adults: the role of key health indicators and risk and protective factors. *Gerontol*. 2013;53(4):664–675. https://doi.org/10.1093/geront/gns123.
45. Yarns BC, Abrams JM, Meeks TW, Sewell DD. The mental health of older LGBT adults. *Curr Psychiatry Rep*. June 2016;18(6):60. https://doi.org/10.1007/s11920-016-0697-y.
46. Ownby RL, Crocco E, Acevedo A, John V, Loewenstein D. Depression and risk for Alzheimer disease: systematic review, meta-analysis, and metaregression analysis. *Arch Gen Psychiatry*. May 2006;63(5):530–538. https://doi.org/10.1001/archpsyc.63.5.530.
47. Barnes DE, Yaffe K, Byers AL, McCormick M, Schaefer C, Whitmer RA. Midlife vs late-life depressive symptoms and risk of dementia: differential effects for Alzheimer disease and vascular dementia. *Arch Gen Psychiatr*. 2012;69(5):493–498. https://doi.org/10.1001/archgenpsychiatry.2011.1481.

48. Fredriksen-Goldsen KI, Jen S, Bryan AEB, Goldsen J. Cognitive impairment, Alzheimer's disease, and other dementias in the lives of lesbian, gay, bisexual and transgender (LGBT) older adults and their caregivers: needs and competencies. *J Appl Gerontol.* 2018;37(5): 545–569. https://doi.org/10.1177/0733464816672047.
49. Fredriksen-Goldsen KI, Kim H-J, Emlet CA, et al. *The Aging and Health Report.* Seattle, WA: Institute for Multigenerational Health; 2011.
50. Fredriksen-Goldsen KI, Cook-Daniels L, Kim H-J, et al. Physical and mental health of transgender older adults: an at-risk and underserved population. *Gerontologist.* 2014;54(3): 488–500. https://doi.org/10.1093/geront/gnt021.
51. Rovio S, Kåreholt I, Helkala EL, et al. Leisure-time physical activity at midlife and the risk of dementia and Alzheimer's disease. *Lancet Neurol.* November 2005;4(11):705–711. https://doi.org/10.1016/s1474-4422(05)70198-8.
52. Pacholko AG, Wotton CA, Bekar LK. Poor diet, stress, and inactivity converge to form a "perfect storm" that drives Alzheimer's disease pathogenesis. *Report. Neurodegener Dis.* 2019;19, 60(+).
53. Erickson KI, Weinstein AM, Lopez OL. Physical activity, brain plasticity, and Alzheimer's disease. *Arch Med Res.* 2012;43(8):615–621.
54. DeFina LF, Willis BL, Radford NB, et al. The association between midlife cardiorespiratory fitness levels and later-life dementia: a cohort study. *Ann Intern Med.* 2013;158(3): 162–168.
55. Booth FW, Roberts CK, Laye MJ. Lack of exercise is a major cause of chronic diseases. *Compr Physiol.* April 2012;2(2):1143–1211. https://doi.org/10.1002/cphy.c110025.
56. Hötting K, Röder B. Beneficial effects of physical exercise on neuroplasticity and cognition. *Neurosci Biobehav Rev.* 2013;37(9):2243–2257.
57. Dao AT, Zagaar MA, Levine AT, Salim S, Eriksen JL, Alkadhi KA. Treadmill exercise prevents learning and memory impairment in Alzheimer's disease-like pathology. *Curr Alzheimer Res.* June 2013;10(5):507–515. https://doi.org/10.2174/1567205011310050006.
58. Adlard PA, Perreau VM, Pop V, Cotman CW. Voluntary exercise decreases amyloid load in a transgenic model of Alzheimer's disease. *J Neurosci.* April 27, 2005;25(17):4217–4221. https://doi.org/10.1523/jneurosci.0496-05.2005.
59. Nomaguchi KM, Bianchi SM. Exercise time: gender differences in the effects of marriage, parenthood, and employment. *J Marriage Fam.* 2004;66(2):413–430. https://doi.org/10.1111/j.1741-3737.2004.00029.x.
60. Parker K, Rhee Y. Alzheimer's disease warning signs: gender and education influence modifiable risk factors-A pilot survey study. *J Am Coll Nutr.* September 24, 2020:1–6. https://doi.org/10.1080/07315724.2020.1812451.
61. Hayden KM, Zandi PP, Lyketsos CG, et al. Vascular risk factors for incident Alzheimer disease and vascular dementia: the Cache County study. *Alzheimer Dis Assoc Disord.* April–June 2006;20(2):93–100. https://doi.org/10.1097/01.wad.0000213814.43047.86.
62. Baker LD, Frank LL, Foster-Schubert K, et al. Effects of aerobic exercise on mild cognitive impairment: a controlled trial. *Arch Neurol.* 2010;67(1):71–79. https://doi.org/10.1001/archneurol.2009.307.
63. Middleton L, Kirkland S, Rockwood K. Prevention of CIND by physical activity: different impact on VCI-ND compared with MCI. *J Neurol Sci.* June 15, 2008;269(1–2):80–84. https://doi.org/10.1016/j.jns.2007.04.054.
64. Goldstein JM, Holsen L, Handa R, Tobet S. Fetal hormonal programming of sex differences in depression: linking women's mental health with sex differences in the brain across the lifespan. *Front Neurosci.* 2014;8:247. https://doi.org/10.3389/fnins.2014.00247.

65. Bromberger JT, Kravitz HM, Chang YF, Cyranowski JM, Brown C, Matthews KA. Major depression during and after the menopausal transition: Study of Women's Health Across the Nation (SWAN). *Psychol Med.* September 2011;41(9):1879–1888. https://doi.org/10.1017/s003329171100016x.
66. Goveas JS, Espeland MA, Woods NF, Wassertheil-Smoller S, Kotchen JM. Depressive symptoms and incidence of mild cognitive impairment and probable dementia in elderly women: the Women's Health Initiative Memory Study. *J Am Geriatr Soc.* January 2011; 59(1):57–66. https://doi.org/10.1111/j.1532-5415.2010.03233.x.
67. Kim S, Kim MJ, Kim S, et al. Gender differences in risk factors for transition from mild cognitive impairment to Alzheimer's disease: a CREDOS study. *Compr Psychiatr.* October 2015;62:114–122. https://doi.org/10.1016/j.comppsych.2015.07.002.
68. Sachdev PS, Lipnicki DM, Crawford J, et al. Risk profiles for mild cognitive impairment vary by age and sex: the Sydney Memory and Ageing study. *Am J Geriatr Psychiatr.* October 2012;20(10):854–865. https://doi.org/10.1097/JGP.0b013e31825461b0.
69. Elbejjani M, Fuhrer R, Abrahamowicz M, et al. Hippocampal atrophy and subsequent depressive symptoms in older men and women: results from a 10-year prospective cohort. *Am J Epidemiol.* August 15, 2014;180(4):385–393. https://doi.org/10.1093/aje/kwu132.
70. Elbejjani M, Fuhrer R, Abrahamowicz M, et al. Depression, depressive symptoms, and rate of hippocampal atrophy in a longitudinal cohort of older men and women. *Psychol Med.* July 2015;45(9):1931–1944. https://doi.org/10.1017/s0033291714003055.
71. Caregiving for a person with Alzheimer's disease or a related dementia. Centers for Disease Control and Prevention. Updated October 30, 2019. https://www.cdc.gov/aging/caregiving/alzheimer.htm.
72. Gaugler JE, Kane RL, Kane RA. Family care for older adults with disabilities: toward more targeted and interpretable research. *Int J Aging Hum Dev.* 2002;54(3):205–231.
73. Schulz R, Quittner A. Caregiving through the life-span: overview and future directions. *Health Psychol.* 1998;17(2):107–111.
74. Prince M, Acosta D, Ferri CP, et al. Dementia incidence and mortality in middle-income countries, and associations with indicators of cognitive reserve: a 10/66 Dementia Research Group population-based cohort study. *Lancet.* July 7, 2012;380(9836):50–58. https://doi.org/10.1016/s0140-6736(12)60399-7.
75. Rabarison KM, Bouldin ED, Bish CL, McGuire LC, Taylor CA, Greenlund KJ. The economic value of informal caregiving for persons with dementia: results from 38 states, the District of Columbia, and Puerto Rico, 2015 and 2016 BRFSS. *Am J Publ Health.* 2018; 108(10):1370–1377.
76. Kasper JD. *Disability and care needs of older Americans by dementia status: an analysis of the 2011 national health and aging trends study.* 2014.
77. Jones RW, Lebrec J, Kahle-Wrobleski K, et al. Disease progression in mild dementia due to Alzheimer disease in an 18-month observational study (GERAS): the impact on costs and caregiver outcomes. *Dementia Geriatr Cogn Dis Extra.* 2017;7(1):87–100.
78. Xiong C, Biscardi M, Astell A, et al. Sex and gender differences in caregiving burden experienced by family caregivers of persons with dementia: a systematic review. *PLoS One.* 2020;15(4):e0231848. https://doi.org/10.1371/journal.pone.0231848.
79. Ma M, Dorstyn D, Ward L, Prentice S. Alzheimers' disease and caregiving: a meta-analytic review comparing the mental health of primary carers to controls. *Aging Ment Health.* November 2018;22(11):1395–1405. https://doi.org/10.1080/13607863.2017.1370689.
80. Vitaliano PP, Murphy M, Young HM, Echeverria D, Borson S. Does Caring for a spouse with dementia promote cognitive decline? A hypothesis and proposed mechanisms. *J Am Geriatr Soc.* 2011;59(5):900–908. https://doi.org/10.1111/j.1532-5415.2011.03368.x.

81. Allen AP, Curran EA, Duggan Á, et al. A systematic review of the psychobiological burden of informal caregiving for patients with dementia: focus on cognitive and biological markers of chronic stress. *Neurosci Biobehav Rev*. February 2017;73:123–164. https://doi.org/10.1016/j.neubiorev.2016.12.006.
82. Cohen SA, Kunicki ZJ, Drohan MM, Greaney ML. Exploring changes in caregiver burden and caregiving intensity due to COVID-19. *Gerontol Geriatr Med*. 2021;7. https://doi.org/10.1177/2333721421999279, 2333721421999279-2333721421999279.
83. Wang Q, Davis PB, Gurney ME, Xu R. COVID-19 and dementia: analyses of risk, disparity, and outcomes from electronic health records in the US. *Alzheimer's Dementia*. 2021;17(8):1297–1306. https://doi.org/10.1002/alz.12296.
84. Barguilla A, Fernández-Lebrero A, Estragués-Gázquez I, et al. Effects of COVID-19 pandemic confinement in patients with cognitive impairment. *Front Neurol*. 2020;11: 589901. https://doi.org/10.3389/fneur.2020.589901.
85. Macchi ZA, Ayele R, Dini M, et al. Lessons from the COVID-19 pandemic for improving outpatient neuropalliative care: a qualitative study of patient and caregiver perspectives. *Palliat Med*. July 2021;35(7):1258–1266. https://doi.org/10.1177/02692163211017383.
86. Borges-Machado F, Barros D, Ribeiro Ó, Carvalho J. The effects of COVID-19 home confinement in dementia care: physical and cognitive decline, severe neuropsychiatric symptoms and increased caregiving burden. *Am J Alzheimers Dis Other Demen*. January–December 2020;35. https://doi.org/10.1177/1533317520976720, 1533317520976720.
87. Carpinelli Mazzi M, Iavarone A, Musella C, et al. Time of isolation, education and gender influence the psychological outcome during COVID-19 lockdown in caregivers of patients with dementia. *Eur Geriatr Med*. 2020;11(6):1095–1098. https://doi.org/10.1007/s41999-020-00413-z.
88. Tam MT, Dosso JA, Robillard JM. The impact of a global pandemic on people living with dementia and their care partners: analysis of 417 lived experience reports. *J Alzheimers Dis*. 2021;80(2):865–875. https://doi.org/10.3233/JAD-201114.
89. Hwang Y, Connell LM, Rajpara AR, Hodgson NA. Impact of COVID-19 on dementia caregivers and factors associated with their anxiety symptoms. *Am J Alzheimers Dis Other Demen*. January–December 2021;36. https://doi.org/10.1177/15333175211008768, 15333175211008768.
90. Hughes MC, Liu Y, Baumbach A. Impact of COVID-19 on the health and well-being of informal caregivers of people with dementia: a rapid systematic review. *Gerontol Geriatr Med*. 2021;7. https://doi.org/10.1177/23337214211020164, 23337214211020164.

Women and Alzheimer's disease: a focus on sex

16

Emma Schindler[1] *and P. Hemachandra Reddy*[2,3,4,5]
[1]University of Miami Miller School of Medicine, Miami, FL, United States; [2]Department of Internal Medicine, Texas Tech University Health Sciences Center, Lubbock, TX, United States; [3]Nutritional Sciences Department, Texas Tech University, Lubbock, TX, United States; [4]Department of Pharmacology and Neuroscience, School of Medicine, Texas Tech University Health Sciences Center, Lubbock, TX, United States; [5]Department of Neurology, Texas Tech University Health Sciences Center, Lubbock, TX, United States

Abstract

Approximately, two-thirds of individuals with Alzheimer's disease (AD) are women. Though previously attributed to differences in lifespan, accumulating evidence suggests that the reasons for the higher prevalence of AD in women are multifactorial and related to differences in risk factors, biomarkers, and neuropathology. Sex also contributes to significant disease heterogeneity, which has important implications for prevention and treatment. This chapter discusses the evidence for sex differences in AD, with an emphasis on disease presentation, biomarkers, pathophysiology, progression, and risk. Women tend to present later in the disease course and with different clinical features, progress faster, and are disproportionately affected by the *APOE*-ϵ4 risk allele and AD neuropathologic changes. Lifetime estrogen exposure, pregnancy, and menopause also affect a woman's risk for cognitive decline later in life. Despite such differences, women are dramatically underrepresented in pharmacologic randomized control trials, leading to significant gaps in knowledge regarding the most effect AD treatment strategies for women. Both researchers and providers need to be aware of sex differences in AD risk, presentation, and outcomes to develop sex-specific prevention and treatment strategies, as well as provide optimum healthcare to women as they age.

Keywords: Alzheimer's disease; COVID-19; Estrogen; Genetics; Pregnancy; Sex; Women.

1. Introduction

Alzheimer's disease (AD) is a progressive neurodegenerative disease that disproportionately affects women. Nearly two-thirds of the 6.2 million affected individuals in the United States are women.[1] At age 65 years, the lifetime risk for developing AD in women is 21.1%, whereas in males, the lifetime risk is 11.6%.[1] Sex differences in AD were previously thought to stem almost exclusively from differences in longevity. Because age is the single biggest risk factor for AD onset and women

COVID-19 in Alzheimer's Disease and Dementia. https://doi.org/10.1016/B978-0-443-15256-6.00001-5

tend to live longer than men, women have a higher probability of developing AD.[1] Another explanation could be that there is significant survivor bias in studies of older adults, as men are more likely to die in middle age of other comorbidities, such as cardiovascular disease. By this logic, older men tend to have better cardiovascular risk profiles and thus a decreased risk of dementia.[1,2] However, recent studies examining structural, functional, and metabolic differences between the male and female aging brain have revealed that the reasons for sex differences in AD prevalence are multifaceted—longevity is only a part of the story. Differences in sex hormones, genetic architecture, and gender-related psychosocial risk factors are also implicated.

This chapter reviews current knowledge on sex differences in AD, while highlighting areas requiring further study. Sex is defined as a biological classification as male or female due to differences in sex chromosomes, genitalia, and gonadal hormones.[3,4]

2. Clinical presentation

The prevalence of AD is higher among women and presents differently in woman in comparison to men. Providers need to understand these differences to prevent delayed diagnosis. Women with AD are more likely to demonstrate the classic amnestic cognitive phenotype, followed by a faster progression of clinical manifestations and brain atrophy after a suspected diagnosis.[5] Men with AD are more likely to demonstrate a nonamnestic clinical picture compared to women despite the presence of characteristic neuropathology. Men also tend to present with cognitive symptoms at an earlier age than do women and have a shorter disease course.[5] In women with AD, sleep disturbances, depression, emotional lability, reclusiveness, delusions, and mania are more common than in men with AD.[5–8] On the other hand, men are more likely to present with apathy, agitation, aggression, and socially inappropriate behavior.[5,6,9,10] Women overall tend to have higher levels of mood and behavioral dysfunction than males.[11]

3. Disease progression

There is a known female advantage in verbal memory throughout the lifespan that is partially explained by sex differences in levels of estradiol.[12–16] This advantage persists even when the degree of AD disease burden (hippocampal atrophy, amyloid beta (Aβ) deposition, and brain hypometabolism) is similar in men.[14,15,17] However, the verbal memory advantage dissipates at high levels of AD disease burden.[17] After a suspected diagnosis, cognitive decline accelerates faster in women than in men in the first year of disease onset[18] and two times faster after 8 years.[19] Similarly, analysis from the French National Alzheimer Database has shown that the rate of progression from MCI to AD is 2%–3% higher in females than in males after adjusting for education.[20] Another study found females to have lower Mini-Mental State Examination (MMSE) scores at the time of AD diagnosis after adjusting for age and education, indicating overall worse cognitive status.[21]

Taken together, these findings led researchers to hypothesize that the female verbal memory advantage represents a "cognitive reserve" that protects against and/or masks the effects of AD neuropathology in early disease stages.[6,22] The downside of this reserve is that it can delay diagnosis—perhaps explaining why women tend to have a higher disease burden at diagnosis, and why cognition declines more rapidly thereafter.[6,18,19,23,24]

4. Neuropathology

The degree of AD neuropathology is often measured via biomarkers of Aβ deposition [Aβ in cerebrospinal fluid (CSF) and ligand uptake on positron emission tomography (PET)], neurofibrillary tangle (NFT) tau pathology [CSF phosphorylated tau (p-tau) and NFT-tau ligand uptake on PET], and markers of neurodegeneration [decreased glucose metabolism on PET, magnetic resonance imaging (MRI) of atrophy in temporoparietal cortex, and CSF tau].[22] Studies of AD biomarkers have found no significant sex differences in Aβ burden among cognitively normal, mildly cognitively impaired, and individuals with AD.[13] Similarly, there are no significant sex differences in CSF tau pathology among cognitively normal, prodromal AD, or individuals with AD.[25] There is minimal evidence of an increased NFT burden in women; however, this has not been corroborated by other studies.[26,27]

Although there are few, if any, sex differences in AD neuropathology, there are definite differences in the association between neuropathology and clinical symptomatology. A report by Buckley et al. supports the finding that Aβ burden does not differ by sex, but they found that females with higher levels of Aβ experienced greater cognitive decline than males with similar Aβ levels. Sex may therefore modify the risk for AD-related cognitive decline.[28] Females with MCI also demonstrate higher rates of global atrophy and cognitive and clinical decline than males.[18,19,29] In particular, clinical progression,[15] overall brain atrophy,[14,17] and hippocampal atrophy[17] have been shown to proceed faster in females than in males with MCI. However, reports of sex differences in hippocampal volume of patients with AD are not consistent.[14,30,31] Barnes and colleagues demonstrated that each unit of increase in AD pathology (measured as the burden of neuritic plaques, diffuse plaques, and NFT in four brain regions) conferred more than a 20-fold increase in the odds of developing dementia in females, compared to a three-fold increase in males.[26] The presence of CSF Aβ and CSF tau has also been associated with more hippocampal atrophy and a higher rate of cognitive decline in females than in males, especially among those with MCI.[32]

5. Genetics

The *APOE* gene is the most significant risk gene for AD. The ε4 allele is the main genetic risk factor for late-onset AD. Apolipoprotein E (ApoE) is a protein encoded by the *APOE* gene on chromosome 19 involved in cholesterol transport in multiple tissues.[33] It is the most prevalent lipoprotein in the brain, where it plays a role in membrane remodeling, synaptic turnover, and dendritic reorganization.[33] ApoE helps break down amyloid beta in the brain by trafficking amyloid to lysosomes; faulty APOE is associated with AD.[34,35]

There are three common *APOE* alleles: ε2, ε3, and ε4.[33] The *APOE*-ε4 allele is the main genetic risk factor for late-onset AD and accounts for nearly 50% of genetically attributable risk.[36] Carrying one or more ε4 alleles increases an individual's risk for developing AD, as the ApoE-ε4 isoform binds poorly to amyloid beta and impairs its clearance.[22,37] *APOE*-ε4 homozygotes have an eight-to-twelve-fold increased risk of developing AD compared to *APOE*-ε3 homozygotes and are diagnosed at a mean age of 68 years.[38–40] *APOE*-ε4 heterozygotes have three times the risk of developing AD compared to *APOE*-ε3 homozygotes and are diagnosed at a mean age of 76 years. In non-*APOE*-ε4 carriers, the average age of onset is 84 years.[38–40] Conversely, the ε2 allele is neuroprotective against AD, as it is associated with increased neural plasticity and synaptic replacement.[41–43]

Some evidence suggests that AD risk in *APOE*-ε4 carriers is modified by sex, with female *APOE*-ε4 carriers having a higher risk of developing AD than male carriers.[33,44–48] Neu and colleagues add to this finding by suggesting that sex differences in *APOE*-ε4-associated AD risk only occur in certain age groups.[42] Their meta-analyses ($N = 58{,}000$) found no difference between male and female risk for developing MCI or AD among *APOE*-ε4 carriers aged 55 to 85 years.[42] However, there was an increased risk of AD among female *APOE*-ε4 carriers compared to male carriers between the ages of 65 and 75 years (female odds ratio (OR): 4.37; 95% CI: 3.82−5.00; male OR: 3.14; 95% CI: 2.68−3.67; $P = .002$).[42] There was also an increased risk of MCI among female *APOE*-ε4 carriers aged 55 to 70 (female OR: 1.43; 95% CI: 1.19−1.73; male OR: 1.07; 95% CI, 0.87−1.30; $P = .05$). Of note, there was no sex difference in the risks for transitioning from MCI to AD.[42] Taken together, their findings suggest that *APOE*-ε4 carriers have similar odds of developing MCI or AD regardless of sex, but that female carriers have an increased risk of developing MCI or AD at a younger age.[42]

The findings of Neu et al. and previous studies beg the question of *why* female *APOE*-ε4 carriers appear to have greater odds of developing MCI and AD, even if only in certain age groups. Some attribute this to differences in AD neuropathology between male and female *APOE*-ε4 carriers. First, AD biomarkers in *APOE*-ε4 carriers have been shown to differ between males and females. Multiple studies report elevated CSF tau pathology in female *APOE*-ε4 carriers compared to males.[23,49,50] On the other hand, CSF Aβ has been found to be reduced in males (suggesting a higher level of Aβ deposition) compared to females.[23] These differences are maintained across cognitively normal, mildly cognitively impaired, and suspected AD groups.[23,49]

In addition, the association between CSF tau and Aβ pathology is reportedly stronger in *APOE*-ε4 carrier females than males, suggesting that the neurodegeneration pathway differs between male and female *APOE*-ε4 carriers.[23,50] Independent of *APOE* genotype, cognitively normal and mildly cognitively impaired females performed better than males on memory tasks despite increased tau pathology. Females may have a cognitive reserve that protects the brain from damage, delaying cognitive decline. However, once a diagnosis is suspected, this advantage seems to dissipate, as females demonstrate a faster rate of cognitive decline.[23]

Others have suggested that menopause mediates the effects of *APOE*-ε4 status on the development of AD in female carriers. One study found that inducing menopause in *APOE*-ε4 mice significantly compounded *APOE*-ε4-associated deficits in recognition. Findings suggest that *APOE*-ε4 carrier status may confer greater sensitivity to neurologic deficits associated with menopause.[51] In addition, neurologic changes in *APOE*-ε4 carriers differ at time points throughout the menopause transition; one neuroimaging study found that perimenopausal and postmenopausal female *APOE*-ε4 carriers had greater Aβ deposition in the brain than male *APOE*-ε4 carriers.[52]

Though its frequency in the general population is quite low (0%–20%), the neuroprotective effects of the *APOE*-ε2 allele may be more pronounced in women than in men.[42,43] One prospective cohort study found that *APOE*-ε2 had a protective effect on verbal memory and fluency in healthy females aged 50–91, but not in males. *APOE*-ε2 female carriers also demonstrated reduced lipidic and inflammatory biomarkers, some of which were found to mediate the *APOE*-ε2 neuroprotective effects, pointing to a possible role for cardiovascular risk factors and inflammation in the differential protection observed in females.[43] In their meta-analysis, Neu et al. also noted a protective effect of the *APOE*-ε2 allele in females, as *APOE*-ε2 reduced the risk of developing AD in females more than in males (female OR: 0.51; 95% CI: 0.43–0.61; male OR: 0.71; 95% CI: 0.60–0.85).[42]

APOE is not the only gene conferring risk for AD. There are over 20 susceptibility loci, and the genetic basis of AD is thought to be highly polygenic.[51,53] Emerging evidence suggests that the genetic architecture of males and females may differ concerning AD risk. Sex differences in genetic predictors of AD biomarkers have been observed, including a stronger association between Serpin genes and amyloidosis in females than in males. There may also be a female-specific role for *OSTN* and *CLDN16* genes in tau neuropathology.[54] Stratifying polygenic hazard scores by sex reveal differences in genetic architecture between males and females independent of APOE carrier status. Loci such as *BIN1*, *MS4A6A*, *DNAJA2*, and *FERMT2* confer higher risk in females than males. *FERMT2* associated with amyloid deposition[55]; *DNAJA2* involved in tau aggregation.[56] Sex-dependent effects highlight differences in neuropathologic pathways between men and women.[57] In light of these findings, a more recent study employing a polygenic framework to estimate sex differences in single nucleotide polymorphism (SNP) heritability of AD found no significant difference between the genetic architecture of males and females.[58] The interaction between sex and genetic risk for AD is most likely multifactorial and difficult to consistently detect.

6. Endogenous estrogen exposure

Estrogen acts on widely distributed estrogen receptors (ERs) in the brain[59,60] to regulate cognition, body temperature, anxiety, feeding, and sexual behavior (Fig. 16.1).[61] Estrogen plays a role in learning and memory,[62,63] and some suggest that relative expression of estrogen receptors as they interact with estradiol in the hippocampus

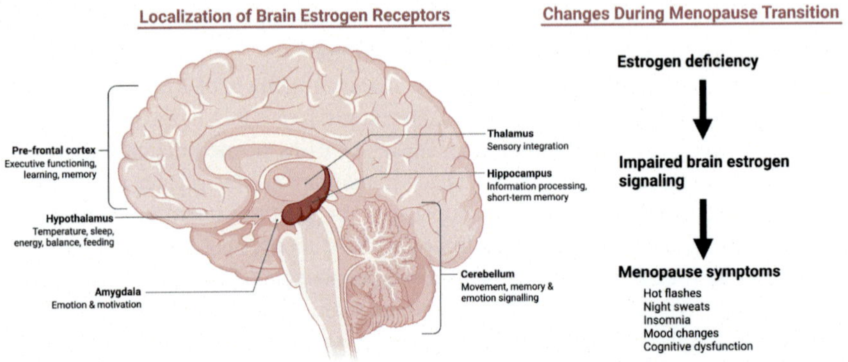

Figure 16.1 Localization of estrogen receptors in the brain and summary of associated changes attributed to decreased brain estrogen signaling during the menopause transition.

influences memory.[64] Estrogen also regulates mitochondrial function in the central nervous system (CNS), including metabolism, biogenesis, and apoptosis.[65,66] Cumulative exposure to estrogen across the lifespan is believed to modify a woman's risk of dementia.[4,67] Lifetime estrogen exposure comes predominantly from endogenous sources, between menarche and menopause, but may be influenced by exogenous sources like oral contraceptives and hormone therapy.[4]

6.1 Age of menarche, menopause, and reproductive span

Older age of menarche may be associated with worse cognitive outcomes.[68–70] However, study findings are mixed, and some have found no association between age of menarche and late-life cognition.[71–74] Similarly, an earlier age of menopause may be associated with worse cognitive outcomes,[68,69,74,75] but findings are not consistent.[70,72,73,75,76]

A better metric of lifetime exposure to estrogen may be reproductive span (time from menarche to menopause).[4,71] One study found a link between longer reproductive span and improved late-life cognition but did not evaluate dementia risk.[77] Others have shown a positive association between shorter reproductive span and dementia risk,[68,69] yet another study reports a positive association between longer reproductive span and AD risk.[71] Longer reproductive span has also been linked to increased AD risk in female carriers of the *APOE*-ε4 risk allele (see "Genetics") in particular.[78] It may be difficult to isolate the effect of reproductive span on late-life cognition due to other variables that affect lifetime estrogen exposure, such as pregnancy and exogenous estrogen sources.

6.2 Natural menopause

The loss of sex hormones, especially estrogen, during aging and menopause is associated with impaired mitochondrial function, neuroinflammation, synaptic decline, cognitive decline, and the risk of developing age-related neurodegenerative disorders (Fig. 16.1).[67] Menopause is a hormonal, neurologic, and bioenergetic transition state that culminates in the complete cessation of menses.[4,22,79] The average age of onset is 51 years, and changes observed during menopause are predominantly mediated by estrogen loss.[79]

As a uniquely female experience, menopause represents a sex-specific driver of cognitive decline. Decreased verbal memory skills have been demonstrated in women going through menopause.[80] These deficits are associated with changes in hippocampal function stemming from loss of estradiol.[12,81] However, some studies have shown these changes to be temporary. Two studies report worse verbal memory in perimenopause but not post menopause.[80,82] Additionally, Mosconi et al.'s recent neuroimaging study of women at different phases of menopause investigated the effect of menopause on gray matter volume, white matter volume, and glucose metabolism.[52] They report significant changes in all biomarkers in brain regions implicated in higher-order cognition across each phase of menopause. Interestingly, changes in most biomarkers stabilized or recovered post menopause. Reduction in gray matter volume in perimenopause was recovered in the postmenopausal group in brain regions implicated in cognitive aging; production of brain ATP also increased in the postmenopausal group, which correlated with preservation of cognitive function. Findings suggest that the brain may be able to compensate for declining estrogen levels and ER activity during menopause.[83] The brain may be able to adapt to low levels of estrogen.[83–85] Importantly, comparison to age-matched males revealed that these effects can be attributed to endocrine aging, rather than chronological aging, highlighting the role of menopause in changes to female cognition independent of age.[52]

Nevertheless, multiple cross-sectional studies demonstrate verbal memory deficits post menopause but not during perimenopause.[12,81,86] Further, a decrease in estrogen during menopause is believed to be a risk factor for developing AD.[87–89] Regarding AD-specific biomarkers, animal models of menopause have shown estrogen depletion to be associated with amyloid-beta (Aβ) plaque accumulation.[90,91]

Bioenergetic changes associated with menopause may be related to cognitive decline thereafter. Estrogen receptors are widely expressed in CNS mitochondrial cells, and changes in estrogen levels affect mitochondrial function.[67] Loss of estrogen during menopause prompts a shift in fuel source from glucose to ketones[52]; continued reliance on ketones has been shown to cause mitochondrial dysfunction, white matter catabolism, and apoptosis in animal models.[65,92] In humans, it has been proposed that this shift may facilitate brain remodeling, loss of hippocampal synaptic spines, and neurodegeneration.[83,90,92–96] Previous neuroimaging findings support this theory, as perimenopausal and postmenopausal women have demonstrated decreased glucose metabolism in conjunction with platelet mitochondrial dysfunction, compared to premenopausal women.[97] In light of this data, Mosconi et al.'s recent finding of increased cerebral blood flow and brain ATP production in postmenopausal women may represent a compensatory response to glucose hypometabolism and a means to promote continued ketone metabolism.[52] Importantly, brain hypometabolism is a characteristic finding in AD.[67,97–99] In addition, mitochondrial dysfunction is pertinent to the mechanism of brain aging and neurologic disorders.[100–102] Oxidative modifications and mitochondrial DNA mutations have been reported in normal brain aging,[103–105] and these changes have been shown to be augmented in neurodegenerative diseases, including.[106,107]

Other symptoms of menopause—such as mood changes, sleep disturbances, and hot flashes, may be risk factors for cognitive decline in and of themselves. First, the prevalence of mood disorders, including depression and anxiety, increases during menopause.[108] Studies have shown that depression and anxiety during menopause predict poorer processing speed, executive function, and verbal memory.[109–111]

Second, sleep disorders are more prevalent in women, especially after menopause.[112] Sleep is critical for the clearance of Aβ, which is produced while awake and cleared during slow-wave sleep.[113] Studies have shown that poor sleep quality, short sleep duration, and high sleep fragmentation in older adults are linked to increased Aβ, cognitive decline, and increased risk of developing AD.[114,115] Sleep apnea tends to be more common among men, but the incidence among women rises after menopause.[99] Those with sleep apnea demonstrate an earlier age of cognitive decline compared to those without sleep apnea.[116] The cascade hypothesis suggests that vasomotor symptoms (hot flashes and night sweats) of menopause may cause sleep disturbances that contribute to cognitive decline.[117,118] Peri- and postmenopausal women who wake up more frequently after falling asleep have been found to have more white matter disease than those without.[119]

Additionally, hot flash frequency and severity are associated with subjective cognitive deficits during menopause.[120] Studies of objective deficits in verbal memory and processing speed, however, report inconsistent findings.[110,121,122] A few studies report a significant association between hot flashes, measured via skin conductance, and poorer verbal memory.[117,123] In addition, hot flashes during sleep are associated with more white matter hyperintensities in middle-aged women.[124] Despite these findings, studies evaluating the link between menopause symptoms and AD biomarkers still need to be conducted.

6.3 Surgical menopause

During natural menopause, hormone levels fluctuate and then decline gradually.[125] Conversely, women who undergo surgical menopause—removal of the uterus and/or one or both ovaries before the natural onset of natural menopause—experience an abrupt drop in hormone levels. After bilateral oophorectomy, both estrogen and progesterone levels decline sharply. Undergoing hysterectomy, with preservation of ovaries, leads to hormone dysregulation by disrupting the ovarian blood supply.[126,127] There are multiple long-term health risks associated with surgical menopause. These include cardiovascular disease, mood dysregulation, and poor bone health.[128,129] Both cardiovascular disease and mood disorders are risk factors for AD.[130,131]

Regarding cognition, most studies report a negative impact of surgical menopause on cognition later in life.[75,76,78,132] Surgical menopause has been associated with a higher risk of cognitive decline and dementia.[133,134] Women who have undergone surgical menopause have been found to have deficits in working, verbal, semantic, and visual memory, as well as a faster decline in overall cognition, semantic memory, and episodic memory later in life.[134–136] Further, increased AD neuropathology has been identified in women who have undergone surgical menopause. This effect was mediated by age at surgery, with a younger age being associated with more rapid cognitive decline and more neuropathologic changes.[134] Women with a history of oophorectomy have almost twice the risk of cognitive impairment or dementia; this risk increases with younger age of oophorectomy.[133] Women with a history of hysterectomy have an increased risk of dementia that is independent of the age of surgery.[69] A history of a hysterectomy has also been linked to an increased risk of early-onset dementia. This risk increases with a younger age of surgery.[137] Bilateral oophorectomy is associated with medial temporal lobe atrophy, whereas hysterectomy has been associated with reduced brain glucose metabolism but no difference in gray or white matter volume.[138,139] Compared to women with the same reproductive span who experienced natural menopause, women who underwent surgical menopause still demonstrate a greater risk of AD neuropathology.[134]

There is evidence that the abrupt drop in hormones at the time of surgery may be implicated in the worse cognitive outcomes observed in women who have undergone surgical menopause, as starting hormone replacement therapy immediately after surgery may be neuroprotective.[134,140] However, the sudden drop in estrogen cannot be the sole explanation for cognitive decline after surgical menopause, as women who had hysterectomies with preserved ovaries or unilateral oophorectomies (and therefore no comparably abrupt drop in estrogen) still had an increased risk of cognitive decline.[141]

7. Exogenous estrogen exposure

7.1 Oral contraceptive pills

Most oral contraceptive pills (OCPs) utilize a combination of estrogen and progestin, though formulations differ in type and amount of hormones, as well as dose.[142] There is some evidence that OCPs benefit cognition, particularly executive functioning, visuospatial abilities, and verbal memory.[74,143–145] One nationwide cohort study found an association between a history of OCP use and reduced risk of all-cause dementia among postmenopausal women.[68] However, multiple studies have found no association between OCP use and cognition[75,146–148]; thus, the true effect of OCP use on late-life cognition is unclear.

7.2 Hormone replacement therapy

Hormone replacement therapy (HRT) is utilized to combat the symptoms of menopause via exogenous replacement of waning sex hormones. HRT may consist of estrogen alone or a combination of estrogen and progesterone. Because estrogen is neuroprotective, it was initially hypothesized that HRT would have a protective effect on cognition. Indeed, early observational studies demonstrated that HRT was associated with decreased AD risk.[146,149,150]

However, the landmark Women's Health Initiative Memory Study (WHIMS) found that women randomized to receive Estrogen-Progestin HRT had twice the risk of developing dementia after 4 years compared to the placebo group.[151] Women who received estrogen alone also had an increased risk of dementia after a 5 year follow-up visit.[152] This seminal work led to changes in HRT prescribing practices due to perceived risk; however, the study later received criticism regarding generalizability, as all participants were aged 65+ years, whereas the average age of menopause onset is 51 years.[4,67,79,151,152] No reports since WHIMS have found a comparably dramatic association between HRT and AD risk. Subsequent reports on the association between HRT and cognitive outcomes vary substantially and continue to be debated among scholars and clinicians.

Multiple randomized control trials have shown no increased risk of cognitive decline with HRT use within 3 years of the last period.[153–156] One of these studies found no cognitive differences after 5 years between women who began HRT 6 versus 10 years after their last period nor a difference in risk of death from AD between groups.[157] A few observational studies report similar findings of minimal to no protective association between HRT and AD onset or death.[158,159]

Some studies support a neuroprotective effect of exogenous estrogen, although it is evident that this effect may depend on the timing of initiation. A nationwide case-control study ($n = 230,580$) in Finland by Imtiaz et al. found that short-term HRT use (<10 years) was associated with a modest increase in AD risk, whereas long-term HRT use (>10 years) was weakly associated with a decrease in AD risk. However, the authors suggest that these findings may not be clinically significant.[160] A recent meta-analysis found a significant association between hormone therapy and AD risk (OR 1.08, 95% CI 1.03–1.14, I^2: 69%).[161] However, this effect was only observed within the first 5 years of using HRT, and the direction of the association appears to reverse after 5 years of use.[161] A Korean nationwide cohort study ($n = 4,696,633$) found that HRT use among postmenopausal women decreased the risk of dementia by approximately 15%, regardless of the duration of use. However, no data regarding the timing of initiation relative to menopause were available.[68] In contrast, another nationwide study ($n = 84,739$) in Finland demonstrated that long-term HRT use (>10 years) was associated with a 9%–17% increase in the risk of developing AD.[162]

Findings from neuroimaging studies in postmenopausal women with and without a history of HRT also vary. Women who started HRT before menopause and stayed on HRT for at least 3 years have been shown to have greater hippocampal volume than those who did not use HRT.[163] Similarly, Boyle et al. report larger gray and white matter volumes in women with a history of HRT, in brain regions relevant to cognitive function (including the frontal, temporal, and parietal lobes) regardless of duration and timing of HRT use.[63] In contrast, Kantarci et al. found reduced whole brain volume, increased ventricle expansion, and greater white matter hyperintensities in postmenopausal women after 48 months of HRT compared to placebo.[164] Interestingly, cognition did not differ between groups. Another imaging study found that estrogen alone was neuroprotective against white matter aging, whereas combination estrogen–

progesterone therapy was associated with accelerated white matter aging compared to women with no history of HRT.[165]

There are a few possible explanations for the inconsistencies across studies. First, there is some evidence to suggest that the timing of HRT administration relative to menopause onset may determine whether it has a protective or detrimental effect on cognition. The "critical window" hypothesis posits that initiating HRT soon after menopause is neuroprotective, whereas beginning HRT after neurons have been deprived of estrogen for a significant period is deleterious.[166] Indeed, a few studies suggest that starting HRT early in the menopause transition may decrease AD risk.[167,168] Similarly, multiple studies have found that HRT exerts a deleterious effect on brain function and can increase AD risk if initiated years after menopause.[65,169–173] Review of previous animal studies also supports the critical window hypothesis; exogenous estrogen administered at the time that endogenous estrogen ceased to function has been shown to benefit brain health, whereas delaying exogenous estrogen administration has not been associated a neuroprotective effect.[174] The critical window hypothesis may also explain why the WHIMS trial found HRT to double the risk of developing AD. Participants were randomized to the HRT or placebo group decades after menopause onset—well outside the narrow window in which HRT may benefit cognition. Nevertheless, there is still one previously mentioned study that refutes the critical window hypothesis, as Henderson et al. found no difference in cognitive outcomes between women who started HRT 6 and 10 years after menopause.[157]

Second, the "healthy cell" hypothesis proposes that exogenous estrogen is only beneficial if administered to healthy neurons. Otherwise, HRT may exacerbate neuron damage.[62,65,88,175,176] According to this hypothesis, the effect of HRT on cognition depends on baseline cognitive status and degree of neuropathology, which may not be consistently and similarly controlled for across studies. Another major limitation of existing studies is variation in the chemical composition of HRT used across participants, even in the same study. It may be that the type and proportions of hormones may modulate the effect on cognition, and this may contribute to inconsistencies across studies. Future studies ought to consider the timing of HRT initiation and duration of use, as well as control for the chemical composition of HRT and baseline cognition and neuropathology to elucidate the true effect, if any, of HRT on cognitive outcomes among postmenopausal women.Whereas the long-term cognitive risk or benefit of using HRT after natural menopause has been debated for decades, fewer studies have focused specifically on cognitive outcomes with HRT use after surgical menopause. A few studies suggest that initiating HRT after surgical menopause may mediate the negative effects of surgical menopause on cognition. Women who started HRT after oophorectomy have been shown to perform similarly on working memory tasks to premenopausal controls.[135] Women who received estrogen therapy from the time of oophorectomy until age 50 years had no increased risk of cognitive impairment or dementia than age-matched women who had not undergone surgical menopause.[133] In addition, initiating HRT within 5 years of surgical menopause, and continuing for 10 years, has been associated with less of a decline in semantic and episodic memory and visuospatial abilities than observed in women who did not use HRT after surgery.[134] There was no association between HRT and AD neuropathology, suggesting that HRT may be protective independent of pathologic brain changes.[134,140]

7.3 Selective estrogen receptor modulators

Selective estrogen receptor modulators (SERMs) are medications that may act as estrogen receptor agonists or antagonists and are used in the treatment of osteoporosis and breast cancer.[4] One study found that Raloxifene reduced the risk of MCI, but not AD, in postmenopausal women, while another study found that Raloxifene did not affect dementia risk.[177,178]

8. Pregnancy

8.1 Uncomplicated pregnancies

Pregnancy itself is potentially neuroprotective due to hormonal and immune system changes that occur during pregnancy.[4] First, pregnancy is associated with an exponential increase in estrogen and progesterone, both of which are thought to be neuroprotective.[71,179] Second, changes in immunoregulation during the first trimester of pregnancy have been associated with reduced AD risk.[180] A study by Fox et al. found that the number of first trimesters was associated with a significant decrease in the risk of developing AD later in life.[180] The number of regulatory T cells (T_{Regs}) increases dramatically in early pregnancy, leading to downregulation of type 1 inflammation, and remains stable in late pregnancy and perhaps through the duration of the lifespan.[180] The opposite trend—fewer T_{Regs} and upregulation of Type 1 inflammation—has been documented in AD.[181] It is therefore plausible that immune changes during pregnancy could protect against AD. Importantly, this benefit was not observed during the third trimester, suggesting a benefit of gravidity that may be independent of parity.[180]

Like Fox et al., another study found the number of incomplete pregnancies was associated with lower AD risk, and the number of completed pregnancies was associated with greater AD risk.[182] But rather than tout the differential protective effects of hormone and immune system changes in early pregnancy over late pregnancy, researchers posited that repeated, extreme upregulation of estradiol over multiple pregnancies impairs cognition later in life. They also hypothesized that the sudden drop in estrogen (and the spike in cortisol) at the time of delivery contributes to later cognitive decline—an effect that is compounded by each completed pregnancy.[182] Indeed, it has been shown that women with a parity of two have better cognition than women with a parity of 0–1 or greater than three.[183–185] These findings support the idea that the neuroprotective effect of pregnancy may dissipate at increasing numbers of completed pregnancies.

However, the link between higher parity and AD risk is somewhat tenuous, as grand multiparity ($\geq$5 completed pregnancies) has been associated with decreased brain volume but not Aβ deposition or white matter lesions.[186] The impact of gravidity and parity on the brain and cognition appears to be multifactorial. Future studies are warranted to better understand AD risk among mothers.

8.2 Adverse pregnancy outcomes

Adverse pregnancy outcomes may affect late-life cognition. Adverse pregnancy outcomes (such as preeclampsia, gestational hypertension, fetal growth restriction, stillbirth, preterm birth, miscarriage, and gestational diabetes) increase maternal risk of cardiovascular disease thereafter.[187–190] Cardiovascular disease is a risk factor for cognitive decline later in life.[183,191,192]

There is some association between hypertensive disorders of pregnancy (such as gestational hypertension, preeclampsia, and eclampsia) and cognitive decline later in life.[4,193] A history of a hypertensive pregnancy disorder has been linked to slower processing speed in older adulthood[194] as well as poor working memory and verbal learning 15 years after pregnancy.[195] One study even found increased brain atrophy and white matter lesion volume decades after pregnancy among women with histories of hypertensive pregnancy disorders, compared to women with uncomplicated pregnancies.[194] In addition, impaired amyloid processing and protein misfolding have been reported in women with preeclampsia, though a direct link to the development of AD has not been elucidated.[196] However, other studies have found no relationship between hypertensive pregnancy disorders and late-life cognition.[191,197] Some suggest an association between hypertensive pregnancy disorders and risk for vascular dementia,[192,198] though others have found no association between hypertensive pregnancy disorders and dementia.[183,199]

9. Vascular risk

Sex differences in AD prevalence and susceptibility may arise from sex differences in cardiovascular and cerebrovascular risk factors. Cerebrovascular events (stroke and microvascular infarcts) and cardiovascular risk factors (hypertension, hyperlipidemia, and diabetes) occur more frequently in men before age 60 years. However, the prevalence in women is equal to or surpasses the prevalence in men after menopause and/or age 60.[200–203]

The 2015 Mayo Clinic Study of Aging (MCSA) identified sex-specific predictors of short-term conversion from normal cognition to MCI in individuals aged 70–89.[130] Specific predictors for women included smoking, dyslipidemia, diabetes, and hypertension, whereas specific predictors for men included obesity and marital status of never married or widowed. Low educational attainment, history of alcohol misuse, stroke, and atrial fibrillation were predictors in both women and men.[130] The MCSA did not evaluate conversion from MCI to dementia; however, another study found that hypertension in middle-age was associated with increased risk of dementia among women only, despite the higher prevalence of hypertension in middle-aged men.[204]

Though diabetes has been characterized as a female-specific predictor of MCI, especially after age 60 years, one study found an increased risk of MCI in men aged 70–89 with type 2 diabetes compared to women.[205] A multicenter cohort study conducted in France found that men aged 65+ with MCI were more likely to have diabetes mellitus, stroke, and high BMI than women with MCI.[206] However, women have a greater risk for diabetic complications of myocardial infarction, depression, and coronary heart disease than men—all of which are AD risk factors.[22,207] The higher prevalence of these diabetic complications in women suggests that a diagnosis of diabetes in middle age may confer a greater risk for developing AD compared to men.[130,208]

Review of the existing literature suggests that individual risk factors for AD interact with sex differently at different stages of progression—such as the progression from normal cognition to MCI or AD and progression from MCI to AD. Some researchers suggest that sex-specific risk profiles may stem from survival bias rather than actual differences in risk. Men are more likely to die of cardiovascular disease in middle age than women, so those who survive to older ages may have better vascular health than women.[209]

10. Pharmacology

There are currently five drugs approved by the Food and Drug Administration (FDA) to treat AD. These include three cholinesterase inhibitors (donepezil, rivastigmine, and galantamine), one N-methyl-D-aspartate (NMDA) receptor antagonist (memantine), and one recently approved monoclonal Aβ antibody (aducanumab).[210,211]

Though current knowledge is limited, some evidence suggests that certain cholinesterase inhibitors affect males and females differently. One study found that rivastigmine treatment in individuals with MCI delayed progression to AD in women only,[212] highlighting a potential sex-specific benefit of early initiation. However, cholinesterase inhibitor treatment for advanced dementia has shown a greater benefit in men,[213–217] except for one study that found the opposite.[218]

Another study found that women treated with cholinesterase inhibitors survived longer than men, suggesting that sex differences in antidementia effects differ from the effect on overall survival.[219] Sex differences in the effects of cholinesterase inhibitors may stem from sexual dimorphism of the cholinergic system[220] or sex differences in susceptibility to AD pathology.[27,221] Cholinesterase inhibitors may also interact with sex hormones, contributing to different pharmacokinetics in males and females.[222,223] This theory is supported by a study that found the effects of donepezil and rivastigmine to be modulated by estrogen receptor 1 (*ESR1*) genotype.[218] Sex and genotype may also interact to determine drug efficacy. Though tacrine has since been withdrawn from the market, female APOE-e4 carriers had a lesser response to tacrine than placebo and women with other APOE genotypes.[224] In men, the treatment effect was not modulated by APOE genotype.[224] Future research exploring sex differences in pharmacokinetics and the efficacy of existing AD treatments is critical, as this knowledge has the potential to guide sex-specific treatment protocols for AD.

Despite known sex differences in AD neuropathology and pharmacokinetics, drug trials infrequently stratify their analyses by sex, potentially masking differences in adverse events and efficacy.[225] A 2017 review of sex differences in 48 clinical trials of donepezil, galantamine, rivastigmine, and memantine found only two studies that analyzed sex differences in safety and efficacy.[214] These two studies found no difference in effect for donepezil.[214] Another systematic review of clinical trials found that most studies presented no data on sex differences in adverse events associated with cholinesterase inhibitors.[226] These knowledge gaps highlight a need for systematic examination of sex differences in the outcomes, adverse effects, and efficacy of cholinesterase inhibitor treatment.[215,225,227,228]

Failure to consider the differential drug effects in males and females may stem from decades-long gender biases in pharmaceutical clinical trial research.[229–232] White men were historically considered the "normal" study population, whereas women were excluded from trials because their fluctuating hormones were thought to make them confounding subjects.[229] Although the NIH now mandates the inclusion of female subjects in all NIH-funded works, as well as requires that approved grants plan to analyze sex as a variable, decades of biased work still permeate the literature—particularly in the field of pharmacology.[5] Further, implicit sex biases in healthcare and clinical research contribute to low enrollment of women in clinical trials, thus achieving a gender-balanced study population is challenging.[233–238]

AD clinical trials face unique challenges regarding enrollment and retention. Overall, the aging population is difficult to recruit; 20%–30% of older adults decline trial participation, compared to 5%–10% of younger adults.[239] There is also concern whether individuals with AD can maintain a clinical trial schedule or survive through the end of the study.[210] Older women with AD are particularly difficult to enroll, they are more likely than men to live alone and may not have a representative to provide consent on their behalf.[240]

Only 50% of participants in phase III trials of candidate AD drugs semagacestat, verubecestat, solanezumab, and bapineuzumab were women.[241–244] However, this does not reflect the true disease epidemiology, as over 60% of individuals with AD are women. Gender differences in enrollment and retention in pharmaceutical clinical trials may reflect inequality in the care that women with AD receive. For example, an analysis of the Swedish Dementia Registry found that women were less likely to undergo lumbar puncture and MRI. This effect was also driven by the higher age of female patients compared to male patients.[245] Concerted efforts to recruit more women in AD clinical trials are necessary but insufficient to fully understand sex differences in the effects of current and prospective AD treatments. The relevance of future studies hinges on both increasing the number of women enrolled in clinical trials and publishing sex-specific analyses.

11. COVID-19 pandemic

The COVID-19 pandemic is an evolving global health threat with far-reaching consequences that include high levels of morbidity and mortality, economic hardship, and tolls on mental health and quality of life.[246] Although men infected with SARS-CoV-2 tend to have worse outcomes and are 2.4 times more likely to die of COVID-19 than women,[247,248] COVID-19 infection in pregnancy puts women at a uniquely elevated risk for poor outcomes. A large multinational study demonstrated an elevated risk for preeclampsia (recurrence risk (RR) 1.76; 95% CI 1.27−2.43), preterm labor (RR 1.59; 95% CI 1.30−1.94), severe COVID-19 (RR 3.38; 95% CI 1.63−7.01), intensive care unit admission (RR 5.04; 95% CI 3.13−8.10), and mortality (RR 22.3; 95% CI 2.88−172) among pregnant women infected with SARS-CoV-2;[249] these findings have been corroborated by similar studies.[250−252]

People with AD represent a particularly vulnerable population due to increased susceptibility to severe disease and unmet care needs driven by lockdowns and social distancing practices.[1,253] Because women comprise two-thirds of individuals AD, it is plausible that the effects of the COVID-19 pandemic on those living with AD are disproportionately felt by women.[1] No studies have explored sex differences in COVID-19 mortality among individuals with AD. However, there has been a marked increase in AD mortality during the COVID-19 pandemic that is more pronounced among women.[254] Preliminary data from the National Institute of Health reports a 10% increase in AD mortality in 2020 compared to previous years that was most significant among women.[254] Several factors may contribute to AD progression, including loss of social and cognitive stimulation due to social distancing practices, difficulty accessing healthcare services and the limitations of telehealth, and increasing rates of psychiatric comorbidities like anxiety and depression due to the stress and uncertainty of the evolving pandemic.[255−257] Women have higher rates of anxiety and depression compared to men; these disparities may be exacerbated among women with AD who are navigating the stresses associated with the COVID-19 pandemic and contribute to worsening neuropsychiatric outcomes.[258,259]

12. Conclusion

This chapter integrates current knowledge of sex differences in AD presentation, biomarkers, pathophysiology, progression, and risk. Providers need to be aware of these differences in order to understand their patients' unique risk profiles, identify early signs of disease, and employ the appropriate treatment strategies. This has become especially important since the start of the COVID-19 pandemic, as women with AD have become increasingly likely to develop worsening of disease. Finally, the underrepresentation of women in drug randomized control trials is particularly concerning and contributes to significant gaps in knowledge regarding effective therapeutics for women with AD. Efforts to diversify enrollment and reporting in future studies are necessary and sex differences in AD need to be considered in the development of future therapies.

References

1. 2021 Alzheimer's disease facts and figures. *Alzheimer Dementia*. March 2021;17(3): 327−406. https://doi.org/10.1002/alz.12328.
2. Seshadri S, Wolf PA, Beiser A, et al. Lifetime risk of dementia and Alzheimer's disease: the impact of mortality on risk estimates in the Framingham Study. *Neurology*. 1997;49(6): 1498−1504.

3. Wizemann TM, Pardue ML, eds. *Exploring the Biological Contributions to Human Health: Does Sex Matter?* The National Academies Collection: Reports funded by National Institutes of Health; 2001.
4. Peterson A, Tom SE. A lifecourse perspective on female sex-specific risk factors for later life cognition. *Curr Neurol Neurosci Rep*. 2021;21(9):46. https://doi.org/10.1007/s11910-021-01133-y.
5. Salminen LE, Tubi MA, Bright J, Thomopoulos SI, Wieand A, Thompson PM. Sex is a defining feature of neuroimaging phenotypes in major brain disorders. *Hum Brain Mapp*. 2021. https://doi.org/10.1002/hbm.25438.
6. Ferretti MT, Iulita MF, Cavedo E, et al. Sex differences in Alzheimer disease — the gateway to precision medicine. *Nat Rev Neurol*. 2018;14(8):457–469. https://doi.org/10.1038/s41582-018-0032-9.
7. Teri L, Borson S, Kiyak HA, Yamagishi M. Behavioral disturbance, cognitive dysfunction, and functional skill. Prevalence and relationship in Alzheimer's disease. *J Am Geriatr Soc*. February 1989;37(2):109–116. https://doi.org/10.1111/j.1532-5415.1989.tb05868.x.
8. Spalletta G, Musicco M, Padovani A, et al. Neuropsychiatric symptoms and syndromes in a large cohort of newly diagnosed, untreated patients with Alzheimer disease. *Am J Geriatr Psychiatry*. November 2010;18(11):1026–1035. https://doi.org/10.1097/JGP.0b013e3181d6b68d.
9. Ott BR, Tate CA, Gordon NM, Heindel WC. Gender differences in the behavioral manifestations of Alzheimer's disease. *J Am Geriatr Soc*. May 1996;44(5):583–587. https://doi.org/10.1111/j.1532-5415.1996.tb01447.x.
10. Kitamura T, Kitamura M, Hino S, Tanaka N, Kurata K. Gender differences in clinical manifestations and outcomes among hospitalized patients with behavioral and psychological symptoms of dementia. *J Clin Psychiatr*. December 2012;73(12):1548–1554. https://doi.org/10.4088/JCP.11m07614.
11. Hollingworth P, Hamshere ML, Moskvina V, et al. Four components describe behavioral symptoms in 1,120 individuals with late-onset Alzheimer's disease. *J Am Geriatr Soc*. September 2006;54(9):1348–1354. https://doi.org/10.1111/j.1532-5415.2006.00854.x.
12. Rentz DM, Weiss BK, Jacobs EG, et al. Sex differences in episodic memory in early midlife: impact of reproductive aging. *Menopause*. April 2017;24(4):400–408. https://doi.org/10.1097/gme.0000000000000771.
13. Jack Jr CR, Wiste HJ, Weigand SD, et al. Age, sex, and APOE ε4 effects on memory, brain structure, and β-amyloid across the adult life span. *JAMA Neurol*. May 2015;72(5): 511–519. https://doi.org/10.1001/jamaneurol.2014.4821.
14. Sundermann EE, Biegon A, Rubin LH, et al. Better verbal memory in women than men in MCI despite similar levels of hippocampal atrophy. *Neurology*. April 12, 2016;86(15): 1368–1376. https://doi.org/10.1212/wnl.0000000000002570.
15. Sundermann EE, Maki PM, Rubin LH, Lipton RB, Landau S, Biegon A. Female advantage in verbal memory: evidence of sex-specific cognitive reserve. *Neurology*. November 1, 2016;87(18):1916–1924. https://doi.org/10.1212/wnl.0000000000003288.
16. McCarrey AC, An Y, Kitner-Triolo MH, Ferrucci L, Resnick SM. Sex differences in cognitive trajectories in clinically normal older adults. *Psychol Aging*. March 2016;31(2): 166–175. https://doi.org/10.1037/pag0000070.
17. Sundermann EE, Biegon A, Rubin LH, et al. Does the female advantage in verbal memory contribute to underestimating Alzheimer's disease pathology in women versus men? *J Alzheimers Dis*. 2017;56(3):947–957. https://doi.org/10.3233/JAD-160716.

18. Holland D, Desikan RS, Dale AM, McEvoy LK. Higher rates of decline for women and apolipoprotein E epsilon4 carriers. *AJNR Am J Neuroradiol*. December 2013;34(12): 2287–2293. https://doi.org/10.3174/ajnr.A3601.
19. Lin KA, Choudhury KR, Rathakrishnan BG, Marks DM, Petrella JR, Doraiswamy PM. Marked gender differences in progression of mild cognitive impairment over 8 years. *Alzheimers Dement (N Y)*. September 1, 2015;1(2):103–110. https://doi.org/10.1016/j.trci.2015.07.001.
20. Tifratene K, Robert P, Metelkina A, Pradier C, Dartigues JF. Progression of mild cognitive impairment to dementia due to AD in clinical settings. *Neurology*. July 28, 2015;85(4): 331–338. https://doi.org/10.1212/wnl.0000000000001788.
21. Pradier C, Sakarovitch C, Le Duff F, et al. The mini mental state examination at the time of Alzheimer's disease and related disorders diagnosis, according to age, education, gender and place of residence: a cross-sectional study among the French National Alzheimer database. *PLoS One*. 2014;9(8):e103630. https://doi.org/10.1371/journal.pone.0103630.
22. Nebel RA, Aggarwal NT, Barnes LL, et al. Understanding the impact of sex and gender in Alzheimer's disease: a call to action. *Alzheimers Dement*. 2018;14(9):1171–1183. https://doi.org/10.1016/j.jalz.2018.04.008.
23. Duarte-Guterman P, Albert AY, Barha CK, Galea LAM. Alzheimer's Dis Neuroimaging I. Sex influences the effects of APOE genotype and Alzheimer's diagnosis on neuropathology and memory. *Psychoneuroendocrinology*. July 2021:129105248. https://doi.org/10.1016/j.psyneuen.2021.105248.
24. Irvine K, Laws KR, Gale TM, Kondel TK. Greater cognitive deterioration in women than men with Alzheimer's disease: a meta analysis. *J Clin Exp Neuropsychol*. 2012;34(9): 989–998. https://doi.org/10.1080/13803395.2012.712676.
25. Mattsson N, Lönneborg A, Boccardi M, Blennow K, Hansson O. Clinical validity of cerebrospinal fluid Aβ42, tau, and phospho-tau as biomarkers for Alzheimer's disease in the context of a structured 5-phase development framework. *Neurobiol Aging*. April 2017;52: 196–213. https://doi.org/10.1016/j.neurobiolaging.2016.02.034.
26. Barnes LL, Wilson RS, Bienias JL, Schneider JA, Evans DA, Bennett DA. Sex differences in the clinical manifestations of Alzheimer disease pathology. *Arch Gen Psychiatr*. 2005; 62(6):685–691. https://doi.org/10.1001/archpsyc.62.6.685.
27. Salehi A, Gonzalez Martinez V, Swaab DF. A sex difference and no effect of ApoE type on the amount of cytoskeletal alterations in the nucleus basalis of Meynert in Alzheimer's disease. *Neurobiol Aging*. November–December 1998;19(6):505–510. https://doi.org/10.1016/s0197-4580(98)00106-7.
28. Buckley RF, Mormino EC, Amariglio RE, et al. Sex, amyloid, and APOE epsilon4 and risk of cognitive decline in preclinical Alzheimer's disease: findings from three well-characterized cohorts. *Alzheimers Dement*. September 2018;14(9):1193–1203. https://doi.org/10.1016/j.jalz.2018.04.010.
29. Hua X, Hibar DP, Lee S, et al. Sex and age differences in atrophic rates: an ADNI study with n=1368 MRI scans. *Neurobiol Aging*. 2010;31(8):1463–1480. https://doi.org/10.1016/j.neurobiolaging.2010.04.033.
30. Apostolova LG, Dinov ID, Dutton RA, et al. 3D comparison of hippocampal atrophy in amnestic mild cognitive impairment and Alzheimer's disease. *Brain*. November 2006; 129(Pt 11):2867–2873. https://doi.org/10.1093/brain/awl274.
31. Perlaki G, Orsi G, Plozer E, et al. Are there any gender differences in the hippocampus volume after head-size correction? A volumetric and voxel-based morphometric study. *Neurosci Lett*. June 6 2014;570:119–123. https://doi.org/10.1016/j.neulet.2014.04.013.

32. Koran MEI, Wagener M, Hohman TJ. For the Alzheimer's Neuroimaging I. Sex differences in the association between AD biomarkers and cognitive decline. *Brain Imaging Behav*. 2017/02/01 2017;11(1):205–213. https://doi.org/10.1007/s11682-016-9523-8.
33. Riedel BC, Thompson PM, Brinton RD. Age, APOE and sex: triad of risk of Alzheimer's disease. *J Steroid Biochem Mol Biol*. 2016;160:134–147. https://doi.org/10.1016/j.jsbmb.2016.03.012.
34. LaDu MJ, Munson GW, Jungbauer L, et al. Preferential interactions between ApoE-containing lipoproteins and Aβ revealed by a detection method that combines size exclusion chromatography with non-reducing gel-shift. *Biochim Biophys Acta*. February 2012;1821(2):295–302. https://doi.org/10.1016/j.bbalip.2011.11.005.
35. Tai LM, Mehra S, Shete V, et al. Soluble apoE/Aβ complex: mechanism and therapeutic target for APOE4-induced AD risk. *Mol Neurodegener*. January 4 2014;9:2. https://doi.org/10.1186/1750-1326-9-2.
36. Raber J, Huang Y, Ashford JW. ApoE genotype accounts for the vast majority of AD risk and AD pathology. *Neurobiol Aging*. May–June 2004;25(5):641–650. https://doi.org/10.1016/j.neurobiolaging.2003.12.023.
37. Verghese PB, Castellano JM, Garai K, et al. ApoE influences amyloid-β (Aβ) clearance despite minimal apoE/Aβ association in physiological conditions. *Proc Natl Acad Sci U S A*. 2013;110(19):E1807–E1816. https://doi.org/10.1073/pnas.1220484110.
38. Liu CC, Liu CC, Kanekiyo T, Xu H, Bu G. Apolipoprotein E and Alzheimer disease: risk, mechanisms and therapy. *Nat Rev Neurol*. February 2013;9(2):106–118. https://doi.org/10.1038/nrneurol.2012.263.
39. Loy CT, Schofield PR, Turner AM, Kwok JBJ. Genetics of dementia. *Lancet*. 2014;383(9919):828–840. https://doi.org/10.1016/S0140-6736(13)60630-3.
40. Michaelson DM. Apoe ε4: the most prevalent yet understudied risk factor for Alzheimer's disease. *Alzheimers Dementia*. 2014;10(6):861–868. https://doi.org/10.1016/j.jalz.2014.06.015.
41. Suri S, Heise V, Trachtenberg AJ, Mackay CE. The forgotten APOE allele: a review of the evidence and suggested mechanisms for the protective effect of APOE ϵ2. *Neurosci Biobehav Rev*. December 2013;37(10 Pt 2):2878–2886. https://doi.org/10.1016/j.neubiorev.2013.10.010.
42. Neu SC, Pa J, Kukull W, et al. Apolipoprotein E genotype and sex risk factors for Alzheimer disease: a meta-analysis. *JAMA Neurol*. 2017;74(10):1178–1189. https://doi.org/10.1001/jamaneurol.2017.2188.
43. Lamonja-Vicente N, Dacosta-Aguayo R, Lopez-Oloriz J, et al. Sex-specific protective effects of APOE epsilon 2 on cognitive performance. *J Gerontol Series A Biol Sci Med Sci*. January 2021;76(1):41–49. https://doi.org/10.1093/gerona/glaa247.
44. Farrer LA, Cupples LA, Haines JL, et al. Effects of age, sex, and ethnicity on the association between apolipoprotein E genotype and Alzheimer disease: a meta-analysis. *JAMA*. 1997;278(16):1349–1356.
45. Altmann A, Tian L, Henderson VW, Greicius MD. Sex modifies the APOE-related risk of developing Alzheimer disease. *Ann Neurol*. April 2014;75(4):563–573. https://doi.org/10.1002/ana.24135.
46. Kim S, Kim MJ, Kim S, et al. Gender differences in risk factors for transition from mild cognitive impairment to Alzheimer's disease: a CREDOS study. *Compr Psychiatr*. October 2015;62:114–122. https://doi.org/10.1016/j.comppsych.2015.07.002.
47. Heise V, Filippini N, Trachtenberg AJ, Suri S, Ebmeier KP, Mackay CE. Apolipoprotein E genotype, gender and age modulate connectivity of the hippocampus in healthy adults. *Neuroimage*. 2014;98:23–30.

48. Damoiseaux JS, Seeley WW, Zhou J, et al. Gender modulates the APOE ε4 effect in healthy older adults: convergent evidence from functional brain connectivity and spinal fluid tau levels. *J Neurosci.* 2012;32(24):8254−8262.
49. Liu M, Paranjpe MD, Zhou X, et al. Sex modulates the ApoE ε4 effect on brain tau deposition measured by 18F-AV-1451 PET in individuals with mild cognitive impairment. Research Paper. *Theranostics.* 2019;9(17):4959−4970. https://doi.org/10.7150/thno.35366.
50. Hohman TJ, Dumitrescu L, Barnes LL, et al. Sex-specific association of apolipoprotein E with cerebrospinal fluid levels of tau. *JAMA Neurol.* 2018;75(8):989−998. https://doi.org/10.1001/jamaneurol.2018.0821.
51. Pontifex MG, Martinsen A, Saleh RNM, et al. APOE4 genotype exacerbates the impact of menopause on cognition and synaptic plasticity in APOE-TR mice. *Faseb J.* May 2021; 35(5):e21583. https://doi.org/10.1096/fj.202002621RR.
52. Mosconi L, Berti V, Dyke J, et al. Menopause impacts human brain structure, connectivity, energy metabolism, and amyloid-beta deposition. *Sci Rep.* 2021;11(1):10867. https://doi.org/10.1038/s41598-021-90084-y.
53. Tan CH, Bonham LW, Fan CC, et al. Polygenic hazard score, amyloid deposition and Alzheimer's neurodegeneration. *Brain.* 2019;142(2):460−470. https://doi.org/10.1093/brain/awy327.
54. Deming Y, Dumitrescu L, Barnes LL, et al. Sex-specific genetic predictors of Alzheimer's disease biomarkers. *Acta Neuropathol.* 2018;136(6):857−872. https://doi.org/10.1007/s00401-018-1881-4.
55. Chapuis J, Flaig A, Grenier-Boley B, et al. Genome-wide, high-content siRNA screening identifies the Alzheimer's genetic risk factor FERMT2 as a major modulator of APP metabolism. *Acta Neuropathol.* 2017;133(6):955−966.
56. Mok S-A, Condello C, Freilich R, et al. Mapping interactions with the chaperone network reveals factors that protect against tau aggregation. *Nat Struct Mol Biol.* 2018;25(5):384−393.
57. Fan CC, Banks SJ, Thompson WK, et al. Sex-dependent autosomal effects on clinical progression of Alzheimer's disease. *Brain.* 2020;143(7):2272−2280. https://doi.org/10.1093/brain/awaa164.
58. Wang H, Lo MT, Rosenthal SB, et al. Similar genetic architecture of Alzheimer's disease and differential APOE effect between sexes. *Front Aging Neurosci.* May 2021:13674318. https://doi.org/10.3389/fnagi.2021.674318.
59. Barth C, Villringer A, Sacher J. Sex hormones affect neurotransmitters and shape the adult female brain during hormonal transition periods. Review. *Front Neurosci.* February 20, 2015;(37):9. https://doi.org/10.3389/fnins.2015.00037.
60. Hara Y, Waters EM, McEwen BS, Morrison JH. Estrogen effects on cognitive and synaptic health over the lifecourse. *Physiol Rev.* July 2015;95(3):785−807. https://doi.org/10.1152/physrev.00036.2014.
61. Do Rego JL, Seong JY, Burel D, et al. Neurosteroid biosynthesis: enzymatic pathways and neuroendocrine regulation by neurotransmitters and neuropeptides. *Front Neuroendocrinol.* August 2009;30(3):259−301. https://doi.org/10.1016/j.yfrne.2009.05.006.
62. Gillies GE, McArthur S. Estrogen actions in the brain and the basis for differential action in men and women: a case for sex-specific medicines. *Pharmacol Rev.* 2010;62(2):155−198. https://doi.org/10.1124/pr.109.002071.
63. Boyle CP, Raji CA, Erickson KI, et al. Estrogen, brain structure, and cognition in postmenopausal women. *Hum Brain Mapp.* 2021;42(1):24−35. https://doi.org/10.1002/hbm.25200.

64. Bean LA, Ianov L, Foster TC. Estrogen receptors, the hippocampus, and memory. *Neuroscientist*. 2014;20(5):534–545. https://doi.org/10.1177/1073858413519865.
65. Brinton RD. The healthy cell bias of estrogen action: mitochondrial bioenergetics and neurological implications. *Trends Neurosci*. October 2008;31(10):529–537. https://doi.org/10.1016/j.tins.2008.07.003.
66. Klinge CM. Estrogens regulate life and death in mitochondria. *J Bioenerg Biomembr*. August 2017;49(4):307–324. https://doi.org/10.1007/s10863-017-9704-1.
67. Zárate S, Stevnsner T, Gredilla R. Role of estrogen and other sex hormones in brain aging. Neuroprotection and DNA repair. *Front Aging Neurosci*. 2017;9:430. https://doi.org/10.3389/fnagi.2017.00430.
68. Yoo JE, Shin DW, Han K, et al. Female reproductive factors and the risk of dementia: a nationwide cohort study. *Eur J Neurol*. 2020/08/01 2020;27(8):1448–1458. https://doi.org/10.1111/ene.14315.
69. Gilsanz P, Lee C, Corrada MM, Kawas CH, Quesenberry CP, Whitmer RA. Reproductive period and risk of dementia in a diverse cohort of health care members. *Neurology*. 2019;92(17):e2005–e2014. https://doi.org/10.1212/wnl.0000000000007326.
70. Shimizu Y, Sawada N, Iwasaki M, et al. Reproductive history and risk of cognitive impairment in Japanese women. *Maturitas*. October 2019;128:22–28. https://doi.org/10.1016/j.maturitas.2019.06.012.
71. Najar J, Östling S, Waern M, et al. Reproductive period and dementia: a 44-year longitudinal population study of Swedish women. *Alzheimers Dement*. August 2020;16(8):1153–1163. https://doi.org/10.1002/alz.12118.
72. Prince MJ, Acosta D, Guerra M, et al. Reproductive period, endogenous estrogen exposure and dementia incidence among women in Latin America and China; A 10/66 population-based cohort study. *PLoS One*. 2018;13(2):e0192889. https://doi.org/10.1371/journal.pone.0192889.
73. Fox M, Berzuini C, Knapp LA. Cumulative estrogen exposure, number of menstrual cycles, and Alzheimer's risk in a cohort of British women. *Psychoneuroendocrinology*. December 2013;38(12):2973–2982. https://doi.org/10.1016/j.psyneuen.2013.08.005.
74. Song X, Wu J, Zhou Y, et al. Reproductive and hormonal factors and risk of cognitive impairment among Singapore Chinese women. *Am J Obstet Gynecol*. 2020;223(3):e410.e1–e410.e23.
75. McLay RN, Maki PM, Lyketsos CG. Nulliparity and late menopause are associated with decreased cognitive decline. *J Neuropsychiatry Clin Neurosci*. 2003;15(2):161–167. https://doi.org/10.1176/jnp.15.2.161.
76. Kuh D, Cooper R, Moore A, Richards M, Hardy R. Age at menopause and lifetime cognition: findings from a British birth cohort study. *Neurology*. May 8, 2018;90(19):e1673–e1681. https://doi.org/10.1212/wnl.0000000000005486.
77. Matyi JM, Rattinger GB, Schwartz S, Buhusi M, Tschanz JT. Lifetime estrogen exposure and cognition in late life: the Cache County Study. *Menopause*. 2019;26(12):1366.
78. Geerlings MI, Ruitenberg A, Witteman JC, et al. Reproductive period and risk of dementia in postmenopausal women. *JAMA*. March 21, 2001;285(11):1475–1481. https://doi.org/10.1001/jama.285.11.1475.
79. McKinlay SM, Brambilla DJ, Posner JG. The normal menopause transition. *Maturitas*. January 1992;14(2):103–115. https://doi.org/10.1016/0378-5122(92)90003-m.
80. Epperson CN, Sammel MD, Freeman EW. Menopause effects on verbal memory: findings from a longitudinal community cohort. *J Clin Endocrinol Metab*. 2013;98(9):3829–3838. https://doi.org/10.1210/jc.2013-1808.

81. Jacobs EG, Weiss BK, Makris N, et al. Impact of sex and menopausal status on episodic memory circuitry in early midlife. *J Neurosci*. 2016;36(39):10163–10173. https://doi.org/10.1523/JNEUROSCI.0951-16.2016.
82. Greendale GA, Huang MH, Wight RG, et al. Effects of the menopause transition and hormone use on cognitive performance in midlife women. *Neurology*. 2009;72(21):1850–1857. https://doi.org/10.1212/WNL.0b013e3181a71193.
83. Brinton RD, Yao J, Yin F, Mack WJ, Cadenas E. Perimenopause as a neurological transition state. *Nat Rev Endocrinol*. 2015;11(7):393–405. https://doi.org/10.1038/nrendo.2015.82.
84. Deecher DC, Dorries K. Understanding the pathophysiology of vasomotor symptoms (hot flushes and night sweats) that occur in perimenopause, menopause, and postmenopause life stages. *Arch Wom Ment Health*. 2007;10(6):247–257. https://doi.org/10.1007/s00737-007-0209-5.
85. Rossmanith WG, Ruebberdt W. What causes hot flushes? The neuroendocrine origin of vasomotor symptoms in the menopause. *Gynecol Endocrinol*. 2009;25(5):303–314. https://doi.org/10.1080/09513590802632514.
86. Jacobs EG, Weiss B, Makris N, et al. Reorganization of functional networks in verbal working memory circuitry in early midlife: the impact of sex and menopausal status. *Cerebr Cortex*. 2017;27(5):2857–2870. https://doi.org/10.1093/cercor/bhw127.
87. Sherwin BB. Estrogen and cognitive aging in women. *Trends Pharmacol Sci*. November 2002;23(11):527–534. https://doi.org/10.1016/s0165-6147(02)02093-x.
88. Brinton RD. Impact of estrogen therapy on Alzheimer's disease: a fork in the road? *CNS Drugs*. 2004;18(7):405–422. https://doi.org/10.2165/00023210-200418070-00001.
89. Pike CJ, Carroll JC, Rosario ER, Barron AM. Protective actions of sex steroid hormones in Alzheimer's disease. *Front Neuroendocrinol*. July 2009;30(2):239–258. https://doi.org/10.1016/j.yfrne.2009.04.015.
90. Yao J, Irwin RW, Zhao L, Nilsen J, Hamilton RT, Brinton RD. Mitochondrial bioenergetic deficit precedes Alzheimer's pathology in female mouse model of Alzheimer's disease. *Proc Natl Acad Sci USA*. 2009;106(34):14670–14675. https://doi.org/10.1073/pnas.0903563106.
91. Yue X, Lu M, Lancaster T, et al. Brain estrogen deficiency accelerates Aβ plaque formation in an Alzheimer's disease animal model. *Proc Natl Acad Sci U S A*. 2005;102(52):19198–19203. https://doi.org/10.1073/pnas.0505203102.
92. Ding F, Yao J, Rettberg JR, Chen S, Brinton RD. Early decline in glucose transport and metabolism precedes shift to ketogenic system in female aging and Alzheimer's mouse brain: implication for bioenergetic intervention. *PLoS One*. 2013;8(11):e79977. https://doi.org/10.1371/journal.pone.0079977.
93. Ding F, Yao J, Zhao L, Mao Z, Chen S, Brinton RD. Ovariectomy induces a shift in fuel availability and metabolism in the hippocampus of the female transgenic model of familial Alzheimer's. *PLoS One*. 2013;8(3):e59825. https://doi.org/10.1371/journal.pone.0059825.
94. Yin F, Yao J, Sancheti H, et al. The perimenopausal aging transition in the female rat brain: decline in bioenergetic systems and synaptic plasticity. *Neurobiol Aging*. 2015;36(7):2282–2295. https://doi.org/10.1016/j.neurobiolaging.2015.03.013.
95. Berti V, Murray J, Davies M, et al. Nutrient patterns and brain biomarkers of Alzheimer's disease in cognitively normal individuals. *J Nutr Health Aging*. 2015;19(4):413–423. https://doi.org/10.1007/s12603-014-0534-0.
96. Mosconi L, Murray J, Davies M, et al. Nutrient intake and brain biomarkers of Alzheimer's disease in at-risk cognitively normal individuals: a cross-sectional neuroimaging pilot study. *BMJ Open*. 2014;4(6):e004850. https://doi.org/10.1136/bmjopen-2014-004850.

97. Mosconi L, Berti V, Quinn C, et al. Perimenopause and emergence of an Alzheimer's bioenergetic phenotype in brain and periphery. *PLoS One*. 2017;12(10):e0185926. https://doi.org/10.1371/journal.pone.0185926.
98. Ott A, Breteler MM, van Harskamp F, Stijnen T, Hofman A. Incidence and risk of dementia. The rotterdam study. *Am J Epidemiol*. March 15, 1998;147(6):574–580. https://doi.org/10.1093/oxfordjournals.aje.a009489.
99. Bixler EO, Vgontzas AN, Lin HM, et al. Prevalence of sleep-disordered breathing in women: effects of gender. *Am J Respir Crit Care Med*. March 2001;163(3 Pt 1):608–613. https://doi.org/10.1164/ajrccm.163.3.9911064.
100. Barja G. Free radicals and aging. *Trends Neurosci*. October 2004;27(10):595–600. https://doi.org/10.1016/j.tins.2004.07.005.
101. Cantuti-Castelvetri I, Lin MT, Zheng K, et al. Somatic mitochondrial DNA mutations in single neurons and glia. *Neurobiol Aging*. November–December 2005;26(10):1343–1355. https://doi.org/10.1016/j.neurobiolaging.2004.11.008.
102. Kujoth GC, Bradshaw PC, Haroon S, Prolla TA. The role of mitochondrial DNA mutations in mammalian aging. *PLoS Genet*. February 23, 2007;3(2):e24. https://doi.org/10.1371/journal.pgen.0030024.
103. Melov S. Modeling mitochondrial function in aging neurons. *Trends Neurosci*. October 2004;27(10):601–606. https://doi.org/10.1016/j.tins.2004.08.004.
104. Beal MF. Mitochondria take center stage in aging and neurodegeneration. *Ann Neurol*. October 2005;58(4):495–505. https://doi.org/10.1002/ana.20624.
105. Vermulst M, Bielas JH, Kujoth GC, et al. Mitochondrial point mutations do not limit the natural lifespan of mice. *Nat Genet*. April 2007;39(4):540–543. https://doi.org/10.1038/ng1988.
106. Gabbita SP, Lovell MA, Markesbery WR. Increased nuclear DNA oxidation in the brain in Alzheimer's disease. *J Neurochem*. November 1998;71(5):2034–2040. https://doi.org/10.1046/j.1471-4159.1998.71052034.x.
107. Sanders LH, McCoy J, Hu X, et al. Mitochondrial DNA damage: molecular marker of vulnerable nigral neurons in Parkinson's disease. *Neurobiol Dis*. October 2014;70:214–223. https://doi.org/10.1016/j.nbd.2014.06.014.
108. Bromberger JT, Kravitz HM, Chang YF, Cyranowski JM, Brown C, Matthews KA. Major depression during and after the menopausal transition: study of women's health across the nation (SWAN). *Psychol Med*. September 2011;41(9):1879–1888. https://doi.org/10.1017/s003329171100016x.
109. Rubin LH, Sundermann EE, Cook JA, et al. Investigation of menopausal stage and symptoms on cognition in human immunodeficiency virus-infected women. *Menopause*. September 2014;21(9):997–1006. https://doi.org/10.1097/gme.0000000000000203.
110. Jaff NG, Rubin LH, Crowther NJ, Norris SA, Maki PM. Menopausal symptoms, menopausal stage and cognitive functioning in black urban African women. *Climacteric*. February 2020;23(1):38–45. https://doi.org/10.1080/13697137.2019.1646719.
111. Greendale GA, Wight RG, Huang MH, et al. Menopause-associated symptoms and cognitive performance: results from the study of women's health across the nation. *Am J Epidemiol*. June 1, 2010;171(11):1214–1224. https://doi.org/10.1093/aje/kwq067.
112. Mallampalli MP, Carter CL. Exploring sex and gender differences in sleep health: a society for women's health research report. *J Wom Health*. 2014;23(7):553–562. https://doi.org/10.1089/jwh.2014.4816.
113. Cedernaes J, Osorio RS, Varga AW, Kam K, Schiöth HB, Benedict C. Candidate mechanisms underlying the association between sleep-wake disruptions and Alzheimer's disease. *Sleep Med Rev*. 2017;31:102–111. https://doi.org/10.1016/j.smrv.2016.02.002.

114. Lim AS, Kowgier M, Yu L, Buchman AS, Bennett DA. Sleep fragmentation and the risk of incident Alzheimer's disease and cognitive decline in older persons. *Sleep*. July 1, 2013; 36(7):1027–1032. https://doi.org/10.5665/sleep.2802.
115. Spira AP, Gamaldo AA, An Y, et al. Self-reported sleep and β-amyloid deposition in community-dwelling older adults. *JAMA Neurol*. 2013;70(12):1537–1543. https://doi.org/10.1001/jamaneurol.2013.4258.
116. Osorio RS, Gumb T, Pirraglia E, et al. Sleep-disordered breathing advances cognitive decline in the elderly. *Neurology*. May 12, 2015;84(19):1964–1971. https://doi.org/10.1212/wnl.0000000000001566.
117. Maki PM, Drogos LL, Rubin LH, Banuvar S, Shulman LP, Geller SE. Objective hot flashes are negatively related to verbal memory performance in midlife women. *Menopause*. September–October 2008;15(5):848–856. https://doi.org/10.1097/gme.0b013e31816d815e.
118. Maki PM, Thurston RC. Menopause and brain health: hormonal changes are only part of the story. *Front Neurol*. 2020;11:562275. https://doi.org/10.3389/fneur.2020.562275.
119. Thurston RC, Wu M, Aizenstein HJ, et al. Sleep characteristics and white matter hyperintensities among midlife women. *Sleep*. June 15, 2020;(6):43. https://doi.org/10.1093/sleep/zsz298.
120. Drogos LL, Rubin LH, Geller SE, Banuvar S, Shulman LP, Maki PM. Objective cognitive performance is related to subjective memory complaints in midlife women with moderate to severe vasomotor symptoms. *Menopause*. December 2013;20(12):1236–1242. https://doi.org/10.1097/GME.0b013e318291f5a6.
121. Triantafyllou N, Armeni E, Christidi F, et al. The intensity of menopausal symptoms is associated with episodic memory in postmenopausal women. *Climacteric*. August 2016; 19(4):393–399. https://doi.org/10.1080/13697137.2016.1193137.
122. LeBlanc ES, Neiss MB, Carello PE, Samuels MH, Janowsky JS. Hot flashes and estrogen therapy do not influence cognition in early menopausal women. *Menopause*. March–April 2007;14(2):191–202. https://doi.org/10.1097/01.gme.0000230347.28616.1c.
123. Fogel J, Rubin LH, Kilic E, Walega DR, Maki PM. Physiologic vasomotor symptoms are associated with verbal memory dysfunction in breast cancer survivors. *Menopause*. November 2020;27(11):1209–1219. https://doi.org/10.1097/gme.0000000000001608.
124. Thurston RC, Aizenstein HJ, Derby CA, Sejdic E, Maki PM. Menopausal hot flashes and white matter hyperintensities. *Menopause*. January 2016;23(1):27–32. https://doi.org/10.1097/GME.0000000000000481.
125. Edwards H, Duchesne A, Au AS, Einstein G. The many menopauses: searching the cognitive research literature for menopause types. *Menopause*. 2019;26(1):45–65. https://doi.org/10.1097/GME.0000000000001171.
126. Trabuco EC, Moorman PG, Algeciras-Schimnich A, Weaver AL, Cliby WA. Association of ovary-sparing hysterectomy with ovarian reserve. *Obstet Gynecol*. May 2016;127(5): 819–827. https://doi.org/10.1097/aog.0000000000001398.
127. Moorman PG, Myers ER, Schildkraut JM, Iversen ES, Wang F, Warren N. Effect of hysterectomy with ovarian preservation on ovarian function. *Obstet Gynecol*. December 2011;118(6):1271–1279. https://doi.org/10.1097/AOG.0b013e318236fd12.
128. Rocca WA, Gazzuola-Rocca L, Smith CY, et al. Accelerated accumulation of multimorbidity after bilateral oophorectomy: a population-based cohort study. *Mayo Clin Proc*. November 2016;91(11):1577–1589. https://doi.org/10.1016/j.mayocp.2016.08.002.
129. Wellons M, Ouyang P, Schreiner PJ, Herrington DM, Vaidya D. Early menopause predicts future coronary heart disease and stroke: the Multi-Ethnic Study of Atherosclerosis. *Menopause*. October 2012;19(10):1081–1087. https://doi.org/10.1097/gme.0b013e3182517bd0.

130. Pankratz VS, Roberts RO, Mielke MM, et al. Predicting the risk of mild cognitive impairment in the Mayo clinic study of aging. *Neurology*. April 7, 2015;84(14): 1433–1442. https://doi.org/10.1212/wnl.0000000000001437.
131. Goveas JS, Espeland MA, Woods NF, Wassertheil-Smoller S, Kotchen JM. Depressive symptoms and incidence of mild cognitive impairment and probable dementia in elderly women: the Women's Health Initiative Memory Study. *J Am Geriatr Soc*. January 2011; 59(1):57–66. https://doi.org/10.1111/j.1532-5415.2010.03233.x.
132. Kritz-Silverstein D, Barrett-Connor E. Hysterectomy, oophorectomy, and cognitive function in older women. *J Am Geriatr Soc*. January 2002;50(1):55–61. https://doi.org/10.1046/j.1532-5415.2002.50008.x.
133. Rocca WA, Bower JH, Maraganore DM, et al. Increased risk of cognitive impairment or dementia in women who underwent oophorectomy before menopause. *Neurology*. September 11, 2007;69(11):1074–1083. https://doi.org/10.1212/01.wnl.0000276984.19542.e6.
134. Bove R, Secor E, Chibnik LB, et al. Age at surgical menopause influences cognitive decline and Alzheimer pathology in older women. *Neurology*. January 21, 2014;82(3): 222–229. https://doi.org/10.1212/wnl.0000000000000033.
135. Gervais NJ, Au A, Almey A, et al. Cognitive markers of dementia risk in middle-aged women with bilateral salpingo-oophorectomy prior to menopause. *Neurobiol Aging*. October 2020;94:1–6. https://doi.org/10.1016/j.neurobiolaging.2020.04.019.
136. Kurita K, Henderson VW, Gatz M, et al. Association of bilateral oophorectomy with cognitive function in healthy, postmenopausal women. *Fertil Steril*. September 1, 2016; 106(3):749–756.e2. https://doi.org/10.1016/j.fertnstert.2016.04.033.
137. Phung TK, Waltoft BL, Laursen TM, et al. Hysterectomy, oophorectomy and risk of dementia: a nationwide historical cohort study. *Dement Geriatr Cognit Disord*. 2010; 30(1):43–50. https://doi.org/10.1159/000314681.
138. Zeydan B, Tosakulwong N, Schwarz CG, et al. Association of bilateral salpingo-oophorectomy before menopause onset with medial temporal lobe neurodegeneration. *JAMA Neurol*. January 1 2019;76(1):95–100. https://doi.org/10.1001/jamaneurol.2018.3057.
139. Rahman A, Schelbaum E, Hoffman K, et al. Sex-driven modifiers of Alzheimer risk: a multimodality brain imaging study. *Neurology*. July 14, 2020;95(2):e166–e178. https://doi.org/10.1212/wnl.0000000000009781.
140. Oveisgharan S, Arvanitakis Z, Yu L, Farfel J, Schneider JA, Bennett DA. Sex differences in Alzheimer's disease and common neuropathologies of aging. *Acta Neuropathol*. December 2018;136(6):887–900. https://doi.org/10.1007/s00401-018-1920-1.
141. Ryan J, Scali J, Carrière I, et al. Impact of a premature menopause on cognitive function in later life. *BJOG*. December 2014;121(13):1729–1739. https://doi.org/10.1111/1471-0528.12828.
142. Beltz AM, Hampson E, Berenbaum SA. Oral contraceptives and cognition: a role for ethinyl estradiol. *Horm Behav*. 2015;74:209–217.
143. Karim R, Dang H, Henderson VW, et al. Effect of reproductive history and exogenous hormone use on cognitive function in mid- and late life. *J Am Geriatr Soc*. December 2016;64(12):2448–2456. https://doi.org/10.1111/jgs.14658.
144. Egan KR, Gleason CE. Longer duration of hormonal contraceptive use predicts better cognitive outcomes later in life. *J Womens Health (Larchmt)*. 2012;21(12):1259–1266. https://doi.org/10.1089/jwh.2012.3522.
145. Li FD, He F, Chen TR, et al. Reproductive history and risk of cognitive impairment in elderly women: a cross-sectional study in eastern China. *J Alzheimers Dis*. 2016;49(1): 139–147. https://doi.org/10.3233/jad-150444.

146. Ryan J, Carrière I, Scali J, Ritchie K, Ancelin ML. Life-time estrogen exposure and cognitive functioning in later life. *Psychoneuroendocrinology*. February 2009;34(2): 287–298. https://doi.org/10.1016/j.psyneuen.2008.09.008.
147. Tierney MC, Ryan J, Ancelin ML, et al. Lifelong estrogen exposure and memory in older postmenopausal women. *J Alzheimers Dis*. 2013;34(3):601–608. https://doi.org/10.3233/jad-122062.
148. Heys M, Jiang C, Cheng KK, et al. Life long endogenous estrogen exposure and later adulthood cognitive function in a population of naturally postmenopausal women from Southern China: the Guangzhou Biobank Cohort Study. *Psychoneuroendocrinology*. July 2011;36(6):864–873. https://doi.org/10.1016/j.psyneuen.2010.11.009.
149. Yaffe K, Sawaya G, Lieberburg I, Grady D. Estrogen therapy in postmenopausal women: effects on cognitive function and dementia. *JAMA*. March 4, 1998;279(9):688–695. https://doi.org/10.1001/jama.279.9.688.
150. LeBlanc ES, Janowsky J, Chan BK, Nelson HD. Hormone replacement therapy and cognition: systematic review and meta-analysis. *JAMA*. March 21, 2001;285(11): 1489–1499. https://doi.org/10.1001/jama.285.11.1489.
151. Shumaker SA, Legault C, Rapp SR, et al. Estrogen plus progestin and the incidence of dementia and mild cognitive impairment in postmenopausal women: the Women's Health Initiative Memory Study: a randomized controlled trial. *JAMA*. May 28, 2003;289(20): 2651–2662. https://doi.org/10.1001/jama.289.20.2651.
152. Shumaker SA, Legault C, Kuller L, et al. Conjugated equine estrogens and incidence of probable dementia and mild cognitive impairment in postmenopausal women: women's health initiative memory study. *JAMA*. 2004;291(24):2947–2958. https://doi.org/10.1001/jama.291.24.2947.
153. Manson JE, Aragaki AK, Rossouw JE, et al. Menopausal hormone therapy and long-term all-cause and cause-specific mortality: the women's health initiative randomized trials. *JAMA*. 2017;318(10):927–938. https://doi.org/10.1001/jama.2017.11217.
154. Gleason CE, Dowling NM, Wharton W, et al. Effects of hormone therapy on cognition and mood in recently postmenopausal women: findings from the randomized, controlled KEEPS-cognitive and affective study. *PLoS Med*. 2015;12(6):e1001833. https://doi.org/10.1371/journal.pmed.1001833.
155. Espeland MA, Shumaker SA, Leng I, et al. Long-term effects on cognitive function of postmenopausal hormone therapy prescribed to women aged 50 to 55 years. *JAMA Intern Med*. August 12, 2013;173(15):1429–1436. https://doi.org/10.1001/jamainternmed.2013.7727.
156. Lethaby A, Hogervorst E, Richards M, Yesufu A, Yaffe K. Hormone replacement therapy for cognitive function in postmenopausal women. *Cochrane Database Syst Rev*. 2008; 2008(1):CD003122. https://doi.org/10.1002/14651858.CD003122.pub2.
157. Henderson VW, St John JA, Hodis HN, et al. Cognitive effects of estradiol after menopause: a randomized trial of the timing hypothesis. *Neurology*. August 16, 2016;87(7): 699–708. https://doi.org/10.1212/wnl.0000000000002980.
158. Mikkola TS, Savolainen-Peltonen H, Tuomikoski P, et al. Lower death risk for vascular dementia than for Alzheimer's disease with postmenopausal hormone therapy users. *J Clin Endocrinol Metab*. 2017;102(3):870–877. https://doi.org/10.1210/jc.2016-3590.
159. Imtiaz B, Tuppurainen M, Rikkonen T, et al. Postmenopausal hormone therapy and Alzheimer disease: a prospective cohort study. *Neurology*. 2017;88(11):1062–1068. https://doi.org/10.1212/wnl.0000000000003696.
160. Imtiaz B, Taipale H, Tanskanen A, et al. Risk of Alzheimer's disease among users of postmenopausal hormone therapy: a nationwide case-control study. *Maturitas*. 2017;98: 7–13. https://doi.org/10.1016/j.maturitas.2017.01.002.

161. Wu M, Li M, Yuan J, et al. Postmenopausal hormone therapy and Alzheimer's disease, dementia, and Parkinson's disease: a systematic review and time-response meta-analysis. *Pharmacol Res*. May 2020;155:104693. https://doi.org/10.1016/j.phrs.2020.104693.
162. Savolainen-Peltonen H, Rahkola-Soisalo P, Hoti F, et al. Use of postmenopausal hormone therapy and risk of Alzheimer's disease in Finland: nationwide case-control study. *BMJ*. 2019;364:l665. https://doi.org/10.1136/bmj.l665.
163. Pintzka CWS, Håberg AK. Perimenopausal hormone therapy is associated with regional sparing of the CA1 subfield: a HUNT MRI study. *Neurobiol Aging*. 2015;36(9): 2555–2562. https://doi.org/10.1016/j.neurobiolaging.2015.05.023.
164. Kantarci OH, Lebrun C, Siva A, et al. Primary progressive multiple sclerosis evolving from radiologically isolated syndrome. *Ann Neurol*. February 2016;79(2):288–294. https://doi.org/10.1002/ana.24564.
165. Nabulsi L, Lawrence KE, Santhalingam V, et al. *Exogenous Sex Hormone Effects on Brain Microstructure in Women: A Diffusion MRI Study in the UK Biobank*. International Society for Optics and Photonics; 2020:1158308.
166. Maki PM. Critical window hypothesis of hormone therapy and cognition: a scientific update on clinical studies. *Menopause*. June 2013;20(6):695–709. https://doi.org/10.1097/GME.0b013e3182960cf8.
167. Eberling JL, Wu C, Haan MN, Mungas D, Buonocore M, Jagust WJ. Preliminary evidence that estrogen protects against age-related hippocampal atrophy. *Neurobiol Aging*. 2003/09/01/2003;24(5):725–732. https://doi.org/10.1016/S0197-4580(02)00056-8.
168. Shao H, Breitner JCS, Whitmer RA, et al. Hormone therapy and Alzheimer disease dementia. *New findings from the Cache County Study*. 2012;79(18):1846–1852. https://doi.org/10.1212/WNL.0b013e318271f823.
169. Morrison JH, Brinton RD, Schmidt PJ, Gore AC. Estrogen, menopause, and the aging brain: how basic neuroscience can inform hormone therapy in women. *J Neurosci*. October 11, 2006;26(41):10332–10348. https://doi.org/10.1523/jneurosci.3369-06.2006.
170. Brinton RD. Estrogen-induced plasticity from cells to circuits: predictions for cognitive function. *Trends Pharmacol Sci*. 2009;30(4):212–222. https://doi.org/10.1016/j.tips.2008.12.006.
171. Scott E, Zhang Q-G, Wang R, Vadlamudi R, Brann D. Estrogen neuroprotection and the critical period hypothesis. *Front Neuroendocrinol*. 2012;33(1):85–104. https://doi.org/10.1016/j.yfrne.2011.10.001.
172. Rettberg JR, Yao J, Brinton RD. Estrogen: a master regulator of bioenergetic systems in the brain and body. *Front Neuroendocrinol*. 2014;35(1):8–30. https://doi.org/10.1016/j.yfrne.2013.08.001.
173. Miller VM, Harman SM. An update on hormone therapy in postmenopausal women: mini-review for the basic scientist. *Am J Physiol Heart Circ Physiol*. 2017;313(5): H1013–H1021. https://doi.org/10.1152/ajpheart.00383.2017.
174. Daniel JM. Estrogens, estrogen receptors, and female cognitive aging: the impact of timing. *Horm Behav*. 2013;63(2):231–237. https://doi.org/10.1016/j.yhbeh.2012.05.003.
175. Naftolin F, Malaspina D. Estrogen, estrogen treatment and the post-reproductive woman's brain. *Maturitas*. May 20, 2007;57(1):23–26. https://doi.org/10.1016/j.maturitas.2007.02.005.
176. Sherwin BB, Henry JF. Brain aging modulates the neuroprotective effects of estrogen on selective aspects of cognition in women: a critical review. *Front Neuroendocrinol*. January 2008;29(1):88–113. https://doi.org/10.1016/j.yfrne.2007.08.002.

177. Thompson MR, Niu J, Lei X, et al. Association of endocrine therapy and dementia in women with breast cancer. *Breast Cancer*. 2021;13:219–224. https://doi.org/10.2147/BCTT.S300455.
178. Yaffe K, Krueger K, Cummings SR, et al. Effect of raloxifene on prevention of dementia and cognitive impairment in older women: the Multiple Outcomes of Raloxifene Evaluation (MORE) randomized trial. *Am J Psychiatr*. April 2005;162(4):683–690. https://doi.org/10.1176/appi.ajp.162.4.683.
179. Singh M. Progesterone-induced neuroprotection. *Endocrine*. April 2006;29(2):271–274. https://doi.org/10.1385/endo:29:2:271.
180. Fox M, Berzuini C, Knapp LA, Glynn LM. Women's pregnancy life history and Alzheimer's risk: can immunoregulation explain the link? *Am J Alzheimers Dis Other Demen*. 2018;33(8):516–526. https://doi.org/10.1177/1533317518786447.
181. Lueg G, Gross CC, Lohmann H, et al. Clinical relevance of specific T-cell activation in the blood and cerebrospinal fluid of patients with mild Alzheimer's disease. *Neurobiol Aging*. January 2015;36(1):81–89. https://doi.org/10.1016/j.neurobiolaging.2014.08.008.
182. Jang H, Bae JB, Dardiotis E, et al. Differential effects of completed and incomplete pregnancies on the risk of Alzheimer disease. *Neurology*. August 14, 2018;91(7): e643–e651. https://doi.org/10.1212/wnl.0000000000006000.
183. Nelander M, Cnattingius S, Åkerud H, Wikström J, Pedersen NL, Wikström AK. Pregnancy hypertensive disease and risk of dementia and cardiovascular disease in women aged 65 years or older: a cohort study. *BMJ Open*. January 21, 2016;6(1):e009880. https://doi.org/10.1136/bmjopen-2015-009880.
184. Read SL, Grundy EMD. Fertility history and cognition in later life. *J Gerontol B Psychol Sci Soc Sci*. October 1, 2017;72(6):1021–1031. https://doi.org/10.1093/geronb/gbw013.
185. Saenz JL, Díaz-Venegas C, Crimmins EM. Fertility history and cognitive function in late life: the case of Mexico. *J Gerontol B Psychol Sci Soc Sci*. March 14, 2021;76(4): e140–e152. https://doi.org/10.1093/geronb/gbz129.
186. Jung JH, Lee GW, Lee JH, et al. Multiparity, brain atrophy, and cognitive decline. *Front Aging Neurosci*. 2020;12:159. https://doi.org/10.3389/fnagi.2020.00159.
187. Okoth K, Chandan JS, Marshall T, et al. Association between the reproductive health of young women and cardiovascular disease in later life: umbrella review. *BMJ*. 2020;371: m3502. https://doi.org/10.1136/bmj.m3502.
188. Staff AC, Redman CWG, Williams D, et al. Pregnancy and long-term maternal cardiovascular health. *Hypertension*. 2016;67(2):251–260. https://doi.org/10.1161/HYPERTENSIONAHA.115.06357.
189. Lane-Cordova AD, Khan SS, Grobman WA, Greenland P, Shah SJ. Long-term cardiovascular risks associated with adverse pregnancy outcomes: JACC review topic of the week. *J Am Coll Cardiol*. April 30, 2019;73(16):2106–2116. https://doi.org/10.1016/j.jacc.2018.12.092.
190. Brown MC, Best KE, Pearce MS, Waugh J, Robson SC, Bell R. Cardiovascular disease risk in women with pre-eclampsia: systematic review and meta-analysis. *Eur J Epidemiol*. January 2013;28(1):1–19. https://doi.org/10.1007/s10654-013-9762-6.
191. Dayan N, Kaur A, Elharram M, Rossi AM, Pilote L. Impact of preeclampsia on long-term cognitive function. *Hypertension*. December 2018;72(6):1374–1380. https://doi.org/10.1161/hypertensionaha.118.11320.
192. Basit S, Wohlfahrt J, Boyd HA. Pre-eclampsia and risk of dementia later in life: nationwide cohort study. *BMJ*. 2018;363:k4109. https://doi.org/10.1136/bmj.k4109.

193. Fields JA, Garovic VD, Mielke MM, et al. Preeclampsia and cognitive impairment later in life. *Am J Obstet Gynecol.* 2017;217(1):74.e1–74.e11. https://doi.org/10.1016/j.ajog.2017.03.008.
194. Mielke MM, Milic NM, Weissgerber TL, et al. Impaired cognition and brain atrophy decades after hypertensive pregnancy disorders. *Circ Cardiovasc Qual Outcomes.* 2016; 9(2 Suppl 1):S70–S76. https://doi.org/10.1161/CIRCOUTCOMES.115.002461.
195. Adank MC, Hussainali RF, Oosterveer LC, et al. Hypertensive disorders of pregnancy and cognitive impairment: a prospective cohort study. *Neurology.* February 2, 2021;96(5): e709–e718. https://doi.org/10.1212/WNL.0000000000011363.
196. Buhimschi IA, Nayeri UA, Zhao G, et al. Protein misfolding, congophilia, oligomerization, and defective amyloid processing in preeclampsia. *Sci Transl Med.* 2014;6(245): 245ra92. https://doi.org/10.1126/scitranslmed.3008808.
197. Harville EW, Guralnik J, Romero M, Bazzano LA. Reproductive history and cognitive aging: the bogalusa heart study. *Am J Geriatr Psychiatr.* February 2020;28(2):217–225. https://doi.org/10.1016/j.jagp.2019.07.002.
198. Andolf E, Bladh M, Möller L, Sydsjö G. Prior placental bed disorders and later dementia: a retrospective Swedish register-based cohort study. *BJOG An Int J Obstet Gynaecol.* 2020; 127(9):1090–1099. https://doi.org/10.1111/1471-0528.16201.
199. Andolf EG, Sydsjö GC, Bladh MK, Berg G, Sharma S. Hypertensive disorders in pregnancy and later dementia: a Swedish National Register Study. *Acta Obstet Gynecol Scand.* April 2017;96(4):464–471. https://doi.org/10.1111/aogs.13096.
200. Cordonnier C, Sprigg N, Sandset EC, et al. Stroke in women - from evidence to inequalities. *Nat Rev Neurol.* September 2017;13(9):521–532. https://doi.org/10.1038/nrneurol.2017.95.
201. Madsen TE, Khoury J, Alwell K, et al. Sex-specific stroke incidence over time in the Greater Cincinnati/Northern Kentucky stroke study. *Neurology.* September 5, 2017; 89(10):990–996. https://doi.org/10.1212/wnl.0000000000004325.
202. Gibson CL. Cerebral ischemic stroke: is gender important? *J Cereb Blood Flow Metab Suppl.* September 2013;33(9):1355–1361. https://doi.org/10.1038/jcbfm.2013.102.
203. Longstreth Jr WT, Sonnen JA, Koepsell TD, Kukull WA, Larson EB, Montine TJ. Associations between microinfarcts and other macroscopic vascular findings on neuropathologic examination in 2 databases. *Alzheimer Dis Assoc Disord.* July–September 2009; 23(3):291–294. https://doi.org/10.1097/WAD.0b013e318199fc7a.
204. Gilsanz P, Mayeda ER, Glymour MM, et al. Female sex, early-onset hypertension, and risk of dementia. *Neurology.* October 31, 2017;89(18):1886–1893. https://doi.org/10.1212/wnl.0000000000004602.
205. Roberts RO, Knopman DS, Geda YE, et al. Association of diabetes with amnestic and nonamnestic mild cognitive impairment. *Alzheimers Dement.* 2014;10(1):18–26. https://doi.org/10.1016/j.jalz.2013.01.001.
206. Artero S, Ancelin ML, Portet F, et al. Risk profiles for mild cognitive impairment and progression to dementia are gender specific. *J Neurol Neurosurg Psychiatry.* September 2008;79(9):979–984. https://doi.org/10.1136/jnnp.2007.136903.
207. Kautzky-Willer A, Harreiter J, Pacini G. Sex and gender differences in risk, pathophysiology and complications of type 2 diabetes mellitus. *Endocr Rev.* June 2016;37(3): 278–316. https://doi.org/10.1210/er.2015-1137.
208. Azad NA, Al Bugami M, Loy-English I. Gender differences in dementia risk factors. *Gend Med.* June 2007;4(2):120–129. https://doi.org/10.1016/s1550-8579(07)80026-x.
209. Chêne G, Beiser A, Au R, et al. Gender and incidence of dementia in the Framingham Heart Study from mid-adult life. *Alzheimers Dement.* March 2015;11(3):310–320. https://doi.org/10.1016/j.jalz.2013.10.005.

210. Olson NL, Albensi BC. Race- and sex-based disparities in Alzheimer's disease clinical trial enrollment in the United States and Canada: an indigenous perspective. *J Alzheimers Dis Rep*. 2020;4(1):325–344. https://doi.org/10.3233/ADR-200214.
211. Sevigny J, Chiao P, Bussière T, et al. The antibody aducanumab reduces Aβ plaques in Alzheimer's disease. *Nature*. September 1, 2016;537(7618):50–56. https://doi.org/10.1038/nature19323.
212. Ferris S, Lane R, Sfikas N, Winblad B, Farlow M, Feldman HH. Effects of gender on response to treatment with rivastigmine in mild cognitive impairment: a post hoc statistical modeling approach. *Gend Med*. July 2009;6(2):345–355. https://doi.org/10.1016/j.genm.2009.06.004.
213. MacGowan SH, Wilcock GK, Scott M. Effect of gender and apolipoprotein E genotype on response to anticholinesterase therapy in Alzheimer's disease. *Int J Geriatr Psychiatr*. September 1998;13(9):625–630. https://doi.org/10.1002/(sici)1099-1166(199809)13:9<625::aid-gps835>3.0.co;2-2.
214. Canevelli M, Quarata F, Remiddi F, et al. Sex and gender differences in the treatment of Alzheimer's disease: a systematic review of randomized controlled trials. *Pharmacol Res*. January 2017;115:218–223. https://doi.org/10.1016/j.phrs.2016.11.035.
215. Haywood WM, Mukaetova-Ladinska EB. Sex influences on cholinesterase inhibitor treatment in elderly individuals with Alzheimer's disease. *Am J Geriatr Pharmacother*. September 2006;4(3):273–286. https://doi.org/10.1016/j.amjopharm.2006.09.009.
216. Davis ML, Barrett AM. Selective benefit of donepezil on oral naming in Alzheimer's disease in men compared to women. *CNS Spectr*. April 2009;14(4):175–176. https://doi.org/10.1017/s1092852900020174.
217. Buccafusco JJ, Jackson WJ, Stone JD, Terry AV. Sex dimorphisms in the cognitive-enhancing action of the Alzheimer's drug donepezil in aged Rhesus monkeys. *Neuropharmacology*. March 2003;44(3):381–389. https://doi.org/10.1016/s0028-3908(02)00378-7.
218. Scacchi R, Gambina G, Broggio E, Corbo RM. Sex and ESR1 genotype may influence the response to treatment with donepezil and rivastigmine in patients with Alzheimer's disease. *Int J Geriatr Psychiatr*. June 2014;29(6):610–615. https://doi.org/10.1002/gps.4043.
219. Wattmo C, Londos E, Minthon L. Risk factors that affect life expectancy in Alzheimer's disease: a 15-year follow-up. *Dement Geriatr Cognit Disord*. 2014;38(5–6):286–299. https://doi.org/10.1159/000362926.
220. Rhodes ME, Rubin RT. Functional sex differences ('sexual diergism') of central nervous system cholinergic systems, vasopressin, and hypothalamic-pituitary-adrenal axis activity in mammals: a selective review. *Brain Res Brain Res Rev*. August 1999;30(2):135–152. https://doi.org/10.1016/s0165-0173(99)00011-9.
221. Counts SE, Che S, Ginsberg SD, Mufson EJ. Gender differences in neurotrophin and glutamate receptor expression in cholinergic nucleus basalis neurons during the progression of Alzheimer's disease. *J Chem Neuroanat*. October 2011;42(2):111–117. https://doi.org/10.1016/j.jchemneu.2011.02.004.
222. Wang RH, Bejar C, Weinstock M. Gender differences in the effect of rivastigmine on brain cholinesterase activity and cognitive function in rats. *Neuropharmacology*. January 28, 2000;39(3):497–506. https://doi.org/10.1016/s0028-3908(99)00157-4.
223. Smith CD, Wright LK, Garcia GE, Lee RB, Lumley LA. Hormone-dependence of sarin lethality in rats: sex differences and stage of the estrous cycle. *Toxicol Appl Pharmacol*. September 15, 2015;287(3):253–257. https://doi.org/10.1016/j.taap.2015.06.010.
224. Farlow MR, Lahiri DK, Poirier J, Davignon J, Schneider L, Hui SL. Treatment outcome of tacrine therapy depends on apolipoprotein genotype and gender of the subjects with

Alzheimer's disease. *Neurology*. March 1998;50(3):669–677. https://doi.org/10.1212/wnl.50.3.669.
225. Martinkova J, Quevenco F-C, Karcher H, et al. Proportion of women and reporting of outcomes by sex in clinical trials for alzheimer disease: a systematic review and meta-analysis. *JAMA Netw Open*. 2021;4(9):e2124124. https://doi.org/10.1001/jamanetworkopen.2021.24124.
226. Mehta N, Rodrigues C, Lamba M, et al. Systematic review of sex-specific reporting of data: cholinesterase inhibitor example. *J Am Geriatr Soc*. October 2017;65(10): 2213–2219. https://doi.org/10.1111/jgs.15020.
227. Hampel H, Mesulam MM, Cuello AC, et al. The cholinergic system in the pathophysiology and treatment of Alzheimer's disease. *Brain*. July 1, 2018;141(7):1917–1933. https://doi.org/10.1093/brain/awy132.
228. Giacobini E, Pepeu G. Sex and gender differences in the brain cholinergic system and in the response to therapy of Alzheimer disease with cholinesterase inhibitors. *Curr Alzheimer Res*. 2018;15(11):1077–1084. https://doi.org/10.2174/1567205015666180613111504.
229. Liu KA, Mager NAD. Women's involvement in clinical trials: historical perspective and future implications. *Pharm Pract*. January–March 2016;14(1):708. https://doi.org/10.18549/PharmPract.2016.01.708.
230. Medeiros AM, Silva RH. Sex differences in Alzheimer's disease: where do we stand? *J Alzheimers Dis*. 2019;67(1):35–60. https://doi.org/10.3233/jad-180213.
231. Burke JF, Brown DL, Lisabeth LD, Sanchez BN, Morgenstern LB. Enrollment of women and minorities in NINDS trials. *Neurology*. 2011;76(4):354–360. https://doi.org/10.1212/WNL.0b013e3182088260.
232. Yakerson A. Women in clinical trials: a review of policy development and health equity in the Canadian context. *Int J Equity Health*. 2019;18(1):56. https://doi.org/10.1186/s12939-019-0954-x.
233. Chadwick AJ, Baruah R. Gender disparity and implicit gender bias amongst doctors in intensive care medicine: a 'disease' we need to recognise and treat. *J Intensive Care Soc*. 2020;21(1):12–17. https://doi.org/10.1177/1751143719870469.
234. Chapman EN, Kaatz A, Carnes M. Physicians and implicit bias: how doctors may unwittingly perpetuate health care disparities. *J Gen Intern Med*. November 2013;28(11): 1504–1510. https://doi.org/10.1007/s11606-013-2441-1.
235. Daugherty SL, Blair IV, Havranek EP, et al. Implicit gender bias and the use of cardiovascular tests among cardiologists. *J Am Heart Assoc*. November 2017;29(12):6. https://doi.org/10.1161/jaha.117.006872.
236. Hansen M, Schoonover A, Skarica B, Harrod T, Bahr N, Guise JM. Implicit gender bias among US resident physicians. *BMC Med Educ*. October 29, 2019;19(1):396. https://doi.org/10.1186/s12909-019-1818-1.
237. Kannan V, Wilkinson KE, Varghese M, et al. Count me in: using a patient portal to minimize implicit bias in clinical research recruitment. *J Am Med Inf Assoc*. August 1, 2019;26(8–9):703–713. https://doi.org/10.1093/jamia/ocz038.
238. Salles A, Awad M, Goldin L, et al. Estimating implicit and explicit gender bias among health care professionals and surgeons. *JAMA Netw Open*. July 3, 2019;2(7):e196545. https://doi.org/10.1001/jamanetworkopen.2019.6545.
239. Ridda I, MacIntyre CR, Lindley RI, Tan TC. Difficulties in recruiting older people in clinical trials: an examination of barriers and solutions. *Vaccine*. January 22, 2010;28(4): 901–906. https://doi.org/10.1016/j.vaccine.2009.10.081.
240. Stahl ST, Beach SR, Musa D, Schulz R. Living alone and depression: the modifying role of the perceived neighborhood environment. *Aging Ment Health*. October 2017;21(10): 1065–1071. https://doi.org/10.1080/13607863.2016.1191060.

241. Doody RS, Thomas RG, Farlow M, et al. Phase 3 trials of solanezumab for mild-to-moderate Alzheimer's disease. *N Engl J Med*. January 23, 2014;370(4):311–321. https://doi.org/10.1056/NEJMoa1312889.
242. Doody RS, Raman R, Farlow M, et al. A phase 3 trial of semagacestat for treatment of Alzheimer's disease. *N Engl J Med*. July 25, 2013;369(4):341–350. https://doi.org/10.1056/NEJMoa1210951.
243. Salloway S, Sperling R, Fox NC, et al. Two phase 3 trials of bapineuzumab in mild-to-moderate Alzheimer's disease. *N Engl J Med*. January 23, 2014;370(4):322–333. https://doi.org/10.1056/NEJMoa1304839.
244. Egan MF, Kost J, Tariot PN, et al. Randomized trial of verubecestat for mild-to-moderate Alzheimer's disease. *N Engl J Med*. May 3, 2018;378(18):1691–1703. https://doi.org/10.1056/NEJMoa1706441.
245. Religa D, Spångberg K, Wimo A, Edlund AK, Winblad B, Eriksdotter-Jönhagen M. Dementia diagnosis differs in men and women and depends on age and dementia severity: data from SveDem, the Swedish Dementia Quality Registry. *Dement Geriatr Cognit Disord*. 2012;33(2–3):90–95. https://doi.org/10.1159/000337038.
246. Cohen SA, Kunicki ZJ, Drohan MM, Greaney ML. Exploring changes in caregiver burden and caregiving intensity due to COVID-19. *Gerontol Geriatr Med*. 2021;7. https://doi.org/10.1177/2333721421999279, 2333721421999279.
247. Griffith DM, Sharma G, Holliday CS, et al. Men and COVID-19: a biopsychosocial approach to understanding sex differences in mortality and recommendations for practice and policy interventions. *Prev Chronic Dis*. July 16, 2020;17:E63. https://doi.org/10.5888/pcd17.200247.
248. Sharma G, Volgman AS, Michos ED. Sex differences in mortality from COVID-19 pandemic: are men vulnerable and women protected? *JACC Case Rep*. July 15, 2020; 2(9):1407–1410. https://doi.org/10.1016/j.jaccas.2020.04.027.
249. Villar J, Ariff S, Gunier RB, et al. Maternal and neonatal morbidity and mortality among pregnant women with and without COVID-19 infection: the INTERCOVID multinational cohort study. *JAMA Pediatr*. 2021;175(8):817–826. https://doi.org/10.1001/jamapediatrics.2021.1050.
250. Ghelichkhani S, Jenabi E, Jalili E, Alishirzad A, Shahbazi F. Pregnancy outcomes among SARS-CoV-2-infected pregnant women with and without underlying diseases: a case-control study. *J Med Life*. July–August 2021;14(4):518–522. https://doi.org/10.25122/jml-2021-0157.
251. Kalafat E, Prasad S, Birol P, et al. An internally validated prediction model for critical COVID-19 infection and intensive care unit admission in symptomatic pregnant women. *Am J Obstet Gynecol*. September 25, 2021. https://doi.org/10.1016/j.ajog.2021.09.024.
252. Karimi-Zarchi M, Schwartz DA, Bahrami R, et al. A meta-analysis for the risk and prevalence of preeclampsia among pregnant women with COVID-19. *Turk J Obstet Gynecol*. September 27, 2021;18(3):224–235. https://doi.org/10.4274/tjod.galenos.2021.66750.
253. Wang Q, Davis PB, Gurney ME, Xu R. COVID-19 and dementia: analyses of risk, disparity, and outcomes from electronic health records in the US. *Alzheimer's Dementia*. 2021;17(8):1297–1306. https://doi.org/10.1002/alz.12296.
254. Glei DA. The us midlife mortality crisis continues: increased excess cause-specific mortality during 2020. *MedRxiv*. October 23, 2021. https://doi.org/10.1101/2021.05.17.21257241.

255. Barguilla A, Fernández-Lebrero A, Estragués-Gázquez I, et al. Effects of COVID-19 pandemic confinement in patients with cognitive impairment. *Front Neurol.* 2020;11: 589901. https://doi.org/10.3389/fneur.2020.589901.
256. Macchi ZA, Ayele R, Dini M, et al. Lessons from the COVID-19 pandemic for improving outpatient neuropalliative care: a qualitative study of patient and caregiver perspectives. *Palliat Med.* July 2021;35(7):1258–1266. https://doi.org/10.1177/02692163211017383.
257. Borges-Machado F, Barros D, Ribeiro Ó, Carvalho J. The effects of COVID-19 home confinement in dementia care: physical and cognitive decline, severe neuropsychiatric symptoms and increased caregiving burden. *Am J Alzheimers Dis Other Demen.* January–December 2020;35. https://doi.org/10.1177/1533317520976720, 1533317520976720.
258. Seedat S, Scott KM, Angermeyer MC, et al. Cross-national associations between gender and mental disorders in the world health organization world mental health surveys. *Arch Gen Psychiatr.* July 2009;66(7):785–795. https://doi.org/10.1001/archgenpsychiatry.2009.36.
259. Carpinelli Mazzi M, Iavarone A, Musella C, et al. Time of isolation, education and gender influence the psychological outcome during COVID-19 lockdown in caregivers of patients with dementia. *Eur Geriatr Med.* 2020;11(6):1095–1098. https://doi.org/10.1007/s41999-020-00413-z.

Effect of COVID-19 on Alzheimer's and dementia measured through ocular indications

17

Harrison Marsh[1], *Stephen Rossettie*[2] *and Albin John*[3]
[1]Department of Ophthalmology, School of Medicine, Texas Tech University Health Sciences Center, Lubbock, TX, United States; [2]School of Medicine, Texas Tech University Health Sciences Center, Lubbock, TX, United States; [3]Department of Neurology, Texas Tech University Health Sciences Center, Lubbock, TX, United States

Abstract

Screening for early detection of Alzheimer's disease (AD) through a comprehensive eye exam appears to be promising and could potentially provide a more sensitive, inexpensive way to visualize early signs of AD for early detection in large populations. Optical coherence tomography (OCT), as well as retinal imaging techniques such as Doppler and fluorescence lifetime imaging ophthalmoscopy (FLIO), can detect signs of early AD such as vascular changes or accumulations of Tau proteins and beta-amyloid proteins. In the age of COVID-19, this screening opportunity is threatened by increased no-show rates leading to decreased early detection of AD. Through the combination of COVID-19 neuroinflammation potentially augmenting AD neurodegeneration, as well as missed opportunity in the use of early ophthalmic detection, the pandemic may have significantly worsened the trajectory of AD.

Keywords: Alzheimer's disease; COVID-19; Neurodegeneration; No-show rates; Ophthalmic screening; Pandemic; Retinal imaging

1. Introduction

The surplus of inflammation due to the immune response to COVID-19 can potentially accelerate the progression of inflammatory changes in the brain as well as increase neurodegeneration.[1] As a result, early detection and management of AD and other neurocognitive diseases are increasingly important. Although COVID-19 has been shown to cause ocular changes such as conjunctivitis,[2] the ocular changes that can be seen in AD in the absence of COVID-19 could be the key in identifying early signs of AD following the pandemic.

The process of screening and diagnosing AD is incredibly expensive, invasive, and inefficient.[3,4] Even with over a 100 years of research, there is still no cure or conclusive method to diagnose AD before death. Currently, there are several treatments, including

COVID-19 in Alzheimer's Disease and Dementia. https://doi.org/10.1016/B978-0-443-15256-6.00017-9

memantine, donepezil, and a new antibody that was recently FDA approved (aducanumab) that targets amyloid beta. However, the available therapies are only effective in minimizing symptoms and potentially slowing the progression of the disease.[5,6] Improvement in early disease detection through new screening methods could greatly improve long-term outcomes of these treatment options, ultimately delaying the onset of more severe symptoms of dementia in AD patients.[7]

Neuroimaging and cerebral spinal fluid (CSF) analysis provide high specificity and accuracy in diagnosing AD.[3] However, these tests are expensive, invasive, and cannot be effectively used on a large scale. If the goal is to detect neurodegenerative changes related to AD earlier, a more practical method must be considered in order to screen a larger portion of the population at risk.

The focus of this chapter is to highlight how screening and disease management of AD through the utilization of ocular indications can provide a noninvasive, inexpensive method to effectively detect early signs of AD and improve treatment outcomes in these patients.

2. Ocular indications for early screening of Alzheimer's disease

Although the pathology of AD typically occurs in the brain, signs of AD have also been reported to affect the eye.[7,8] Through the eye, one is able to rapidly and through noninvasive imaging methods, visualize neural tissue and vasculature.[7,9] This neural snapshot is possible due to the common embryological origin, neuronal cell layers, neurotransmitter systems, glial cells, blood barriers, and microvasculature shared by both the cerebral cortex and retina. Additionally, the optic nerve axons form direct connections between the brain and retina, facilitating the transport of amyloid precursor protein (APP) created in retinal ganglion cells (RGCs).[10] As a result, the retina can portray a variety of pathologies in patients with AD progression. Some ocular pathologies that are seen in AD patients include degeneration of retinal ganglion cells, retinal nerve fiber layer thinning, and vascular changes.[7,11]

Visual acuity, color vision, and contrast sensitivity have all been shown to be significantly lower in patients with AD compared to healthy controls.[12] These visual changes can likely be explained by structural pathologies in the retina. All stages of AD have been associated with microvascular pathology in the retina.[13] Patients with mild AD have shown significant macular thinning in the central region. Macular thickening has been associated with moderate AD, a finding also common in glaucoma. Given the anatomical and etiological similarities between glaucoma and AD, one study has even raised the question of whether AD could be a cerebral form of glaucoma.[14] In addition to the common finding of retinal nerve fiber layer (RNFL) thinning, the ganglion cell layer and the other plexiform layer have also shown thinning in AD patients when compared to healthy controls. However, the outer nuclear layer has been associated with thickening in these patients as well as significant thinning of the nasal, subfoveal, and inferior sectors of the choroid in patients with mild AD.[12]

A study by Haoshen Shi and associates in 2020 found that certain vascular changes that appear prominently in early brain pathogenesis can predict the cognitive decline in AD patients. They showed that in mild cognitively impaired (MCI) and AD patients, vascular Aβ deposition in the neurosensory retina was associated with an identified deficiency in platelet-derived growth factor receptor beta (PDGFRβ) and pericyte loss.[15] However, the mechanisms of vascular changes in the retina, their association with vascular amyloidosis, the loss of blood−retinal barrier (BRB) integrity, and pericyte loss remains unclear. Overall, their study on mice showed that the identification of retinal capillary degeneration and compromised BRB integrity in the early disease stages of AD dependent on age and Alzheimer's pathology could contribute to the development of novel targets that could aid in both the therapy, and diagnosis of AD.[16]

3. Methods of visualizing the retina

Classical imaging techniques can be used in the screening process of AD. Ocular fundus or retinal photography are two common techniques for investigating the retina that can be employed during routine checks at eye clinics. Retinal photography is capable of detecting a variety of retinal vascular signs, including arteriolar signs in the retina or qualitative retinopathy, vascular caliber changes in the retina, as well as geometrical pattern changes of the retina.[17] Several studies have recorded qualitative retinal vascular signs as well as quantitative vascular measures, including retinopathy, vascular reduction, increasing vessel width variability, decreased complexity of branching characteristic, reduced efficiency of the branching geometry and less tortuous venules, are associated with decreased cognitive performance based on different neuropsychological tests.[18] A more recent pilot study identified peripheral biomarkers by using ultrawide field retinal imaging, including a large increase in number of drusen, significant increase in venular width gradient, as well as a significant decrease in arterial fractal dimension, for AD and its progression over 2 years.[17,19]

Tau proteins and beta-amyloid proteins in the brain have long been understood to strongly correlate to AD, and more recently studies have found that increased levels of *retinal* amyloid-beta peptides correlate with levels found in brain tissue.[20] Retinal imaging techniques such as Doppler and fluorescence lifetime imaging ophthalmoscopy (FLIO) are also noninvasive and inexpensive methods that can detect the accumulation of these proteins.[21] A research group has trialed using the naturally occurring compound curcumin, which gives turmeric its color and flavor—to fluoresce retinal amyloid, which works due to its extreme affinity to amyloid beta, the protein that makes up the plaques in AD.[22,23] Other studies have supported the usefulness of retinal amyloid imaging (RAI), as these proteins can begin accumulating in the retina as early as 20 years before any significant symptoms occur.[10,24]

Optical coherence tomography (OCT), which are easy to perform with other ocular tests, shows the potential of utilizing examinations of the retina in AD detection. OCT is a noninvasive diagnostic technique that renders an in vivo cross-sectional view of the retina. OCT utilizes a concept known as inferometry to create a cross-sectional map of the retina that is accurate to within at least 10–15 microns. The noninvasive and direct measurement of morphology in the retina using OCT has proven functionality in therapeutic and diagnostic applications in a variety of central nervous system (CNS) diseases, including, MS, Parkinson's, and AD.[25] Several previous studies have evaluated the peripapillary retinal nerve fiber layer (RNFL) thickness measurements assessed by OCT, and all of them were able to demonstrate that most of RNFL parameters were reduced in patients with AD.[26] The degree of thinning of the retina's ganglion cell layer measured using OCT has the most predictive power for an AD diagnosis when used to distinguish between patients with clinically diagnosed Alzheimer's and health controls of a similar age.[27] Ophthalmic findings, especially those found via OCT, provide a key opportunity to potentially detect AD in the preclinical stage. However, for AD to be detected early and treated during the prodromal phase, these findings would have to be accompanied by a high degree of clinical suspicion to correlate a decline in cognition with changes in the eye, specifically the retina, ocular movements, and pupil.[21,28]

The retinal nerve fiber layers is a routine measurement in comprehensive eye exams that can be performed by an inexpensive OCT scan to aid in the screening for pathology related to glaucoma and age-related macular degeneration.[29] This scan can be completed in a matter of minutes and is capable of examining pathologic changes that have been shown to be related to AD such as peripapillary retinal nerve fiber layer (pRNFL) and internal molecular layer thinning (Fig. 17.1). More specifically,

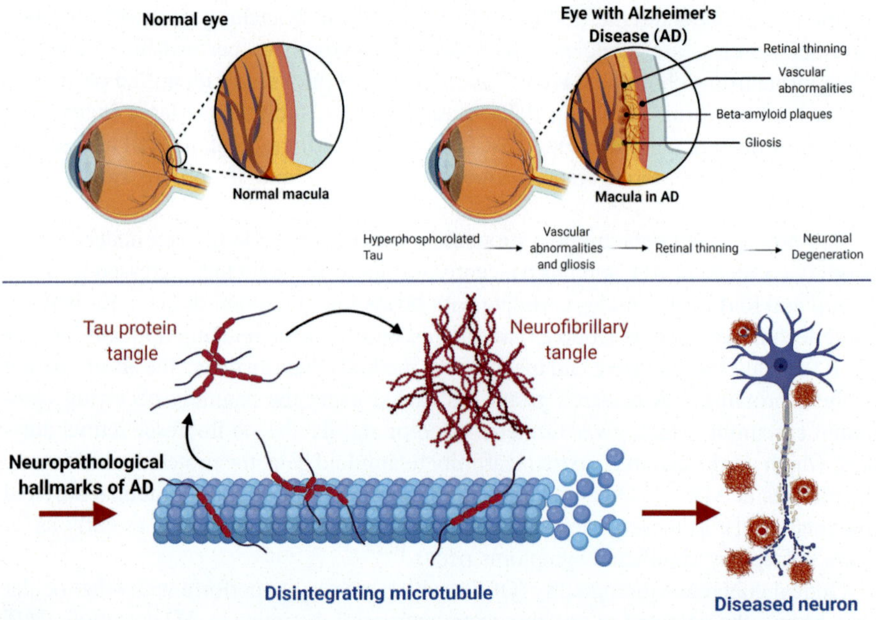

Figure 17.1

significant thickness reduction has been shown in AD patients in the global and temporal superior quadrants of the pRNFL, as well as the peripheral and superior pericentral sector of retinal thickness.[30] This further supports the association of AD with a significant reduction in choroidal thickness. This choroidal thinning could be utilized as another adjunctive biomarker for both diagnosis and follow up of AD.[31] For these reasons, microvascular imaging and imaging of the neuronal structure of the retina by OCT/OCTA would be an excellent option for screening and monitoring AD response to therapies for a large-scale population.

4. Ophthalmology during COVID-19

Unfortunately, the COVID-19 outbreak has led to significantly increased no-show rates, with one study quoting the no-show rate increased from 13% before the pandemic to 33% after it had begun.[32] In addition to an increased risk of sight-threatening conditions, the increased no-show rate is likely to lead to decreased detection of AD, as patients are missing appointments that may otherwise have resulted in early detection.

Although COVID-19 has brought many difficulties, it has certainly sparked an interest and urgency for the use of telemedicine in many fields, including ophthalmology. The use of smartphone-based apps and online consultations has increased, which may offset some of the burdens of the pandemic.[33] Furthermore, the relaxation of regulatory restrictions and increased remote care reimbursement aid in the establishment of ophthalmic telemedicine visits.[34] In the face of another unpredictable crisis within the health system, these technologies and reorganization of ophthalmology systems could offset the disturbances seen in this pandemic.[35]

Despite telemedicine's benefits, the decrease in physical visits may harm the progress of early ocular detection of AD because screening includes OCT and in-person posterior chamber examination by the physician, as discussed in the body of this chapter. Therefore, from an ophthalmic perspective, the COVID-19 pandemic is largely problematic for AD. The combination of increased risk of neurodegeneration due to inflammation as well as decreased early detection due decreased in-person exams is likely to lead to increased progression of undetected AD.

5. Conclusion

The diagnostic potential and the ability to monitor disease progression in AD through the utilization of ocular indications still requires further research and clinical evidence. However, the possibilities of efficient, inexpensive, and noninvasive large-scale screening along with the early, preclinical diagnostic capability of ocular symptoms and biomarkers could potentially revolutionize the AD treatment outcomes while potentially improving the long-term quality of life of those suffering from the disease. As we continue to study AD in the brain, the last decade of optimistic research involving ocular AD indications warrants further investigation of how the retina may reliably reflect the neurological disease. In the age of COVID-19, this possibility is threatened by increased no-show rates leading to decreased early detection of AD. The combination of COVID-19 neuroinflammation potentially augmenting AD neurodegeneration, as well as missed opportunity in the use of early ophthalmic detection, the pandemic may have significantly worsened the trajectory of AD.

References

1. Naughton SX, Raval U, Pasinetti GM. Potential novel role of COVID-19 in Alzheimer's disease and preventative mitigation strategies. *J Alzheim Dis*. 2020;76(1):21–25. https://doi.org/10.3233/jad-200537.
2. Bertoli F, Veritti D, Danese C, et al. Ocular findings in COVID-19 patients: a review of direct manifestations and indirect effects on the eye. *J Ophthalmol*. 2020;2020:1–9. https://doi.org/10.1155/2020/4827304.
3. Khan TK, Alkon DL. Alzheimer's disease cerebrospinal fluid and neuroimaging biomarkers: diagnostic accuracy and relationship to drug efficacy. *J Alzheim Dis*. 2015;46(4): 817–836. https://doi.org/10.3233/jad-150238.
4. Brookmeyer R, Evans DA, Hebert L, et al. National estimates of the prevalence of Alzheimer's disease in the United States. *Alzheimer's Dementia*. 2011;7(1):61–73. https://doi.org/10.1016/j.jalz.2010.11.007.
5. Chen R, Chan PT, Chu H, et al. Treatment effects between monotherapy of donepezil versus combination with memantine for Alzheimer disease: a meta-analysis. *PLoS One*. 2017; 12(8):e0183586. https://doi.org/10.1371/journal.pone.0183586.
6. Schneider L. A resurrection of aducanumab for Alzheimer's disease. *Lancet Neurol*. 2020; 19(2):111–112. https://doi.org/10.1016/s1474-4422(19)30480-6.
7. Frost S, Martins RN, Kanagasingam Y. Ocular biomarkers for early detection of alzheimer's disease. *J Alzheim Dis*. 2010;22(1):1–16. https://doi.org/10.3233/jad-2010-100819.
8. Heaton G, Davis B, Turner L, Cordeiro M. Ocular biomarkers of Alzheimer's disease. *Cent Nerv Syst Agents Med Chem*. 2015;15(2):117–125. https://doi.org/10.2174/1871524915666150319123015.
9. Chang LYL, Lowe J, Ardiles A, et al. Alzheimer's disease in the human eye. Clinical tests that identify ocular and visual information processing deficit as biomarkers. *Alzheimer's Dementia*. 2014;10(2):251–261. https://doi.org/10.1016/j.jalz.2013.06.004.
10. Alber J, Goldfarb D, Thompson LI, et al. Developing retinal biomarkers for the earliest stages of Alzheimer's disease: what we know, what we don't, and how to move forward. *Alzheimer's Dementia*. 2020;16(1):229–243. https://doi.org/10.1002/alz.12006.
11. Hart NJ, Koronyo Y, Black KL, Koronyo-Hamaoui M. Ocular indicators of Alzheimer's: exploring disease in the retina. *Acta Neuropathol*. 2016;132(6):767–787. https://doi.org/10.1007/s00401-016-1613-6.
12. Barnes S, Salobrar-García E, de Hoz R, et al. Changes in visual function and retinal structure in the progression of Alzheimer's disease. *PLoS One*. 2019;14(8). https://doi.org/10.1371/journal.pone.0220535.

13. Zhang Y, Wang Y, Shi C, Shen M, Lu F. Advances in retina imaging as potential biomarkers for early diagnosis of Alzheimer's disease. *Transl Neurodegener.* 2021;10(1). https://doi.org/10.1186/s40035-021-00230-9.
14. Wostyn P, Audenaert K, De Deyn PP. Alzheimer's disease: cerebral glaucoma? *Med Hypotheses.* 2010;74(6):973–977. https://doi.org/10.1016/j.mehy.2009.12.019.
15. Shi H, Koronyo Y, Rentsendorj A, et al. Identification of early pericyte loss and vascular amyloidosis in Alzheimer's disease retina. *Acta Neuropathol.* 2020;139(5):813–836. https://doi.org/10.1007/s00401-020-02134-w.
16. Shi H, Koronyo Y, Fuchs D-T, et al. Retinal capillary degeneration and blood-retinal barrier disruption in murine models of Alzheimer's disease. *Acta Neuropathol Commun.* 2020; 8(1). https://doi.org/10.1186/s40478-020-01076-4.
17. Liao H, Zhu Z, Peng Y. Potential utility of retinal imaging for alzheimer's disease: a review. *Front Aging Neurosci.* 2018;10:188. https://doi.org/10.3389/fnagi.2018.00188.
18. Cheung CY-l, Ikram MK, Sabanayagam C, Wong TY. Retinal microvasculature as a model to study the manifestations of hypertension. *Hypertension.* 2012;60(5):1094–1103. https://doi.org/10.1161/hypertensionaha.111.189142.
19. Csincsik L, MacGillivray Thomas J, Flynn E, et al. Peripheral retinal imaging biomarkers for Alzheimer's disease: a pilot study. *Ophthalmic Res.* 2018;59(4):182–192. https://doi.org/10.1159/000487053.
20. Doustar J, Rentsendorj A, Torbati T, et al. Parallels between retinal and brain pathology and response to immunotherapy in old, late-stage Alzheimer's disease mouse models. *Aging Cell.* 2020;19(11). https://doi.org/10.1111/acel.13246.
21. Mahajan D, Votruba M. Can the retina be used to diagnose and plot the progression of Alzheimer's disease? *Acta Ophthalmol.* 2017;95(8):768–777. https://doi.org/10.1111/aos.13472.
22. Zhang K, Chen M, Du Z-Y, Zheng X, Li D-L, Zhou R-P. Use of curcumin in diagnosis, prevention, and treatment of Alzheimer's disease. *Neural Regen Res.* 2018;13(4):742. https://doi.org/10.4103/1673-5374.230303.
23. Dumitrascu OM, Lyden PD, Torbati T, et al. Sectoral segmentation of retinal amyloid imaging in subjects with cognitive decline. *Alzheimer's Dementia: Diagn, Assess Dis Monit.* 2020;12(1). https://doi.org/10.1002/dad2.12109.
24. Gupta VB, Chitranshi N, den Haan J, et al. Retinal changes in Alzheimer's disease— integrated prospects of imaging, functional and molecular advances. *Prog Retin Eye Res.* 2021;82. https://doi.org/10.1016/j.preteyeres.2020.100899.
25. Cabrera DeBuc D, Gaca-Wysocka M, Grzybowski A, Kanclerz P. Identification of retinal biomarkers in Alzheimer's disease using optical coherence tomography: recent insights, challenges, and opportunities. *J Clin Med.* 2019;8(7). https://doi.org/10.3390/jcm8070996.
26. Cunha LP, Almeida ALM, Costa-Cunha LVF, Costa CF, Monteiro MLR. The role of optical coherence tomography in Alzheimer's disease. *Int J Retina Vitr.* 2016;2(1). https://doi.org/10.1186/s40942-016-0049-4.
27. Yoon SP, Grewal DS, Thompson AC, et al. Retinal microvascular and neurodegenerative changes in Alzheimer's disease and mild cognitive impairment compared with control participants. *Ophthalmol Retina.* 2019;3(6):489–499. https://doi.org/10.1016/j.oret.2019.02.002.
28. Ausó E, Gómez-Vicente V, Esquiva G. Biomarkers for Alzheimer's disease early diagnosis. *J Personalized Med.* 2020;10(3). https://doi.org/10.3390/jpm10030114.

29. Miki A, Medeiros FA, Weinreb RN, et al. Rates of retinal nerve fiber layer thinning in glaucoma suspect eyes. *Ophthalmology*. 2014;121(7):1350–1358. https://doi.org/10.1016/j.ophtha.2014.01.017.
30. Cunha JP, Proença R, Dias-Santos A, et al. OCT in Alzheimer's disease: thinning of the RNFL and superior hemiretina. *Graefes Arch Clin Exp Ophthalmol*. 2017;255(9):1827–1835. https://doi.org/10.1007/s00417-017-3715-9.
31. Gharbiya M, Trebbastoni A, Parisi F, et al. Choroidal thinning as a new finding in alzheimer's disease: evidence from enhanced depth imaging spectral domain optical coherence tomography. *J Alzheim Dis*. 2014;40(4):907–917. https://doi.org/10.3233/jad-132039.
32. Lim LW, Yip LW, Tay HW, et al. Sustainable practice of ophthalmology during COVID-19: challenges and solutions. *Graefes Arch Clin Exp Ophthalmol*. 2020;258(7):1427–1436. https://doi.org/10.1007/s00417-020-04682-z.
33. Wan KH, Lin TPH, Ko C-N, Lam DSC. Impact of COVID-19 on ophthalmology and future practice of medicine. *Asia-Pac J Ophthalmol*. 2020;9(4):279–280. https://doi.org/10.1097/apo.0000000000000305.
34. Saleem SM, Pasquale LR, Sidoti PA, Tsai JC. Virtual ophthalmology: telemedicine in a COVID-19 era. *Am J Ophthalmol*. 2020;216:237–242. https://doi.org/10.1016/j.ajo.2020.04.029.
35. Ferrara M, Romano V, Steel DH, et al. Reshaping ophthalmology training after COVID-19 pandemic. *Eye*. 2020;34(11):2089–2097. https://doi.org/10.1038/s41433-020-1061-3.

Surgical and nonsurgical interventions for Alzheimer's disease

18

P. Hemachandra Reddy[1,2,3,4] *and Albin John*[4]
[1]Department of Internal Medicine, Texas Tech University Health Sciences Center, Lubbock, TX, United States; [2]Nutritional Sciences Department, Texas Tech University, Lubbock, TX, United States; [3]Department of Pharmacology and Neuroscience, School of Medicine, Texas Tech University Health Sciences Center, Lubbock, TX, United States; [4]Department of Neurology, Texas Tech University Health Sciences Center, Lubbock, TX, United States

Abstract

Over the years, many surgical and nonsurgical interventions have been adapted to manage Alzheimer's disease (AD). While many of these tools were developed to primarily treat other neurological conditions, increased understanding of AD pathology has opened up new opportunities to apply established techniques in novel fashions. This chapter discusses neurosurgical interventions for AD especially in the context of the coronavirus pandemic.

Keywords: Alzheimer's disease; Deep brain stimulation; Invasive brain stimulation; Neurology; Neurosurgery; Noninvasive brain stimulation

1. Introduction

The coronavirus pandemic has significantly affected everyone's lives especially those afflicted with the infection and those who are caught in the crossfires of hospital regulations placed to adequately care for those in need. Neurosurgery in particular has been affected by coronavirus restrictions as emergency neurosurgical referrals have dropped by 33.6% and elective operations by 55.6%.[1] In this chapter, we discuss surgical and nonsurgical managements of Alzheimer's disease (AD) that may have been affected by the coronavirus disease 2019 (COVID-19) pandemic.

1.1 Alzheimer's disease—general

Alzheimer's disease (AD) is a neurodegenerative condition that affects more than 50 million people worldwide and places a significant economic burden especially on

COVID-19 in Alzheimer's Disease and Dementia. https://doi.org/10.1016/B978-0-443-15256-6.00012-X

those with advanced AD.[2] Pharmacotherapy for AD has limited benefits in reducing cognitive decline and cell death.[3] An early event in AD is the degeneration of glutamatergic and GABAergic neurons in the hippocampus. Further, degeneration of the cholinergic neurons in the forebrain can lead to the cognitive decline that is characteristic of AD.[4] AD patients can experience a 4%–5% atrophy of the hippocampus every year that can further lead to cognitive deficits.[3,4] Anisotropy, or white matter impairments, can be appreciated using diffusion tensor imaging and can indicate cognitive decline due to impaired synaptic connections.

1.2 Alzheimer's disease—circuit anatomy

The memory circuit is a complex cross-communicating system of memory storage and retrieval. Much of long-term memory storage is located in the medial temporal lobe in structures such as the hippocampus and amygdala, along with other brain structures such as the mammallary bodies and thalamus.[5]

The entorhinal cortex (EC), found on the para-hippocampal gyrus, sends outputs to the hippocampus. These structures are involved in both recognition and spatial memory.[5] The EC has medial and lateral components; the medial component plays a role in encoding object information and attention. Medial EC outputs to the hippocampus are divided into two pathways: the temporoammonic (direct pathway) and the perforant (indirect pathway). The output from the lateral EC is the lateral perforant pathway, and neurons from this pathway have excitatory synapses with pyramidal neurons in the hippocampus.[4] The perforant pathway is an important afferent pathway to the hippocampus from the EC that is composed of glutamatergic fibers. These projects to the CA3 and CA1 subfields in the hippocampus.[5]

Information from the hippocampus travels to the fornix, through the precommissural branches of the fornix, and then to the anterior cingulate cortex.[5] The fornix receives outputs from the hippocampus. It is a white matter tract running through the cerebral hemispheres medially and serves as an additional connecting bridge between the two cerebral hemispheres. Fornices, structures that are part of the Papez circuit, play a role in memory recall, episodic memory formation, and efficient encoding of memories. As a result, damage to the fornices can lead to memory impairment.[6] The fornix is found on the same transverse plane as the foramen of Monroe and the anterior commissure. The precomissural fibers of the fornix come from the hippocampus anterior to the anterior commissure, while the postcommissural fibers exit the subiculum of the hippocampus. These fiber tracts innervate the thalamus and basal forebrain structures.[4] As one ages, the fornices lose their integrity, a change that can result in mild cognitive impairment (MCI) or AD.[4]

Postcommissural fibers branch onto the thalamus and mammallary bodies and eventually synapse onto the cingulate gyrus. The hippocampus and the anterior thalamic nuclei process episodic, spatial, and recognition memory.

2. Invasive brain stimulation

2.1 Deep brain stimulation

Deep brain stimulation (DBS) is a minimally invasive neurosurgical procedure performed to stimulate specific brain regions via implanted electrodes.[5] A small burr hole is made into the skull so that thin electrodes can be inserted deep into specific brain targets. These electrodes then stimulate the brain tissue at various frequencies, voltages, and pulse widths.[7,8] A pulse generator is implanted below the collar bone and the wires from the leads are tunneled underneath the patient's skin to the generator.[6,8] Currently, DBS is a U.S. Food and Drug Administration (FDA)-approved treatment for symptomatic Parkinson's disease with motor improvement as one of the recorded benefits. It has also been used to treat obsessive compulsive disorder, obesity, Tourette's syndrome, and epilepsy[2,4,6] with ongoing expansion of its indications.

DBS was first used in AD patients in 1984 by Turnbull and colleagues when they stimulated the Nucleus basalis of Meynert (NBM). Their unremarkable findings led to a drought in DBS research for the next 26 years until patients receiving DBS for other diseases started reporting Deja-vu-like episodes and improved cognition.[5,6,8] One such patient reported memory flashbacks along with improvements in the California Verbal Learning test administered 12 months later. Hammani and colleagues attributed this change to stimulation of the fornix leading to increased activity in the hippocampus. However, many were skeptical of this anecdote. This led to an increase in case studies investigating DBS that noted improvements in apraxia and memory.

2.1.1 DBS—stimulation locations

Different locations of DBS have been investigated for AD using animal models. In addition to identifying effective target sites, these models noted that DBS could overcome neurogenesis that was suppressed by corticosterone injection. Researchers have also noted improvements in cognition and long-term changes; however, they did not note immediate improvements (<4 weeks after surgery).[5]

A cornerstone of AD pathogenesis is a decrease in cholinergic transmission. As a result, the Nucleus Basalis of Meynert (NBM) was the first target for AD patients with reduced cholinergic transmission and NBM volume.[6] DBS to the NBM has also shown efficacy in patients with early-onset AD.[4] NBM stimulation affects the activity of glutamic acid decarboxylase, which helps synthesize GABA and glutamate.[3] Modulating stimulation frequency has been noted to greatly activate the NBM.[5] Lower frequency stimulation has been shown in anesthetized rats to increase nerve growth factor (NGF) release. NGF plays a role in the survival of cholinergic neurons. These improvements were noted in aged rats but have not been shown in animal models of dementia yet.[5]

Other researchers have investigated stimulation of the entorhinal cortex (EC), a structure that uses the perforant pathway to reach the dentate gyrus. They noted that DBS increased proliferation in the dentate gyrus that resulted in differentiation of the dentate gyrus granule cell morphology. The EC responds well to low-current, high-frequency stimulation of 50 μA in rodents and 0.5−1.5 mA in humans. The neurons in this location have a low activation threshold. Stimulation of the memory circuit using the EC as a target for stimulation improved memory as verified by mice models and a spatial water maze task.[5]

Direct stimulation of the hippocampus may not serve as a good target for DBS stimulation, as 1 ms stimulation at 4−6 mA resulted in deficits and disruption in memory function.[5,9] However, stimulation of the pathways into and out of the hippocampus, such as the fornix, could be an effective target for DBS. The fornix responds best to current density and less to frequency of stimulation. This may be due to its composition of bundles of myelinated fibers that are easily stimulated regardless of frequency.[5] Clinically, DBS has been used in a few MCI patients at the fornix, with noted reduction in memory. Furthermore, 1 year of continuous DBS resulted in less cognitive decline as measured by the AD assessment Scale cognitive subscale (ADAS-cog).[5,10] Hescham and colleagues noted that DBS stimulation of the fornix in both orthodromic and antidromic manners excited large myelinated axons and was capable of reversing memory impairments induced by scopolamine.[5] Other researchers have noted that fornix stimulation led to increased acetylcholine levels in the hippocampus in the first 20 min. Intermittent stimulation may maintain an increased acetylcholine level.[3] There is a circulating hypothesis that stimulation of the fornix could delay or even reverse memory loss. Laxton and his team noted that stimulation of the fornix showed some promise in slowing the progression of AD and in improving the cognitive capabilities of patients.[2,10]

The vagus nerve is another favored target for DBS due to its connection to the locus coeruleus. The locus coeruleus sends out noradrenergic projections throughout the brain. This brain region reduces in volume in AD patients, and the loss of norepinephrine has been correlated with increased AD severity. Norepinephrine (NE) is not only directly involved with cognition and synaptic activity but also involved in prevention of microglial inflammatory actions.[6] Vagus nerve stimulation involves direct stimulation using a pulse generator and electrodes that are placed on the patient's left side. Swedish researchers have noted that at a frequency of 20 Hz, pulse width of 500 μs, and current of 0.25 mA, the stimulation had a 70% response rate and improved cognition in 9 of 10 patients. These improvements were sustained over the next 6 months. After expanding the study pool, researchers noted that vagus nerve stimulation may play a role in maintaining cognitive capacity and preventing memory decline.[6,8]

2.1.2 DBS—trials

Multiple preliminary trials have been conducted to investigate the role of DBS in AD patients. Results from a 12-month DBS phase one trial of six adults with early stage AD noted improved memory, decreased cognitive decline, improved glucose metabolism, and increased hippocampal volume. As a result of the positive outcomes, this trial progressed to a phase two study during which researchers noted that DBS had better outcomes in improving cognition and reducing cognitive decline in individuals over the age of 65 years. Another trial investigated the use of DBS of the hippocampus and hypothalamus in 42 AD patients. The device delivered a continuous 130 Hz 3–3.5 V stimulation (similar stimulation parameter used for Parkinson's patients) for 1 year. Researchers then assessed changes in patient's cognition via questionnaires. They also used PET scans to observe any changes in participant's brain structure. Patients under the age of 65 years did not experience improvements.[3,6,8,11] Those over the age of 65 years with AD, on the other hand, had less cognitive decline compared to those without the device.[11] How long this improvement lasts is still under investigation with preliminary research demonstrating cognitive improvements failing to last longer than 1 year.[3]

The cognitive benefits of DBS continued to drive research in the use of DBS in AD. This work was conducted by two groups: Lozano in Canada and the Germany based Kuhn and colleagues. These two groups focused on two different stimulation targets: Fornix (Lozano) and the Nucleus Basalis of Meynert (Kuhn).[7]

German researchers noted that patients had increased temporal and amygdalohippocampal glucose metabolism after stimulation for about 1 year. Kuhn and colleagues also explored the use of DBS in the Nucleus Basalis of Meynert (NBM) of young patients with low Alzheimer's Disease Assessment Scale–Cognitive Subscale (ADAS-Cog) scores. While one patient experienced steep cognitive decline, the others had a stabilized ADAS-Cog score and improved Mini-mental state examination score after 28 months of stimulation. Research from Hardenacke and team noted some cognitive benefit and slowing of disease progression with the use of DBS in the NBM of younger patients.[8,12]

New groups have joined the investigation, including a research team from the United States, that is investigating stimulation of the frontal lobe as a potential brain target. France also pursued investigation of DBS in AD patients and noted reduced cognitive decline and improvement in cognitive performance in MCI patients.[5,8]

2.1.3 DBS—benefits in AD

As a result of many preliminary trials and experiments, researchers have started to better understand the positive effects of DBS on AD patients. They note that the effects of DBS may include:

1. Reduced synaptic loss
2. Improved spine density
3. Induction of neurogenesis
4. Improved glucose metabolism
5. Enhancement of memory circuits

Of note, the improvements in glucose metabolism due to DBS are particularly important in AD patients who have reduced glucose metabolism and atrophy of the hippocampus. Glucose metabolism is a sign that the brain is active and using energy appropriately. In AD patients, there is reduced glucose metabolism. The improvement in glucose metabolism due to DBS correlates with neuronal activation with studies noting a 2%–5% increase in glucose metabolism especially in the amygdala, hippocampus, and temporal regions.[3,4] AD patients can also experience atrophy of the hippocampus every year. DBS can reduce this atrophy and may even increase hippocampal size by 5%–8%. These effects are much greater than that of physical/mental exercise and other conservative measures, such as socialization and diet (1%–1.5% increase).[4,13] These findings were corroborated in a 2010 clinical trial by the University Health Network, Toronto. These researchers also noted that patients receiving DBS had visions of autobiographical experiences along with less severe memory decline, an increase in glucose metabolism throughout the cerebrum, increased hippocampal volumes, and a reduction in brain atrophy. Importantly, the increase in hippocampal volumes was appreciated in patients who received DBS and not in those who were solely pharmacologically treated.[4]

Overall, DBS is a burgeoning area of research for AD therapy. If DBS can achieve a success rate of at least 3%, with success defined as an immediate improvement in the patient's cognition that is sustained for 1 year, Mirsaeedi and team agree that DBS is a superior treatment to the current standard of care (ex: pharmacologic, lifestyle). Furthermore, if success rates rise to 80%, this treatment could be cost effective. As a result of its potential therapeutic effect in AD, DBS requires a low threshold of success, and the ceiling for achievement is very high.[2] However, the question remains whether enhancing memory will improve quality of life considering AD patients suffer from a myriad of cognitive deficits.[3]

2.2 CSF shunting

The cerebrospinal fluid (CSF) plays a variety of functions in the brain ranging from acid–base buffering to transportation of micronutrients. As one ages, CSF production decreases by over 50%, and turnover of CSF reduces from 4 to 1.5 times per day. The combination of these two age-related declines leads to impaired homeostasis and impaired clearance of toxic metabolites. Decreased clearance of amyloid beta may result in deposition of the fibrils at the arachnoid granulations, locations where CSF is drained. By plugging these drains, amyloid can lead to increased intracranial pressure (ICP).[14] Researchers have postulated that reducing ICP may benefit AD patients by improving cerebral blood flow. As a result, researchers have investigated ventriculoarterial shunting to reduce ICP and improve nutrient supply to relevant brain regions. Furthermore, CSF shunting can help clear the accumulation of tau and soluble amyloid-beta by improving CSF clearance.[15]

Pomeraniec and colleagues investigated this hypothesis by studying whether CSF shunt placement in patients for normal pressure hydrocephalus (NPH) between 1998 and 2013 at the University of Virginia could improve concurrent AD pathology. It was noted that many patients who fit the criteria of NPH had AD.[16] Upon further investigation, NPH may be an early indicator of AD. Researchers studied 142 NPH patients, of which 19% had concomitant AD at the time of CSF shunt implementation.

However, researchers noted that while initial positive outcomes were noted, gait and cognition were not significantly improved. On the other hand, CSF shunting may slow down the accelerated degeneration of the brain due to AD, thus may play a promising role in combatting AD degeneration.[17]

2.3 Omental patch

Researchers postulate that the decreased blood flow noted in many AD patients may be a cause of dementia in AD due to a lack of perfusion, especially in the intraparenchymal regions of the anterior choroidal and anterior perforating arteries, causing neuronal death.[18–20] This correlation was noted in spin-labeled magnetic resonance studies that looked at cerebral blood flow (CBF) in AD patients versus that in age-matched controls. It was noted that there was a significant decreased in blood flow in patients with AD that was further apparent in those with early-onset AD, compared to age-matched controls. Decreasing CBF may be due to reduced cardiac function normally associated with aging. Vascular changes, such as kinking, twisting, and looping of blood vessels may further exacerbate the reduced CBF conundrum.[19]

To improve both CBF volume and velocity, surgeons have tried using an omental transposition from the peritoneal cavity (with blood supply intact) and placing this highly vascular patch on the brain.[20] Blood vessels will form due to the increased amount of VEGF, an angiogenic substance in the body, and penetrate vertically into the brain. From neurotransmitters to nerve growth factors, the omental transposition creates an environment for increased vascularization.[20] Another role of an omental transposition may be in improving the clearance system of the brain. Impaired clearance of amyloid can lead to amyloid buildup in the brain that contributes to AD progression.[19,20]

Initial studies were carried out in animal models. Human trials followed, in which 25 AD patients underwent the transposition. Six patients did not show improvement, 10 patients had minimal changes, and nine patients had significant improvement. Omental patches have been shown to have long-term improvements in cognition such that some patients can return to activities of daily living; however, researchers have also noted that the longer patients had AD, the less effective the procedure was.[19] While still in its infancy, this approach may hold promise and requires further testing with larger sample sizes.[19,20]

2.4 Other invasive surgical interventions for AD

1. Intraventricular infusions—Using intraventricular infusions, surgeons can deliver neuroprotective factors and medications more directly to areas of damage.[15] One such substance that could be delivered is clusterin, or apolipoprotein J. Clusterin binds to Aβ as an extracellular chaperone to improve clearance of Aβ. Researchers noted that intraventricular injection of clusterin in Tg6799 transgenic mice (mice with mutations in Amyloid precursor protein and presenilin 1) improved memory loss and reduced amyloid burden in the mice brains.[21]

2. Gene therapy—Using gene therapy, surgeons can assist in intracerebral delivery of DNA that encodes for neuroprotective and neurogenic factors using a viral vector at sites of atrophy such as the NBM.[15,22] Dr. Rafii and colleagues tried using gene therapy with 49 patients with AD by which an adeno-associated-virus-2 vector was used to deliver the NGF DNA. However, researchers did not note statistically significant cognitive differences between those who received the virus vector and those who did not. In fact, they noted marginal decline in cognitive performance over the course of 24 months (not significant).[22,23]

3. Non-invasive brain stimulation procedures

The field of noninvasive brain stimulation for AD and related dementias is growing, with over 12 randomized controlled trials and proof of principles published. However, application in clinical practice has not yet become routine.

Locations for stimulation include:

- Dorsolateral prefrontal cortex: working memory and improves plasticity of the prefrontal cortex if stimulated.[6]
- Broca, Wernicke, parietal somatosensory association cortex: language, visual, and spatial improvements.[6]
- Inferior frontal gyrus: reflexive reorienting, go/no-go tasks.[6]
- Temporal cortex: mesial temporal lobe atrophy can lead to memory defects.[6]

3.1 Transcranial magnetic stimulation

Transcranial magnetic stimulation (TMS) was first used by Barker and colleagues in 1985 to target the motor cortex. Using Faraday's law of electromagnetic induction, TMS creates a magnetic field across the skull that can generate an electric current for action potential generation. Higher frequencies of stimulation increase cortical excitability, while lower frequencies inhibit it.[6,8] The pulses depolarize the neuronal membranes, leading to action potential generation. This method of stimulation can excite the brain and its effects can last even after the stimulation period.[6,24]

Currently, TMS has been approved for use in major depression and obsessive compulsive disorder. Researchers are now looking at the use of TMS in AD by targeting various brain regions. Italian researchers used TMS aimed at the dorsolateral prefrontal cortex and noted that stimulation led to increased accuracy in naming tasks. After expanding the cohort size, the researchers noted a similar improvement in patients with moderate to severe AD that was not reproduceable in mild AD patients. Similar studies by Egyptian researchers noted that TMS showed promising improvement in the Mini-Mental State Exam (MMSE) after high frequency stimulation over the inferior frontal gyrus.[6,8] Similar results were noted in Egypt and China with long sessions of high frequency TMS showing cognitive improvements in patients. Overall, TMS stimulation of various brain regions demonstrated universally positive results.[6] Additionally, researchers are looking at using a combination therapy of both TMS and cognitive training regiments. This method, termed the NeuroAD protocol, is being pioneered by Neuronix, Ltd. from Israel (ClinicalTrials.gov: NCT01825330).[24]

3.2 Transcranial electrical stimulation

Transcranial electrical stimulation, on the other hand, is passage of weak currents between electrodes placed on the scalp. The most common form, transcranial direct current stimulation, sends a constant current to create electrical gradients that can excite the brain and thus depolarize or hyperpolarize the neurons.[24] These electric currents, between 1 and 2 mA, can improve word recognition and visual recognition memory. However, conflicting results necessitate further exploration of this modality.[6]

3.3 Other NIBS

1. Transcranial alternating current stimulation (tACS)—Clinicians use weak electric fields to modulate brain activity via interactions of the gamma oscillating sinusoidal current delivered by tACS with the brain's own oscillations. tACS may improve memory and problem solving skills if used over the frontal and temporal lobes. tACS can reduce amyloid and tau burden by enhancing synaptic function, increasing neuroprotective factors, and reducing damage.[6,25,26] Use of tACS along with cognitive exercises (exercises that focus on spatial, cognitive, and associative skills) further amplify the cognitive benefit noted with mental exercises alone.[26]
2. Electroconvulsive therapy and magnetic seizure therapy—These therapies are very useful for symptomatic control (suicidality, catatonia) in AD.[6,27] Magnetic seizure therapy can also offer therapy for depression in AD.[28]
3. Noninvasive stimulation of the Vagus nerve—This therapy uses transcutaneous stimulation of the vagus nerve at the neck or ear (on the skin) and has been reported to lead to long term cognitive improvements. Further studies are necessary due to limited sample size.[6,28]

One of the drawbacks of the NIBS techniques is that they are limited to superficial regions of the cortex. While TMS can, with certain coils, achieve deeper stimulation, it loses its accuracy and is not as effective. For NIBS to become a mainstream AD therapy, there is a need for more clinical trials with larger cohorts and standardization of the parameters of stimulation.[24]

3.4 Neurosurgery during COVID-19

Elective neurosurgical procedures have been significantly affected by COVID-19 hospital restrictions.[29] As more hospital staff are needed to care for the increasing COVID-19-infected patient load, fewer operating rooms can be adequately staffed. As a result, many elective surgical procedures were placed on hold.[30] In hospitals that were not significantly affected by COVID-19 patient loads, face-to-face clinic visits and elective surgeries were not curtailed.[31]

Of the invasive surgical intervention presented, DBS in particular is affected by the difficulty of clinical follow-up and care. Patients can present with hardware emergencies such battery failure, electrical malfunction, or lead displacement.[29] They can also develop serious conditions due to interruption of neurostimulation such as DBS withdrawal syndrome, that needs emergent attention.[29] These complications may have been missed or may have experienced a delay in appropriate management due to the pandemic.

It is possible that this observation is simply a statistical anomaly. As time progresses, additional data will be collected to further support or refute our current observation, particularly when looking at the same study period over future years post pandemic. There may be an increase in the rate of revision shunt surgeries as restrictions start to be lifted, implying that these cases were simply delayed. However the cost of delay may also become apparent as state health departments have started reporting increases in possible shunt-related deaths.[29–31]

Additionally, access to physicians for routine clinical care, examinations, intervention sessions, and support with in-person clinic visits has been significantly decreased as a result of the COVID-19 pandemic.[31] However, some neurological practices have reported a 40-fold increase in telemedicine visits with researchers celebrating the success of telemedicine for neurosurgical patients.[32] While both the invasive and noninvasive interventions listed in this chapter are useful therapies in AD patients, the efficacy of virtual visits, remote patient monitoring, and sparse in-clinic visits have hampered proper administration of these tools.[1,31]

4. Conclusion

Throughout the years of AD research, many new interventions, both surgical and nonsurgical have been developed to best care for patients. In the coming years, further study and assessment of these tools will shed greater light on the efficacy of these relatively new tools with managing AD. While surgical and nonsurgical treatments are a burgeoning area of research and development for patients with AD, the coronavirus pandemic has significantly affected access to such interventions. The aftermath of the brief pause on neurosurgical interventions is of great interest and further insights can be gleaned from future studies.

References

1. Ashkan K, Jung J, Velicu AM, et al. Neurosurgery and coronavirus: impact and challenges—lessons learnt from the first wave of a global pandemic. *Acta Neurochir*. 2020; 163(2):317–329. https://doi.org/10.1007/s00701-020-04652-8.
2. Mirsaeedi-Farahani K, Halpern CH, Baltuch GH, Wolk DA, Stein SC. Deep brain stimulation for Alzheimer disease: a decision and cost-effectiveness analysis. *J Neurol*. 2015; 262(5):1191–1197. https://doi.org/10.1007/s00415-015-7688-5.
3. Hescham S, Aldehri M, Temel Y, Alnaami I, Jahanshahi A. Deep brain stimulation for Alzheimer's Disease: an update. *Surg Neurol Int*. 2018;9(1). https://doi.org/10.4103/sni.sni_342_17.

4. Yu D, Yan H, Zhou J, Yang X, Lu Y, Han Y. A circuit view of deep brain stimulation in Alzheimer's disease and the possible mechanisms. *Mol Neurodegener*. 2019;14(1). https://doi.org/10.1186/s13024-019-0334-4.
5. Hescham S, Lim LW, Jahanshahi A, Blokland A, Temel Y. Deep brain stimulation in dementia-related disorders. *Neurosci Biobehav Rev*. 2013;37(10):2666−2675. https://doi.org/10.1016/j.neubiorev.2013.09.002.
6. Fried I. Brain stimulation in Alzheimer's disease. *J Alzheim Dis*. 2016;54(2):789−791. https://doi.org/10.3233/jad-160719.
7. Bittlinger M, Müller S. Opening the debate on deep brain stimulation for Alzheimer disease − a critical evaluation of rationale, shortcomings, and ethical justification. *BMC Med Ethics*. 2018;19(1). https://doi.org/10.1186/s12910-018-0275-4.
8. Chang C-H, Lane H-Y, Lin C-H. Brain stimulation in Alzheimer's disease. *Front Psychiatr*. 2018:9. https://doi.org/10.3389/fpsyt.2018.00201.
9. Lacruz ME, Valentín A, Seoane JJG, Morris RG, Selway RP, Alarcón G. Single pulse electrical stimulation of the hippocampus is sufficient to impair human episodic memory. *Neuroscience*. 2010;170(2):623−632. https://doi.org/10.1016/j.neuroscience.2010.06.042.
10. Laxton AW, Tang-Wai DF, McAndrews MP, et al. A phase I trial of deep brain stimulation of memory circuits in Alzheimer's disease. *Ann Neurol*. 2010;68(4):521−534. https://doi.org/10.1002/ana.22089.
11. Leoutsakos J-MS, Yan H, Anderson WS, et al. Deep brain stimulation targeting the fornix for mild Alzheimer dementia (the ADvance trial): a two year follow-up including results of delayed activation. *J Alzheim Dis*. 2018;64(2):597−606. https://doi.org/10.3233/jad-180121.
12. Hardenacke K, Hashemiyoon R, Visser-Vandewalle V, et al. Deep brain stimulation of the nucleus basalis of Meynert in Alzheimer's dementia: potential predictors of cognitive change and results of a long-term follow-up in eight patients. *Brain Stimul*. 2016;9(5): 799−800. https://doi.org/10.1016/j.brs.2016.05.013.
13. Mendiola-Precoma J, Berumen LC, Padilla K, Garcia-Alcocer G. Therapies for prevention and treatment of Alzheimer's disease. *BioMed Res Int*. 2016;2016:1−17. https://doi.org/10.1155/2016/2589276.
14. Qin Y, Gu JW. A surgical method to improve the homeostasis of CSF for the treatment of Alzheimer's disease. *Front Aging Neurosci*. 2016;8. https://doi.org/10.3389/fnagi.2016.00261.
15. Laxton AW, Stone S, Lozano AM. The neurosurgical treatment of Alzheimer's disease: a review. *Stereotact Funct Neurosurg*. 2014;92(5):269−281. https://doi.org/10.1159/000364914.
16. Golomb J. Alzheimer's disease comorbidity in normal pressure hydrocephalus: prevalence and shunt response. *J Neurol Neurosurg Psychiatr*. 2000;68(6):778−781. https://doi.org/10.1136/jnnp.68.6.778.
17. Pomeraniec IJ, Bond AE, Lopes MB, Jane JA. Concurrent Alzheimer's pathology in patients with clinical normal pressure hydrocephalus: correlation of high-volume lumbar puncture results, cortical brain biopsies, and outcomes. *J Neurosurg*. 2016;124(2): 382−388. https://doi.org/10.3171/2015.2.Jns142318.
18. Rafael H. Neural transplantation. *J Neurosurg*. 2006;104(2):336−337. https://doi.org/10.3171/jns.2006.104.2.336.
19. Goldsmith H. Alzheimer's disease can be treated: why the delay? *Surg Neurol Int*. 2017; 8(1). https://doi.org/10.4103/sni.sni_116_17.
20. Goldsmith HS. Omental transposition in treatment of Alzheimer disease. *J Am Coll Surg*. 2007;205(6):800−804. https://doi.org/10.1016/j.jamcollsurg.2007.06.294.

21. Qi X-M, Wang C, Chu X-K, Li G, Ma J-F. Intraventricular infusion of clusterin ameliorated cognition and pathology in Tg6799 model of Alzheimer's disease. *BMC Neurosci.* 2018; 19(1). https://doi.org/10.1186/s12868-018-0402-7.
22. Honig LS. Gene therapy in Alzheimer disease—it may be feasible, but will it be beneficial? *JAMA Neurol.* 2018;75(7). https://doi.org/10.1001/jamaneurol.2017.4029.
23. Rafii MS, Tuszynski MH, Thomas RG, et al. Adeno-associated viral vector (serotype 2) —nerve growth factor for patients with Alzheimer disease. *JAMA Neurol.* 2018;75(7). https://doi.org/10.1001/jamaneurol.2018.0233.
24. Buss SS, Fried PJ, Pascual-Leone A. Therapeutic noninvasive brain stimulation in Alzheimer's disease and related dementias. *Curr Opin Neurol.* 2019;32(2):292—304. https://doi.org/10.1097/wco.0000000000000669.
25. Xing Y, Wei P, Wang C, et al. TRanscranial AlterNating current Stimulation FOR patients with Mild Alzheimer's disease (TRANSFORM-AD study): protocol for a randomized controlled clinical trial. *Alzheimer Dementia Trans Res Clin Interv.* 2020;6(1). https://doi.org/10.1002/trc2.12005.
26. Moussavi Z, Kimura K, Kehler L, de Oliveira Francisco C, Lithgow B. A novel program to improve cognitive function in individuals with dementia using transcranial alternating current stimulation (tACS) and tutored cognitive exercises. *Front Aging.* 2021;2. https://doi.org/10.3389/fragi.2021.632545.
27. Tampi RR, Tampi DJ, Young J, Hoq R, Resnick K. The place for electroconvulsive therapy in the management of behavioral and psychological symptoms of dementia. *Neurodegener Dis Manag.* 2019;9(6). https://doi.org/10.2217/nmt-2019-0018.
28. Lin Y-C, Wang Y-P. Status of noninvasive brain stimulation in the therapy of Alzheimer's disease. *Chinese Med J.* 2018;131(24):2899—2903. https://doi.org/10.4103/0366-6999.247217.
29. Holla VV, Neeraja K, Surisetti BK, et al. Deep brain stimulation battery exhaustion during the COVID-19 pandemic: crisis within a crisis. *J Mov Disord.* 2020;13(3):218—222. https://doi.org/10.14802/jmd.20073.
30. Laxpati N, Bray DP, Wheelus J, et al. Unexpected decrease in shunt surgeries performed during the shelter-in-place period of the COVID-19 pandemic. *Operative Neurosurgery.* 2021;20(5):469—476. https://doi.org/10.1093/ons/opaa461.
31. Zhang C, Zhang J, Qiu X, et al. Deep brain stimulation for Parkinson's disease during the COVID-19 pandemic: patient perspective. *Front Hum Neurosci.* 2021;15. https://doi.org/10.3389/fnhum.2021.628105.
32. Richards AE, Curley K, Christel L, et al. Patient satisfaction with telehealth in neurosurgery outpatient clinic during COVID-19 pandemic. *Interdisciplinary Neurosurgery.* 2021:23. https://doi.org/10.1016/j.inat.2020.101017.

Index

Note: Page numbers followed by 'f' indicate figures those followed by 't' indicate tables and 'b' indicate boxes.

Printed in the United States
by Baker & Taylor Publisher Services